Mechanical Engineering under Uncertainties

SCIENCES

Mechanics, Field Director – Gilles Pijaudier-Cabot

Numerical Methods in Mechanics, Subject Head – Sylvain Drapier

Mechanical Engineering under Uncertainties

From Classical Approaches to Some Recent Developments

Coordinated by
Christian Gogu

ISTE

WILEY

First published 2021 in Great Britain and the United States by ISTE Ltd and John Wiley & Sons, Inc.

ISTE Ltd
27-37 St George's Road
London SW19 4EU
UK

www.iste.co.uk

John Wiley & Sons, Inc.
111 River Street
Hoboken, NJ 07030
USA

www.wiley.com

Library of Congress Control Number: 2020946721

British Library Cataloguing-in-Publication Data
A CIP record for this book is available from the British Library
ISBN 978-1-78945-010-1

ERC code:
PE1 Mathematics
PE1_13 Probability
PE1_14 Statistics
PE8 Products and Processes Engineering
PE8_3 Civil engineering, architecture, maritime/hydraulic engineering, geotechnics, waste treatment
PE8_7 Mechanical and manufacturing engineering (shaping, mounting, joining, separation)

MIX
Paper from responsible sources
FSC
www.fsc.org
FSC® C013604

Contents

Foreword

Maurice Lemaire
SIGMA Clermont, Clermont-Ferrand, and
Phimeca Engineering, Cournon, France

For centuries, mechanics has been a rational science motivated by astronomers observing the solar system with remarkable predictions of ephemeris, until Henri Poincaré brought out the chaos. For their part, the builders knew that their forecasts could be thwarted by dubious or unforeseen events, of which they were protected from by the so-called safety coefficients introduced by Claude-Henri Navier.

The 20th Century was marked by advances in materials and mechanical system behavior models, and by the considerable development of digital solutions. However, there will always be an uncertain gap (not an error) between the model and the physical reality.

If Andrey Kolmogorov made the theory of probabilities quasi-perfect, the statistical model necessary for its implementation remains, and will remain, uncertain in mechanics, even if the big data of the 21st Century will bring much to the exploration of the Mediocristan (the life around the median), but little to that of the Extremistan (the life at borders), according to the mapping of Nassim-Nicholas Taleb.

Immersed in the uncertainties of its models and its data, the domain of mechanics was initially based on expertise and feedback to justify its new works. Only recently did it realize that probability theory could provide an additional precision, by eliciting the content of its rules and codes. Engineers gradually accept to go from quantities of interest justification by a deterministic calculation, to their probability distributions acceptance stressing the failure.

This paradigm shift is cultural, scientific and technological all at once. It is cultural because it involves a new way of thinking that both scientific and educational communities were unprepared for. Mechanics of materials and structures, statistics (often limited to process control) and probabilities fall under disciplinary fields independent from one another.

It is scientific because it involves new algorithm development, recognizing the uncertain nature of models and parameters by the probability theory or other approaches. It is technological because it involves a new approach to rules, now based on probabilistic content, like the EN Eurocodes did. Finally, it involves thought on the decision: to entrust it to the probabilities in Mediocristan and to the consequences of the failures in Extremistan.

Some visionaries became aware of the importance of variability from the beginning of the 20th Century. A reliability index concept introduced in 1949, by Rzhanitzyn in steel construction, was quickly popularized because of the need to provide answers to new situations, for example in off-shore platforms engineering, and by the development of computer tools, which is now motivating methodological works to be implemented numerically.

Since the early 1980s, several scientific communities in the world have gradually seized the subject, bringing more arguments to the probabilistic approach and giving it a scientific consistency and operational tools exiting academic applications. Allow me to have the pleasure of pointing out that the Clermont-Ferrand team has done its part.

Christian Gogu joined talented authors from university centers and research organizations to contribute to the Mechanics and Uncertainty French School, and I want to congratulate them.

In this book, the reader will find a clarification of the concepts of epistemic and random, because the first requirement is to ensure that the vocabulary is not random. This book's contents provide details about historical methods, as well as the introduction of methods that overcome their shortcomings. It addresses the issue of sensitivity factors, essential for both reliability and optimization. It sheds light on the fundamental uncertainty of the statistical model, which should be considered as fractal. It takes into account the uncertainty of the models and not only that of the parameters. It gives meaning to the notion of optimization with uncertain variables.

However, it leaves open questions on the curse of dimensionality and on the time-variant problem; it will be the topic of the next edition in the coming years. It illustrates the approaches by various technological fields in mechanical engineering for aeronautical and space engineering, and in civil engineering for nuclear engineering. In this book the reader will find both matter to initiation, as well as an introduction to advanced tools.

I hope that this book will be a motivation for researchers, and especially young researchers, to find a scientific field in which to express their skills, where it is necessary to innovate for the scientific and technological progress of humanity.

November 2020

Preface

Christian GOGU
Clément Ader Institute, Paul Sabatier University, Toulouse, France

This preface is intended to thank all those who contributed to this book and to establish the context of each contribution.

First of all, I would like to thank Sylvain Drapier, professor at the École des Mines de Saint-Étienne, whom I had as a professor during my studies and whom I subsequently worked with at the Tribology and Systems Dynamics Laboratory, where I completed my PhD thesis concerning the identification of mechanical properties under uncertainty. In charge of the topic "Numerical modeling in mechanics" for the new *Sciences* encyclopedia launched by ISTE Editions, Sylvain Drapier invited me to be the editor of this work and I thank him for the trust he has placed in me.

I would then like to give my warm thanks to Maurice Lemaire, Professor Emeritus at IFMA/SIGMA Clermont and Scientific Advisor to Phimeca, for agreeing to write the foreword of this book. I met Maurice a long time ago, and came to know him personally first and then professionally. He has pioneered the research on considering uncertainties in mechanical engineering and it is an honor to have his contribution to this book.

Thereafter I would like to thank all the authors of the different chapters in this book, while establishing the context of each of their contributions in order to highlight the structuring and articulation of the book.

Without being exhaustive, this book aims to present a number of fundamental concepts and classical approaches, as well as several recent developments and advanced methods. It is divided into two parts, each with four chapters. The first part is dedicated to the modeling, quantification and propagation of uncertainties. The second part is dedicated to taking uncertainties into account through reliability analyses and optimization under uncertainty.

Chapter 1, which I have written, aims to review a number of conventional uncertainty modeling methods. It also seeks to introduce a number of notations and terminologies. A distinction is specifically made between random and epistemic uncertainties, and the different modeling methods associated with each are presented, illustrated and discussed.

Chapter 2, written by François Willot, illustrates the use of probabilistic tools to characterize environments exhibiting a microheterogeneous structure, a typical example of which is concrete. Such a characterization of "random" microstructures makes it possible to later return to a statistical characterization of macroscopic properties, necessary to characterize the reliability or robustness of the structures.

Chapter 3, written by David Bouhjiti, Julien Baroth and Frédéric Dufour, illustrates methods of uncertainty propagation on large civil engineering structures, in this case containment facilities for nuclear power plants. It discusses the essential notions of variability of material properties and their characterization by random fields. The stochastic finite-element technique, as well as several improvements, is then introduced in order to achieve the propagation of random fields to the quantities of interest.

Chapter 4, co-authored with Sylvain Dubreuil, Nathalie Bartoli and Thierry Lefebvre, presents a new method for characterizing and then reducing uncertainties, due to the use of substitution models based on Gaussian processes within multidisciplinary analysis. This method is illustrated with a classical example of multidisciplinary coupling, in this case, a fluid-structure coupling on an aeronautical wing.

Chapter 5, written by Jean-Marc Bourinet, reviews a large number of methods for calculating the probabilities of rare events, typically failure probabilities. The reader will find an introduction to the fundamental concepts and to these different methods for calculating failure probabilities.

Chapter 6, written by Cécile Mattrand, Pierre Beaurepaire and Nicolas Gayton, reviews a number of relatively recent developments associated with a certain type of method, known as the adaptive method, for calculating the probabilities of failure. These methods are based on the use of kriging surrogate models and their adaptive enrichment, in order to calculate failure probability with the most moderate numerical cost possible.

Chapter 7, written by Vincent Chabridon, Mathieu Balesdent, Guillaume Perrin, Jérôme Morio, Jean-Marc Bourinet and Nicolas Gayton, proposes a new method for the calculation of the (global) sensitivity of the probability of failure, taking into account the uncertainty regarding the parameters of the distributions, characterizing the variabilities of the problem. This method is illustrated by an example in the field of space launchers.

Chapter 8, written by Quentin Mercier and Fabrice Poirion, presents a new method for solving stochastic multiobjective optimization problems and, more particularly, those where the expectation of the different objectives is to be optimized. It is also shown how this formulation can be used to efficiently solve optimization problems under reliability constraints, by reformulating them as multiobjective optimization problems.

PART 1

Modeling, Propagation and Quantification of Uncertainties

1

Uncertainty Modeling

Christian GOGU
Clément Ader Institute, Paul Sabatier University, Toulouse, France

1.1. Introduction

Several decades ago, the scientific and engineering community began to recognize the value of considering uncertainties in the design, optimization and risk analysis of complex systems, such as aircraft, space vehicles or nuclear power plants (Wagner 2003; Lemaire 2014). These uncertainties can manifest themselves in many forms, originating from many different sources. In the field of mechanics, sources of uncertainty can typically be due to:

– sources of uncertainty in the manufactured system: tolerances, variability of material properties, manufacturing methods, or loads and boundary conditions over the lifecycle;

– sources of uncertainty in experimental data: test apparatus and protocols, inaccuracies in measurements, as well as experimental and environmental conditions;

– sources of uncertainty in modeling: modeling approaches and choices, modeling tools, numerical resources available and model calibration based on experimental data.

With respect to these different sources of uncertainty, a distinction is often made between aleatory and epistemic uncertainties (Vose 2008; National Research Council 2009), although this distinction is debatable, as will be discussed in more detail in section 1.2.

Mechanical Engineering under Uncertainties,
coordinated by Christian GOGU. © ISTE Ltd 2021.

Aleatory uncertainty is also referred to as irreducible uncertainty, stochastic uncertainty, inherent uncertainty or type I uncertainty. This uncertainty typically arises from environmental stochasticity, fluctuations in time, variations in space, heterogeneities and other intrinsic differences in a system. It is often referred to as irreducible uncertainty because it cannot be reduced further except by modification of the problem under consideration. On the other hand, it can be better characterized when it is empirically estimated. For example, the characterization of the variability of a material property can be improved by increasing the number of samples used, allowing a better estimate of statistical properties such as the mean and standard deviation.

An example of random uncertainty can be seen in an (unbiased) coin toss (that is, "heads or tails"). The intrinsic characteristics of the toss create uncertainty about the outcome of the toss, with a probability of 0.5 of obtaining tails. Assuming that tails is an undesirable outcome, it would then be desirable to reduce the probability of obtaining tails in order to reduce the uncertainty of the desired outcome (obtaining heads). However, without breaking the rules of the game, that is, without modifying the problem under consideration, it is not possible to reduce this uncertainty, hence the term "irreducible uncertainty". On the other hand, if this uncertainty is characterized empirically on the basis of several tosses, it is obvious that this uncertainty can be better characterized by increasing the number of tosses.

In terms of an engineering example, the amplitude of the gusts that an aircraft will be likely to encounter during its lifetime can be seen as a random or irreducible uncertainty. Indeed, when designing a new model of aircraft, it would be desirable for the engineer to reduce this uncertainty as much as possible in order to reduce the weight of the aircraft structure. Unfortunately, since this uncertainty is essentially related to environmental stochasticity, the engineer has no way to reduce it without changing the design being considered, therefore, the uncertainty is seen as irreducible. However, it could perhaps be reduced by changing the problem under study. For example, the engineer might want to consider installing sensors that allow the aircraft to detect the amplitude of turbulence in its trajectory in advance, thus allowing pilots to undertake avoidance maneuvers. Nevertheless, such a choice would have numerous and serious consequences. (Would such an aircraft be certifiable? Is it more economical to avoid turbulence rather than designing the aircraft to withstand it? Would passenger comfort be satisfactory? etc.) Usually, the problem is thus considered fixed and the irreducible or non-irreducible nature of the uncertainties is estimated on a given problem.

Epistemic uncertainty, also called reducible uncertainty or type II uncertainty, is the result of a lack of knowledge. This type of uncertainty is usually associated with uncertainty in measurements, a small number of experimental data, censorship or the unobservability of phenomena or scientific ignorance, which, in general terms, amounts to a lack of knowledge of one kind or another. It is often referred to as reducible uncertainty, as it can potentially be reduced by additional actions (for example, additional studies), leading to an improvement in knowledge. It should be noted that once epistemic uncertainty has been reduced, random uncertainty may remain, which could then become the predominant uncertainty and would therefore be irreducible.

An example of epistemic uncertainty would be that associated with the estimation of the age (in years) of an individual. Let us assume that we have just heard on the radio that a Nobel Prize winning scientist is going to give an acceptance speech in our town and we would like to know the age of the scientist. There is no mention of that on the radio. At this point, let us say that we can only estimate the age of this person between 30 and 90 years old. Therefore, there is a lot of uncertainty about their age at that point. However, as mentioned earlier, epistemic uncertainty can be reduced through improved knowledge. If we attend the acceptance speech, we will have the opportunity to see this person, which could allow us to reduce the uncertainty about their age to between, let us say, 50 and 60 years of age. If we even go to talk to them after the speech, we may even be able to get additional information that will allow us to further reduce the uncertainty about their age. In this case, the uncertainty could even be reduced to zero if we can find out their date of birth.

A typical example of epistemic uncertainty in engineering problems is measurement uncertainty. Similar to the scientist's age, the quantity that must be measured has a true value that is fixed (considering measurements at the macroscopic, not the quantum scale). Nevertheless, measurement instrumentation usually only allows this quantity to be determined with uncertainty. This uncertainty can be reduced by developing better instruments, using a better knowledge of the measurement phenomena involved, hence the term reducible uncertainty, although, in general, this does not mean that the uncertainty can be reduced to zero.

Probability theory has historically provided the first framework for modeling and quantifying uncertainties. Today, it is generally accepted that aleatory uncertainties can be adequately handled by probability theory. While the probabilistic approach can also be used to model epistemic uncertainties, other alternative representations have been proposed for such uncertainties, such as interval analysis, fuzzy set theory, possibility theory or evidence theory. These alternative approaches address, in particular, the need to quantify uncertainty when little or no data (either numerical or experimental) are available. Attempts to unify all of these approaches under a generalized theory of inaccurate probabilities have also been undertaken (Walley 2000; Klir 2004), without, however, leading to fully satisfactory approaches.

The purpose of this chapter is therefore to provide an overview of some of the different approaches used to represent and quantify uncertainties, both aleatory and epistemic. It is organized as follows. In section 1.2, a discussion about the need to distinguish between epistemic and aleatory uncertainty is presented. In section 1.3, an illustration of the probabilistic modeling approach is provided, including illustrations of some cases where its use may be problematic for representing epistemic uncertainty. In section 1.4, *p-box* theory is presented, which is an extension of probability theory that is designed to better address problematic cases in modeling epistemic uncertainties. In section 1.5, interval analysis is briefly discussed and in section 1.6 fuzzy set theory is addressed. In section 1.7, possibility theory is introduced, while evidence theory (or Dempster–Shafer theory) is presented in section 1.8. Some concluding remarks and discussions are provided in section 1.9.

1.2. The usefulness of separating epistemic uncertainty from aleatory uncertainty

The need to classify uncertainty into epistemic and aleatory has been the subject of much debate and discussion (Hoffman and Hammonds 1994; Apostolakis 1999; Der Kiureghian and Ditlevsen 2009; Lemaire 2014). It therefore seemed useful to create a section dedicated to the challenges and the usefulness of this distinction. This debate arises from the controversy between Niels Bohr and Albert Einstein concerning the nature of randomness observed at the quantum scale. For the former, it was a fundamental randomness, while for the latter, it was linked to an incomplete knowledge of phenomena at these scales. To illustrate his point of view, Einstein stated that "God does not play dice". Subsequent theoretical and experimental work by physicists seems to have proved Niels Bohr right: randomness observed at these scales is intrinsic to quantum nature. Now that the debate at the microscopic scale has been closed, one may raise the same question at the macroscopic level of mechanics, which is the subject of this book. Are there uncertainties at the macroscopic level of an intrinsically aleatory nature that can affect mechanical systems? Or are all uncertainties at the macroscopic level related only to a lack of knowledge?

It is the author's view that macroscopic mechanical systems can be affected by uncertainty of an inherently aleatory nature. This is notably related to two aspects:

free will and the chaotic nature of some systems. Free will refers to the ability of human beings to freely self-determine and thus to be responsible for their actions. An opposite school of thought, articulated by Pierre Simon Laplace through "Laplace's demon", would establish that perfect knowledge of the present and past would make it possible to perfectly predict the future, including with regard to human actions. In such a vision, all uncertainties would obviously be of an epistemic nature, and an improvement in knowledge would make it possible to reduce them. Since all our societies are based, particularly in terms of justice systems, on the responsibility of each individual for their actions, we shall admit here the existence of the faculty of free will. Free will thus leads to a degree of aleatory uncertainty, which cannot be reduced by improvement in knowledge.

An example from the mechanical field is now provided to illustrate the uncertainty associated with free will. Anyone who has seen a construction excavator at work has quickly realized that it is an illusion to try to predict the load that its articulated arms will be subjected to during its lifetime, as the conditions under which the machine is used can be extremely unexpected. The ingenuity (if we look on the positive side) of the operators of the machine means that they will always find new uses for the machine, which the engineer designing the machine would never have thought of. For example, using the bucket to push a truck that the excavator has just loaded and that skids up the slope to get out of the hole, and also using it for different kinds of demolitions. Consequently, it is not possible to consider the uncertainty about the loads being applied on the arms of the machine as an uncertainty of a purely epistemic nature, solely linked to lack of knowledge and therefore reducible by improving knowledge. As a matter of fact, this would mean that an improvement in knowledge would ultimately be able to predict the ways in which operators will decide to use the machine, which is obviously contrary to free will.

Another uncertainty of an intrinsically aleatory nature at the macroscopic level is that related to chaotic systems. Chaos theory has shown that such systems cannot be predicted beyond a more or less long-term horizon. A classic example of a chaotic system is the atmospheric system, which is reflected by the fact that the state of the system (namely what is commonly known as weather) cannot be predicted beyond a certain period of time; this is the famous butterfly effect. Note that it is the system itself that is chaotic and not just our modeling of the system. An improvement of our models will therefore not change anything in the system itself, which will remain chaotic. Because of their chaotic nature, and therefore unpredictability in the more or less long term, chaotic systems are a source of aleatory uncertainty, which cannot be reduced by improvement in knowledge.

In order to illustrate this with a mechanical case, let us again consider the case of the gusts that an aircraft will experience. When designing a new model of aircraft, it is useful for the engineer to know the amplitude of the gusts that the aircraft will experience during its lifetime in order to be able to size the aircraft structure as efficiently as possible. The amplitude of these gusts can then be considered as aleatory, irreducible uncertainty. Indeed, due to the chaotic system that is the atmosphere, it is not possible to predict air movements on a scale of a few meters over a time scale of several decades (that is, the lifetime of the aircraft). Moreover, it should be noted that the chaotic nature of the atmospheric system is further amplified by interaction with human free will. Uncertainty about human response to climate change makes it difficult to predict the state of the atmospheric system on the time and space scales under consideration. The engineer therefore has no means of reducing this uncertainty about the amplitude of gusts by improving knowledge.

After having seen two sources of uncertainty that are inherently aleatory in nature at the macroscopic scale, one may ask whether other uncertainties can be classified as aleatory. While the previous two sources are intrinsically aleatory, the classification of other sources of uncertainty may depend on the situation. It should not be forgotten that the classification between epistemic and aleatory uncertainty introduced in the previous section is specific to the problem under consideration. When we considered the favorable outcome of the "coin toss" as aleatory uncertainty, the rules of the game are fixed (in particular, considering that the toss cannot be biased). If it is considered that biasing the toss is part of the rules of the game, then the uncertainty about the outcome is obviously reducible.

In mechanical problems, the engineer is responsible for articulating the problem, and, in particular, for setting "the rules of the game", namely what can and cannot be changed. The classification is then specific to the problem under consideration and an uncertainty on the same physical quantity can then be classified as being epistemic for certain problems and aleatory for others.

To illustrate this, let us take the example of the uncertainty on the limit stress at failure of a spar of an aircraft wing. Without further details on the problem under consideration, it is not possible to say whether this uncertainty should be classified as epistemic or aleatory. Indeed, in the first case, assume that the problem considered is that of designing a new wing, keeping the same raw material supplier and the existing quality control process, and that we are focusing on the uncertainty in the limit stress at failure of the spar of one of the future aircraft that will be manufactured. Since the raw material batches obtained from the supplier have some variability, which cannot be reduced by staying within the scope of the problem under consideration (that is, without changing supplier and the quality control process), the uncertainty about the limit stress at failure in question can be classified as aleatory, irreducible uncertainty. In the second case, assume that we are studying the limit stress at failure of the spar of a particular aircraft that has already been manufactured. The uncertainty about this limit stress at failure is of interest to ensure that this aircraft will be able to withstand a higher load associated with a particular mission being considered for this aircraft. In this case, the failure limit stress has a single "true" value, but which is not known. The uncertainty associated with the failure limit stress is this time classified as epistemic uncertainty. It can be reduced by improving knowledge, for example, here by testing a sample kept by the manufacturer from the batch of material used for the spar.

In light of this ambivalence, the interest of classifying uncertainties into epistemic and aleatory for a given problem can be questioned. From the author's perspective, the usefulness of this classification is most apparent from an engineering point of view in terms of distinguishing between reducible and irreducible uncertainties. This distinction is important because the solutions to be provided by the engineer to these two types of uncertainties are very different. For the former, the solutions essentially consist of seeking to reduce uncertainties, while for the latter, the solutions essentially consist of designing the systems in their presence (for example, by integrating partial safety factors, redundancies or by explicitly calculating the reliability of the system).

The interest of alternative approaches to modeling epistemic uncertainties is also questionable compared to the probabilistic approach typically used for aleatory uncertainties. Clearly, since the distinction between epistemic and aleatory uncertainty is based on the engineer's decisions regarding the problem to be considered, this, in itself, cannot justify different modeling methods for these two types of uncertainties. Taking the example of the limit stress at failure of a spar, illustrating the dependence of the classification on the problem under consideration, we can elaborate by noting that in the first case probabilistic modeling is typically adopted to model the variability of the stress at failure. It is obvious that when we shift to case 2, there is no reason not to maintain the probabilistic approach to model the epistemic uncertainty before trying to reduce it by way of testing (using Bayesian approaches, for example). There are thus many cases where probabilistic modeling is well suited to modeling epistemic uncertainties.

On the other hand, there are also situations where probabilistic modeling may not be the most appropriate way to model uncertainties. These are typically cases where tests or simulations on the uncertain quantity of interest are difficult, if not impossible, to obtain. In such cases, the available data are not sufficient to construct a suitable probabilistic model. Alternative approaches that are less data-intensive and based on expert opinions have been proposed and will be reviewed in the following sections. Finally, it should be noted that, in practice, for the same quantity, there is often an epistemic (reducible) and an aleatory (irreducible) part of uncertainty, making the analysis even more complex.

1.3. Probability theory

Probability theory is the most widely used approach for the representation of uncertainties. Its theoretical framework is recalled here from an angle that makes it easy to compare it subsequently with alternative uncertainty representation approaches. We shall also present some situations illustrating the limits of probability theory for representing some types of epistemic uncertainties.

1.3.1. *Theoretical context*

Formally, probabilities are defined on a probability space *EP* defined as follows.

DEFINITION 1.1.– *Let* Ω *be a set. A* σ*-algebra or* σ*-field on* Ω *refers to a set T of subsets of* Ω *verifying:*

– *T not empty*: $T \neq \varnothing$;

– *T stable by complementary*: $\forall A \in T, A^C \in T$;

– *T stable by countable union*: $\forall n \in \mathbb{N}, A_n \in T \Rightarrow \bigcup_{n \in \mathbb{N}} A_n \in T$.

DEFINITION 1.2.– *Let* Ω *be a set and T a* σ*-algebra of* Ω*. A probability measure on* (Ω*, T) is a function* $\mathbb{P}: T \to \mathbb{R}$ *satisfying the following three axioms, called Kolmogorov axioms:*

– *the function has values between 0 and 1*: $\forall A \in T, 0 \leq \mathbb{P}(A) \leq 1$;

– *the image of all the events in the universe is 1*: $\mathbb{P}(\Omega) = 1$;

– *the function is* σ*-additive*: $\forall A_n \in T,\ \mathbb{P}\left(\bigcup_n A_n\right) = \sum_n \mathbb{P}(A_n)$.

DEFINITION 1.3.– *Let* Ω *be a set, T a* σ*-algebra of* Ω *and* $\mathbb{P}$ *a probability measure. The triplet (*Ω*, T,* $\mathbb{P}$*) is called a probability space EP.*

Concretely, Ω is called a universe, comprising the set of the possible events (or outcomes) under consideration. T is a subset of the universe, that is, a subset of possible events, having a σ-algebra structure. $\mathbb{P}$ is a probability measure that allows a certain probability to be attributed to any event in the universe.

PROPERTY 1.1.– Let (Ω, T, $\mathbb{P}$) denote a probability space. Then $\forall A \in T, \mathbb{P}(A) + \mathbb{P}(A^C) = 1$.

This property is trivial within the context of probability theory, but we shall see that in alternative theories of representation of uncertainties it appears differently.

DEFINITION 1.4.– *Let (Ω, T, $\mathbb{P}$) be a probability space. Any measurable function from Ω to $\mathbb{R}$ is called a real-valued random variable X. $x = X(\omega)$ is called a realization of the random variable X:*

$$\begin{aligned} X &: \Omega \to \mathbb{R} \\ &\omega \mapsto x \end{aligned} \quad [1.1]$$

DEFINITION 1.5.– *The probability distribution of the real-valued random variable X is the probability measure, denoted $\mathbb{P}_X$, resulting from the transport of the probability measure $\mathbb{P}$ on Ω into a probability measure $\mathbb{P}_X$ on $\mathbb{R}$.*

The probability distribution $\mathbb{P}_X$ is then a function that allows any subset of $\mathbb{R}$ to be associated with an associated probability. Note that there are different types of probability distributions, some examples of which are the binomial, geometric or Bernoulli distributions for discrete random variables and normal (or Gaussian), uniform or gamma distributions for continuous random variables. In the following, we will be interested, essentially, in continuous random variables, which are those most often used in engineering problems.

DEFINITION 1.6.– *The cumulative distribution function (CDF), sometimes referred to as the distribution function (DF), of the probability distribution $\mathbb{P}_X$ is the function F_X defined by:*

$$\begin{aligned} F_X &: \mathbb{R} \to [0,1] \\ &x \mapsto \mathbb{P}_X(]-\infty, x]) \end{aligned} \quad [1.2]$$

The CDF thus associates to any real value x the value of the probability that the random variable of distribution $\mathbb{P}_X$ is less than or equal to the value x.

DEFINITION 1.7.– *The complementary cumulative distribution function (CCDF), sometimes called the complementary cumulative probability function (CCPF), of the probability distribution* $\mathbb{P}_X$ *is the function* $\mathbb{F}_X^C$ *defined by:*

$$\begin{aligned} F_X^C &: \mathbb{R} \to [0,1] \\ x &\mapsto \mathbb{P}_X(]x,+\infty]) \end{aligned} \qquad [1.3]$$

The CCDF is introduced here because it is sometimes used in risk analysis, where it provides a means to know what is the probability that a quantity will exceed a threshold.

An example of CDF and CCDF is provided in Figure 1.1 for a Gaussian distribution of mean 4 and standard deviation 1.

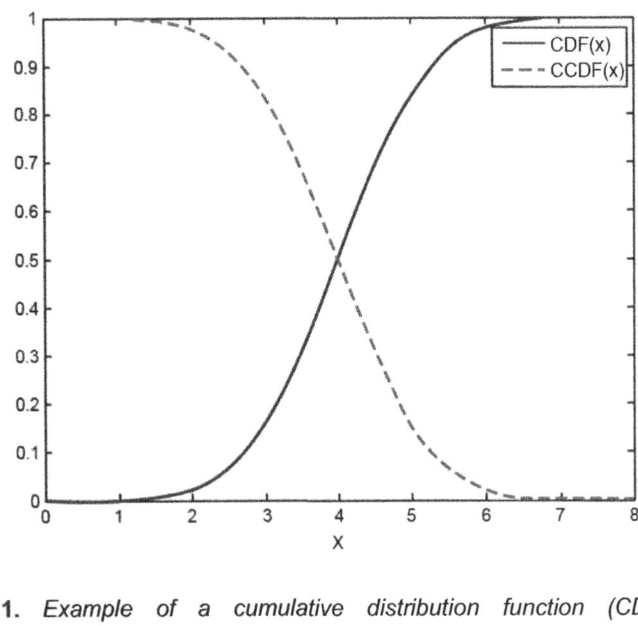

Figure 1.1. *Example of a cumulative distribution function (CDF) and a complementary cumulative distribution function (CCDF) corresponding to a Gaussian probability distribution of mean 4 and standard deviation 1. For a color version of this figure, see www.iste.co.uk/gogu/uncertainties.zip*

DEFINITION 1.8.– *For a random variable of absolutely continuous distribution (by misuse of language, this is referred to as continuous random variable), the probability density function (PDF) is defined by*

$$f_X(x) = \frac{dF_X(x)}{dx} \qquad [1.4]$$

DEFINITION 1.9.– *The expectation of a real-valued continuous random variable is given as*

$$\mathbb{E}[X] = \int_{\mathbb{R}} x f_X(x)dx \qquad [1.5]$$

The expectation of a random variable corresponds to the mean expected value or, by misuse of language, the mean value. It is often denoted as μ_X.

DEFINITION 1.10.– *The variance of a real-valued continuous random variable is given as*

$$Var[X] = \mathbb{E}\left[(X-\mu_X)^2\right] = \int_{\mathbb{R}} (x-\mu_X)^2 f_X(x)dx \qquad [1.6]$$

The variance corresponds to the dispersion of the values of the variable around its mean value. It is often denoted σ_X^2 and its square root σ_X is called the standard deviation. The coefficient of variation (CoV) of the random variable is then defined by $CoV = \sigma_X / |\mu_X|, \forall \mu_X \neq 0$.

DEFINITION 1.11.– *The k-order centered moment of a real-valued continuous random variable is written as*

$$\mathbb{E}\left[(X-\mu_X)^k\right] = \int_{\mathbb{R}} (x-\mu_X)^k f_X(x)dx \qquad [1.7]$$

Note that the second-order centered moment is the variance. The third-order centered moment is usually referred to as the coefficient of skewness, and characterizes the asymmetry of the probability distribution of the random variable. The fourth-order centered moment is usually referred to as kurtosis and characterizes the flattening of the probability distribution of the random variable.

1.3.2. *Probabilistic approach for modeling aleatory uncertainties*

Uncertainty of an aleatory nature can typically be modeled by a random variable with a certain probability distribution. In an engineering context, the probability distributions of the random variable quantities of interest can rarely be described in a fully analytical manner. In this case, these distributions are instead constructed from samples of the quantities of interest, which can be obtained either experimentally or by simulation. These samples can then be used to construct an empirical estimate of the distribution function of each random variable.

DEFINITION 1.12.– *Let* $(x_1, \dots, x_n)$ *denote n samples of a random variable of distribution function* F_X. *The empirical cumulative distribution function (ECDF) is then defined by*:

$$\widehat{F_X^n} : \mathbb{R} \to [0,1] \\ x \mapsto \frac{1}{n}\sum_{i=1}^{n} \mathbf{1}_{x_i \le x} \quad [1.8]$$

where $\mathbf{1}_A$ is the indicator function of event A. It can be shown that the ECDF is an unbiased estimator of the CDF. An illustration of the ECDF obtained from 100 samples from a reduced centered normal distribution is provided in Figure 1.2. The CDF of this normal distribution is also shown.

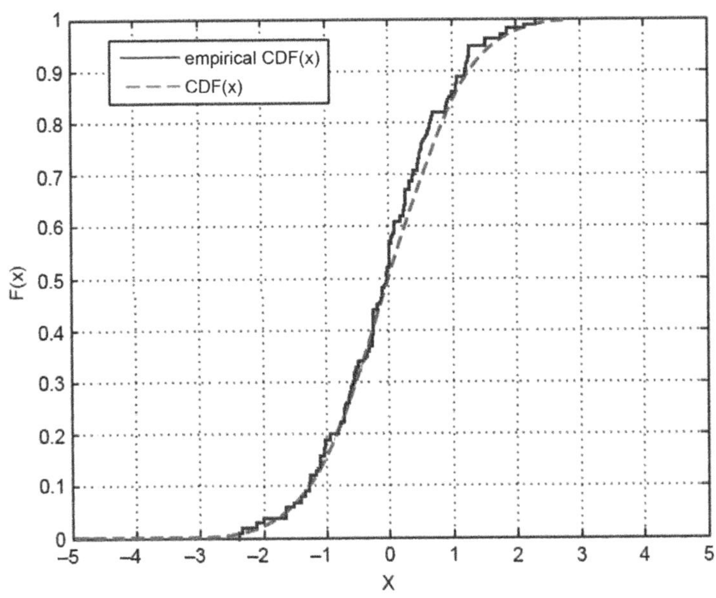

Figure 1.2. *Example of an empirical cumulative distribution function (ECDF, in blue) obtained from 100 samples of a reduced centered normal distribution. The true cumulative distribution function (CDF) is given in dotted magenta. For a color version of this figure, see www.iste.co.uk/gogu/uncertainties.zip*

From these same samples, empirical quantities such as the empirical mean or variance can also be calculated.

DEFINITION 1.13.– *Let* $(x_1, ..., x_n)$ *denote n samples of a random variable. The empirical mean of the samples is defined by*

$$\widehat{\mu_X} = \frac{1}{n}\sum_{i=1}^{n} x_i \qquad [1.9]$$

DEFINITION 1.14.– *Let* $(x_1, ..., x_n)$ *denote n samples of a random variable. The unbiased empirical variance of the samples is defined by*

$$\widehat{\sigma^2} = \frac{1}{n-1}\sum_{i=1}^{n}\left(x_i - \widehat{\mu_X}\right)^2 \qquad [1.10]$$

Hypothesis tests, such as the Kolmogorov–Smirnov test, make it possible, in addition, to verify whether a sample is derived from a certain distribution function. It should be noted that, when these tests do not allow us to determine with a sufficient level of confidence that the sampling follows a certain probability distribution, it is still possible to construct a continuous approximation of the distribution function using kernel estimation techniques (also known as Parzen–Rosenblatt methods) (Silverman 2018, see also section 7.5.1 of this book).

All of these techniques of empirical statistics work quite well when the number of samples is large enough. This is typically the case in the field of mechanics, when tests are carried out on elementary specimens in order to characterize the variability of a property (for example, failure stress and fracture toughness). In this case, up to 100 samples may be available, which is usually sufficient to characterize the variability in question.

There are, however, many cases where the number of samples available is not sufficient to properly characterize the probability distribution related to an aleatory uncertainty. In this case, this means that, in addition to the aleatory uncertainty, there is also an epistemic uncertainty related to the poor knowledge of the probability distribution of the random variable from which the samples are derived. Several options are then available, depending on the situation:

– if it is easy to obtain more samples, the best way to reduce this epistemic uncertainty is to increase the number of samples. This is typically the case for elementary tests on specimens where the number of tests is increased progressively until a satisfactory estimate of the probability distribution is obtained;

– if more samples cannot easily be obtained experimentally, it may be worth considering whether it would be easier to obtain them by simulation. This is typically the case when moving up the test pyramid of a complex system (for example, an aircraft). While it is relatively easy to perform hundreds of tests on elementary specimens at the base of the pyramid, it is not feasible to perform failure tests on a hundred complete aircraft, that is, at the top of the pyramid. Typically, only one fracture test is performed on a complete aircraft, which is obviously totally insufficient to determine any probability distribution thereupon. On the other hand, numerical simulation can then be considered in order to obtain samples in a numerical way. This typically amounts to propagating characterized uncertainties at a lower level of the test pyramid through a numerical model representing the complete system. For this, consider the function g modeling the response of the system $y = g(x_1, \ldots, x_k)$ (namely, the quantity of interest on the complete system) as a function of a number of input variables (namely, the parameters whose uncertainties have been characterized at a lower level). Associating random variables $X_1, \ldots, X_k$ with the parameters $x_1, \ldots, x_k$, the problem of uncertainty propagation amounts to calculating the probability distribution of the random variable Y associated with the response y. When the function g is complex, such as in the case of numerical models (for example, finite elements models) for complex systems, it is not usually possible to characterize the probability distribution of Y in an analytical way. Several methods exist to propagate the uncertainty by sampling, the best-known of which is the Monte Carlo simulation (MCS). It consists of performing N random draws on each of the random variables $X_1, \ldots, X_k$ and calculating the value of the response y for each draw. N samples of Y are thus obtained, which makes it possible to characterize the probability distribution of Y using the methods seen previously. For this approach to be interesting, the cost of obtaining N samples of Y by simulation must obviously be much lower than the cost of obtaining the same number of samples experimentally;

– if it is not easy to obtain more samples, neither numerically nor experimentally, this implies that a large epistemic uncertainty will have to be accepted. The question then arises as to the most appropriate approach to model these epistemic uncertainties.

1.3.3. *Probabilistic approach for modeling epistemic uncertainties*

There are many situations where epistemic uncertainties can be represented strictly within the same probabilistic framework as described in the previous sections.

To illustrate such situations, let us take an example addressed earlier. Consider the limit stress at failure of a wing spar. Let us assume that the variability characterization provided by the raw material supplier allows us to model the variability on the limit stress at failure of the spars of all the aircraft that will be produced by a given probability distribution $\mathbb{P}_X$. Now we assume that we are interested in a specific aircraft among those that have already been produced. The failure limit stress of the spar of this aircraft has a "true" value that is fixed, but which we do not know. The associated uncertainty is therefore epistemic in nature. How can we model this epistemic uncertainty?

The spar in question represents one realization among all the realizations (that is, all the spars that will be produced); the failure limit stress of this spar thus represents a realization of the random variable X characterized by the distribution $\mathbb{P}_X$. It is then absolutely reasonable to consider at first that the epistemic uncertainty that we want to represent can be modeled by a random variable of the same distribution $\mathbb{P}_X$. Since this uncertainty is epistemic, it is however possible to reduce it, for example, by carrying out characterization tests, which could provide a new probability distribution for this epistemic uncertainty, a distribution which will typically be "narrower" than $\mathbb{P}_X$. In any case, the epistemic uncertainty on the failure limit stress of a specific aircraft spar is modeled here all along within a probabilistic framework, a framework that seems to be well suited to this problem.

On the other hand, there are other situations where retaining this probabilistic framework to represent epistemic uncertainties may be more problematic. These are typically cases where the amount of data is limited. This happens quite frequently, since having large sample sizes (experimental or by simulation) is often expensive or simply not feasible in practice. What can be said, for example, about the inherent variability of systems that have been manufactured only a few times, such as the space shuttle, which has been manufactured only five times?

In such cases, the key questions to ask before adopting a probabilistic modeling for these epistemic uncertainties are as follows:

– do we have sufficient information to consider a particular type of probability distribution (normal, uniform, etc.) for this uncertainty?

– if a certain type of probability distribution is assumed even in the absence of sufficient information to justify it, will it always lead to reasonable results for the analysis?

In the following, we present three examples of situations where the probabilistic modeling of epistemic uncertainties can be problematic or lead to counterintuitive results.

Problems can arise when epistemic uncertainties are represented by uniform probability distributions, which are supposed to account for a lack of knowledge about what happens between the upper and lower bounds of the uniform distribution. In the event, an expert can only provide two bounds to characterize the uncertainty on a quantity of interest and the principle of indifference could lead us to believe that modeling this uncertainty using a uniform distribution between these two bounds is a reasonable choice.

However, this can be problematic, as a uniform distribution has relatively strong specific implications in probability theory. In particular, making the assumption of a uniform distribution between two bounds is a much stronger assumption than the assumption that the quantity can be indifferently (that is, without preferential concentration) located anywhere between the two bounds. As a matter of fact, a uniform distribution implies a particular uncertainty structure between the two bounds, which is not without consequence. To illustrate this, consider the following example.

Let us consider a bar with a square cross-section subjected to a specific tensile force. In order to calculate the stress in the bar, we need the cross-sectional area of the bar. However, the dimensions of the bar are not precisely known. They thus comprise an epistemic uncertainty. Assume that the only thing that is known is that the value of the side of the square is between a and b. How should we model these uncertainties? The principle of indifference could lead us to model the side X of the square using a uniform distribution between a and b. However, the same principle of indifference could lead us to model these epistemic uncertainties by considering that it is the area of section X^2 that is uniformly distributed between a^2 and b^2, since ultimately it is the section that we need for the calculation of the stress. Which of these two models should we choose?

The two choices might seem indifferent, but unfortunately, they are not. Indeed, by modeling uncertainty using a uniform distribution on X, when it is propagated to X^2, it will not lead to a uniform distribution on X^2. This is illustrated in Figure 1.3 when X follows a uniform distribution between 0 and 1. The distribution of X^2 is then far from being a uniform distribution, with low values of X^2 much more likely than high values. This dilemma illustrates one of the difficulties that can arise when trying to model certain knowledge gaps by imposing *a priori* choices of probability distributions.

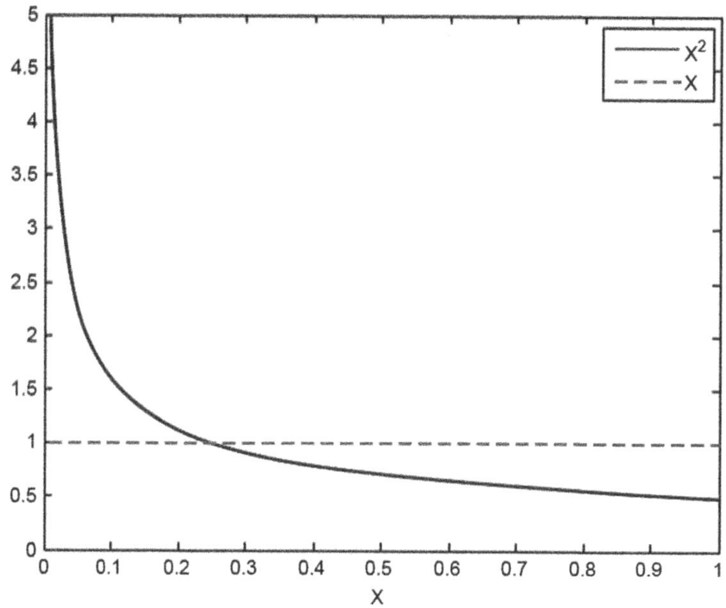

Figure 1.3. *Distribution function of a uniform distribution on X between [0,1] (magenta line) and the corresponding distribution function of X^2 (blue line). For a color version of this figure, see www.iste.co.uk/gogu/uncertainties.zip*

A second example illustrating counterintuitive results that could emerge from inadequate modeling of epistemic uncertainties is based on Nikolaidis *et al.* (2004). Let us consider a structure subjected to vibrational stresses. Due to the resonance phenomenon, some excitation frequencies must be avoided in order to prevent failure of the structure. If we impose that the structure's response (for example, its maximum deflection) must be below a threshold, this delimits a zone of failure, illustrated in Figure 1.4 for a generic frequency response function.

Let us now consider that there is a large epistemic uncertainty on the values of the frequencies at which the structure in service will be excited and that only the bounds a and b on these frequencies are known. In the absence of additional information, the principle of indifference is applied and a uniform distribution in $[a,b]$ is initially considered for the excitation frequencies (see Figure 1.4).

This defines the probability of failure by the area below the distribution within the failure zone. Now let us suppose that the engineer is looking to be more conservative and considers the uncertainty about the excitation frequencies to be higher. Therefore, they consider a wider uniform distribution in $\left[\tilde{a},\tilde{b}\right]$ (see Figure 1.4) to account for the higher epistemic uncertainty. This results in a decrease in the area under the distribution that lies within the failure zone and thus decreases the probability of failure. This example thus illustrates a situation where increasing epistemic uncertainty (that is, increasing the lack of knowledge) decreases the probability of failure. This is somewhat counterintuitive, as one would normally expect that high epistemic uncertainty (that is, a significant lack of knowledge) would lead to unreliable products (namely, having a high probability of failure) and that products can be made more reliable by increasing knowledge (namely, decreasing epistemic uncertainty).

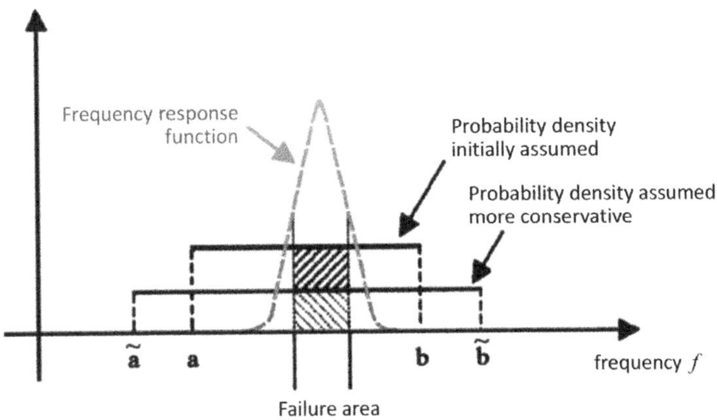

Figure 1.4. *Illustration with a system presenting a resonance of a case where modeling epistemic uncertainties using a uniform probability distribution leads to counterintuitive results. For a color version of this figure, see www.iste.co.uk/gogu/uncertainties.zip*

We can also deduce, from the previous case study, a third example of a counterintuitive situation. Let us consider that the true variability on the excitation frequencies corresponds to a uniform distribution in $[a,b]$. This uncertainty is not exactly known and by attempting to be conservative we assume that the uncertainty is quite large, modeling it by a uniform distribution in $\left[\tilde{a},\tilde{b}\right]$. In the design phase, the engineer can typically look for ways to modify the structure in order to reduce the probability of failure. Suppose the engineer makes a modification for this purpose (for example, a change in dimensions, in materials), which results in moving the failure area somewhere in the interval $\left[\tilde{a},a\right]$. If we consider the real probability distribution over the excitation frequencies f (uniform in $[a,b]$), this change in the design has the consequence of drastically reducing the probability of failure, since the failure area is outside the interval $[a,b]$. However, since the real distribution is not known and a wider distribution (uniform in $\left[\tilde{a},\tilde{b}\right]$) has been considered to be conservative, the consequence of the change in design is totally imperceptible in this case. The probability of failure remains exactly the same as before the change in design, since the region of failure is still located within the bounds $\left[\tilde{a},\tilde{b}\right]$. This third example illustrates situations where the inappropriate modeling of epistemic uncertainties by probability distributions can prevent the proper prediction of tendencies in the variation of the probability of failure due to changes in design.

We have seen here, through three examples, some of the limitations of an entirely probabilistic modeling of some uncertainties of epistemic nature, especially in the presence of a very small amount of data, which does not allow a specific distribution to be associated with random variables. These limitations have motivated the development of additional approaches, which we shall review in the following sections. A first such approach simply consists of an extension of the probability theory: this is the probability box approach.

1.4. Probability box theory (p-boxes)

Probability box (or p-box) theory has been developed to deal with cases where the probability distribution $\mathbb{P}_X$ of a random variable X cannot be accurately determined, thus combining an aleatory component and an epistemic component of uncertainty. Within the framework of probability box theory, the distribution function F_X associated with the unknown distribution $\mathbb{P}_X$ can only be bounded (exactly or at a given level of confidence): the left bound is a distribution function that will be denoted $\overline{F}$ and the right bound is a distribution function denoted $\underline{F}$. This is illustrated in Figure 1.5.

Any distribution function comprised between $\overline{F}$ and $\underline{F}$ can thus represent the probability distribution of X. The spacing between $\overline{F}$ and $\underline{F}$ represents the magnitude of the epistemic uncertainty associated with the lack of knowledge of the distribution of X. If $\overline{F}$ and $\underline{F}$ are superimposed, there is no epistemic uncertainty and the aleatory uncertainty is represented by the distribution function $F_X = \overline{F} = \underline{F}$.

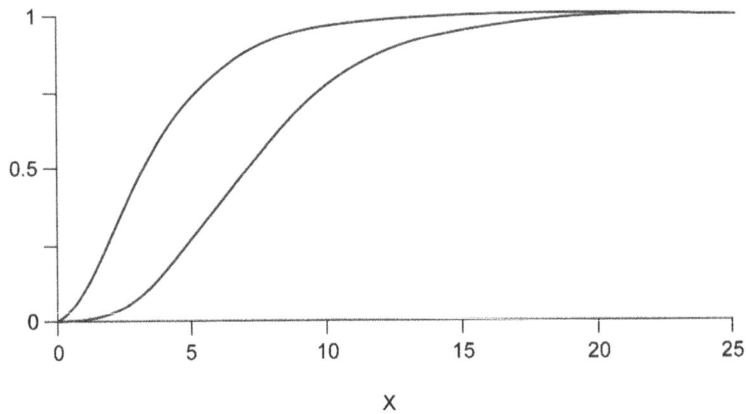

Figure 1.5. *Illustration of a probability box*

A probability box can be interpreted in two different ways: as bounds on the cumulative probability for a given value of x or as bounds on the values of x for a given confidence level. In the example in Figure 1.5, for example, one can read that the probability that x is less than 3 is comprised between 0.05 and 0.3. We can also read that the 95% quantile of X is between $x = 8$ and 17.

From a formal point of view, a probability box can be defined, adopting the formalism of Ferson *et al.* (2015), as follows:

DEFINITION 1.15.– *Let $\overline{F}$ and $\underline{F}$ be non-decreasing functions from $\mathbb{R}$ to [0,1] satisfying the condition $\underline{F}(x) \leq \overline{F}(x), \forall x \in \mathbb{R}$. Let $[\underline{F}, \overline{F}]$ denote the set of all non-decreasing functions F from $\mathbb{R}$ to [0,1] satisfying $\underline{F}(x) \leq F(x) \leq \overline{F}(x)$, $\forall x \in \mathbb{R}$. $[\underline{F}, \overline{F}]$ is then called a "probability box".*

Note that this approach can be seen as an extension of inaccurate probabilities (Walley 1991) to distribution functions and, consequently, to probability distributions. There are different variants existing derived from this concept of a probability box. For example, in addition to bounding the distribution function F with $\overline{F}$ and $\underline{F}$ one can impose additional constraints such as:

– the mean of the probability distributions associated with F must lie within a given interval;

– the variance of the probability distributions associated with F must lie within a given interval;

– the probability laws associated with F must belong to a certain class of distributions.

There are also different types of probability boxes:

– Distributional probability boxes: the probability distribution is assumed to be known and it is only the parameters of the distribution that are not accurately known, thus generating different possible distribution functions (but all of the same type of distribution). Such an approach will be illustrated in Chapter 7 of this book.

– Confidence bands: instead of seeing the bounds of the distribution function in an absolute way, this bounding can be seen at a given level of confidence. When an empirical distribution function is obtained from a sample, many developments (for example, the inequalities of (Dvoretzky *et al.* 1956)) have made it possible to bound the true distribution function from which the samples are derived at a given confidence level. These approaches, known as confidence bands, are obviously closely related to the notion of probability boxes.

– *C-boxes*: this approach can be seen as an extension of the confidence bands making it possible to work on structures modeling all levels of confidence at the same time.

Probability box-based approaches have been applied in many areas of engineering. In the context of mechanics, we can cite the application to the analysis of the buckling load of the Ariane 5 fairing (Oberguggenberger *et al.* 2009). Roy and Balch (2012) applied probability boxes for predicting the thrust of a supersonic nozzle, while Zhang *et al.* (2011) applied them to the analysis of a finite element lattice.

Probability boxes-based approaches are generally well suited for analyses where aleatory and epistemic uncertainties are simultaneously present. Its major disadvantage lies in the difficulty of obtaining bounding functions $\overline{F}$ and $\underline{F}$ in practice, especially in the presence of small amounts of data (a small number of samples or even no samples at all and only the availability of expert opinions, etc.). Indeed, when few samples are available, the bounding that can usually be obtained on the distribution function is too large to be useful in practice. Sometimes there is even a total absence of samples (that is, there have been no formalized experiments that can be traced and exploited) and only expert opinions can provide information on existing uncertainties. How can the bounds $\overline{F}$ and $\underline{F}$ be constructed based on these expert opinions? Various approaches, which we shall review later, and, in particular, the Dempster–Shafer theory, which is conceptually quite close to probability boxes, have sought to overcome these difficulties.

1.5. Interval analysis

As we have seen in previous sections, one of the major problems in modeling epistemic uncertainties resides in the determination of the probability distribution (or its bounding) that a certain physical quantity follows. However, in a large majority of cases, the uncertainty can at least be bounded. A range with very wide bounds can often be obtained by considering constraints of physical nature on the quantities of interest. For example, physical considerations make it possible to limit the Poisson ratio of a homogeneous isotropic material between −1 and 0.5. Even if relatively narrow, this range is not very informative. The opinion of an expert, on the other hand, could allow this range to be further significantly reduced. In our example, an expert could, for example, state that, for the material that we are considering, the Poisson ratio will be between 0.2 and 0.4. Note that, based on this information, we cannot characterize how the uncertainty varies within this range (are the boundary values as plausible as the center values?). Nonetheless, there are situations where a range is the only information available.

Interval analysis can be used to model such cases where the uncertainty is provided by bounds on the quantity of interest. The uncertainty on the quantities $x_1, \ldots, x_n$ is then described by their lower bound $x_1^{lb}, \ldots, x_n^{lb}$ and their upper bound $x_1^{ub}, \ldots, x_n^{ub}$. A lot of research work has then focused on the propagation of uncertainties from $x_1, \ldots, x_n$ to a quantity $y = g(x_1, \ldots, x_n)$, where g is any function modeling the relation between the input and output variables. The problem of determining the lower and upper bound on y is equivalent to solving a global optimization problem. Consequently, many global optimization algorithms can be applied to solve this problem (Hansen and Walster 2003). Note that in cases where the function g is monotonic over the range of variation of the input variables (namely, on the range domain), then the lower and upper bounds on y can be determined very efficiently: due to the monotonicity, y has merely to be evaluated at the vertices of the hypercube constituted by the bounds of the input variables. The lower and upper bounds are then necessarily the minimum and maximum among these points, respectively. This method is often known as the vertex method. Other techniques use Taylor expansions to approximate the bounds on the output variable. Note that sampling techniques, such as Monte-Carlo simulation, can also be used to solve this problem, but this quickly becomes prohibitively time-consuming and methods using global optimization algorithms are usually more efficient. For a review of the different algorithms for efficient interval analysis, the reader may refer to Kreinovich and Xiang (2008).

A major disadvantage of interval approaches is the lack of a measure of uncertainty, similar to probability within the context of probability theory. In interval analysis, uncertainty is characterized by the boundaries of the interval only, but no information is available on the likelihood of different values within that interval. In probabilistic approaches, the likelihood of different values is characterized by the PDF, which defines the probability that the quantity of interest lies within a certain range (for example, an interval). Such a measure is not available in the interval approach.

Interval arithmetic and associated uncertainty propagation methods will not be discussed in more detail because of the lack of an uncertainty measure in interval analysis. This reduces the usefulness of the method for reliability- or robustness-based applications where quantification of the risk associated with various decisions is required. Nevertheless, the interval approach is still useful in situations where only the worst case needs to be considered.

1.6. Fuzzy set theory

While bounds on an interval may be sufficient to quantify epistemic uncertainty within the context of interval analysis, there are many situations where additional information is available, making it possible to refine uncertainty quantification.

Let us consider the case where experts believe that the Poisson ratio of a material is bounded by the interval [0.2;0.4] for a given material. It is very likely that, with their expertise, they will be able to provide a more refined analysis. For example, they can state that most of the time the Poisson ratio of the material will be between 0.3 and 0.35 and only in rare cases it will fall outside this narrower range. This information on the epistemic uncertainty provided by experts can be modeled within the context of fuzzy set theory by a trapezoidal membership function, as illustrated in Figure 1.6. The values in the interval [0.3;0.35] are the most likely values and their corresponding value in terms of the membership function is 1. The membership function then linearly decreases down to the bounds 0.2 and 0.4 provided by the experts. For any value α of the membership function, we can obtain an interval on the quantity of interest, called α-cut. In the example in Figure 1.6, we illustrate the interval for a level $\alpha = 0.6$, denoted 0.6-cut.

Triangular membership functions are also frequently used in fuzzy set theory. They allow a single value to be modeled as the most probable, as well as bounds outside which the quantity of interest cannot be located. More generally, any kind of membership functions can be used, but this is very rare, in practice, because of the difficulty of specifying them for concrete problems.

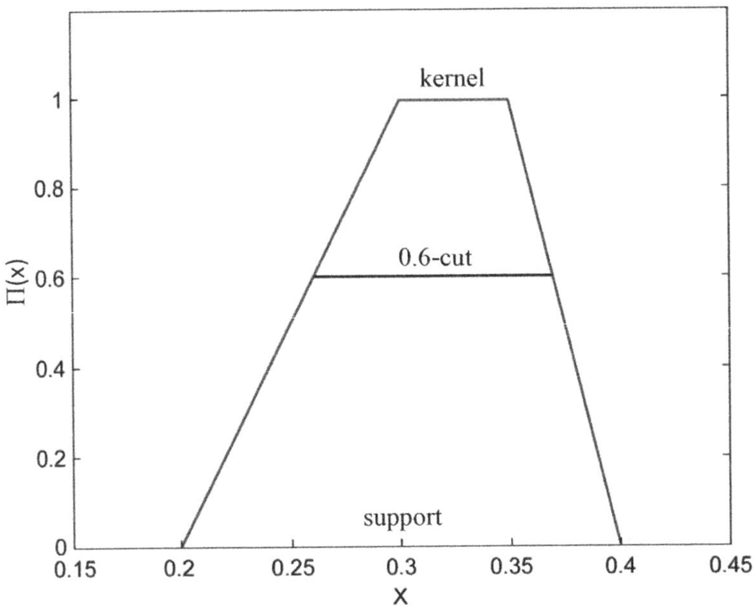

Figure 1.6. *Illustration of a trapezoidal membership function in fuzzy set theory*

As with previous approaches, a fundamental question concerns the propagation of uncertainties through a function g. As for any value α of the membership function, we can associate an interval; this is tantamount to ultimately performing interval analysis propagation at each level α. The methods discussed in the previous section apply in this way. Purely algebraic propagation is also possible for simple operations (additions, multiplications, powers). For a more detailed view of fuzzy set theory, the reader may refer to Zimmermann (2011).

One of the major drawbacks of fuzzy set theory lies, as for interval analysis, in the absence of a measure of uncertainty, which is equivalent to probability in probability theory. Uncertainty propagation is carried out at a level α, but the link with a measure allowing the quantification of the risk associated with this level α is still missing. In order to remedy this, a measure of uncertainty called possibility has been introduced. This has led to possibility theory, which will be presented in the next section.

1.7. Possibility theory

1.7.1. *Theoretical context*

Formally, possibility theory is defined on a possibility space *Epos*, defined as follows:

DEFINITION 1.16.– *Let* Ω *be a set and E the set of the subsets of* Ω*. A function* $\Pi : E \to \mathbb{R}$ *is called a possibility measure (or possibility distribution function [PoDF]) if it satisfies the following axioms:*

– *the function has values between 0 and 1:* $\forall A \in E, 0 \leq \Pi(A) \leq 1$;

– *the image of the empty set is 0:* $\Pi(\varnothing) = 0$;

– *the image of all the events in the universe is 1:* $\Pi(\Omega) = 1$;

– *the function is monotonic:* $\forall A, B \in E,$ *if* $A \subseteq B$ *then* $\Pi(A) \leq \Pi(B)$;

– *the function is subadditive:* $\forall A_n \in E,\ \Pi\left(\bigcup_n A_n\right) = \max_n (\Pi(A_n))$.

DEFINITION 1.17.– *Let* Ω *be a set, called universe, E the set of the subsets of* Ω *and* Π *a measure of possibility. The triplet (*Ω*, E,* Π*) is called the possibility space EPos.*

Note that the possibility measure (or PoDF) Π provides a measure of the likelihood of each element of *E*. The membership functions introduced for fuzzy sets (for example, triangular, trapezoidal functions) are typically used as PoDFs.

DEFINITION 1.18.– *Let* ξ *be an uncertain variable to which the possibility distribution* Π *is associated. Let* ξ_α^i *and* ξ_α^s *be the lower and upper bounds of the α-cut of* ξ*. Let f be a non-negative weighting function* $f : [0,1] \to \mathbb{R}$*, monotonically decreasing and verifying the condition of normalization in the integral sense. The f-weighted possibilistic mean operator is:*

$$\mathbb{E}_{\Pi}^{f}[\xi] = \int_0^1 \frac{\xi_\alpha^i + \xi_\alpha^s}{2} f(\alpha) d\alpha \qquad [1.11]$$

This operator is the equivalent of expectation in the context of probability theory. By introducing various weighting functions, different levels of importance can be assigned to the different α-cuts of the possibility distribution. There are also similar expressions existing for higher order moments in a possibility context.

In order to be able to compare the likelihood of different events, two quantities, the possibility and necessity of an event, are introduced.

DEFINITION 1.19.– *Let* (Ω, E, Π) *be a space of possibility. We call the possibility and necessity of an event* $e \in E$, *respectively:*

$$Pos(e) = Sup\{\Pi(x), x \in e\} \quad [1.12]$$

$$Nec(e) = 1 - Sup\{\Pi(x), x \in e^c\} = 1 - Pos(e^c) \quad [1.13]$$

PROPERTY 1.2.– Let (Ω, E, Π) be a space of possibility and an event $e \in E$. We have the following properties:

$$Nec(e) \leq Pos(e) \quad [1.14]$$

$$Nec(e) + Pos(e^C) = 1 \quad [1.15]$$

$$Pos(e) + Pos(e^C) \geq 1 \quad [1.16]$$

$$Pos(e) < 1 \Rightarrow Nec(e) = 0 \quad [1.17]$$

$$Nec(e) > 0 \Rightarrow Pos(e) = 1 \quad [1.18]$$

More intuitively, we can say that possibility measures the degree to which the facts do not contradict the assumption that an event can happen. If an event has a possibility of 1, it means that there is no reason to believe that the event cannot happen. It does not mean, however, that the event will certainly happen. In order for an event to be certain, both its possibility and its necessity must be equal to 1. On the other hand, if we consider that an event cannot happen, then it must be assigned a possibility of 0.

It should be noted that, similarly to what has been done in probability theory, a cumulative possibility function (CPoF), a cumulative necessity function (CNeF), a cumulative complementary possibility function (CCPoF) and a cumulative complementary necessity function (CCNeF) can be associated with a PoDF. Figure 1.7 illustrates these different functions with an example.

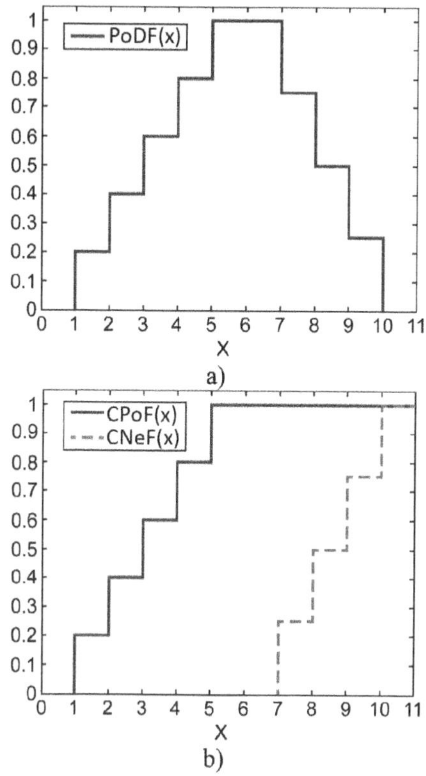

Figure 1.7. *Illustrations of (a) a possibility distribution function (PoDF); (b) the cumulative possibility function (CPoF) and the corresponding cumulative necessity function (CNeF)*

Note that for the example shown in Figure 1.7, the most likely values are those in the interval [5,7], which have a possibility of 1. The larger intervals encompassing this latter then have progressively smaller possibility values, until they reach 0 for the interval [1,10]. Values below 1 or above 10 are thus considered impossible. All the properties of equations [1.13]–[1.17] can be verified on these curves. Note also that the stepwise nature of the cumulative functions is typical of the modeling of epistemic uncertainties using alternative approaches. Indeed, the quantification of uncertainty is, most of the time, based on experts who would assign likelihood values to the different intervals, thus allowing the construction of the PoDF. The more finely the expert can elicit the uncertainty (by assigning likelihood values to many successively larger nested intervals), the smaller the steps will be. Finally, in terms of uncertainty propagation, given that the PoDFs are, most of the time, based on fuzzy set theory membership functions, they also make use of the same propagation techniques discussed in section 1.6. Propagation is thus typically done by interval propagation for different α-cuts.

1.7.2. *Comparison between probability theory and possibility theory*

This section aims to highlight the commonalities and differences between probability and possibility theories. First of all, in terms of axioms, the main difference lies in terms of σ-additivity for probability theory and subadditivity for possibility theory (see the definitions given in sections 1.3.1 and 1.7.1). Let us recall that the probability of the union of disjoint events is equal to the sum of the probabilities of the events. Thus, if $\{A_1, \ldots, A_n\}$ is a partition of the universe, then the sum of the probabilities of the A_i must be equal to 1. There is no similar constraint in terms of possibilities of events A_i. For example, the sum of the probabilities of the events "tomorrow it will snow" and "tomorrow it will not snow" must be equal to 1. On the other hand, if we assign a possibility of 0.7 to "it will snow tomorrow", we must assign a possibility of 1 to the event "it will not snow tomorrow". This is because the maximum possibility in the universe must be equal to 1. Thus, since the possibility of the universe Ω is the maximum possibility of the events in the universe, this implies that the possibility of at least one event in the partition must be equal to 1.

Both probability theory and possibility theory work with distribution functions, called PDF in the probabilistic framework and PoDF within the possibility framework. However, despite visual similarities, these two functions account for significantly different modeling. Figure 1.8 illustrates some of these differences.

We can see the following differences in Figure 1.8:

– the area under the probability density curve provides the probability of the corresponding event while the area under the curve of a PoDF has no significance;

– in the case of (absolutely) continuous random variables the probability of the variable taking a specific value is zero, while the possibility of the same event can be any value between 0 and 1;

– the area under the curve of a PDF must be 1 while its maximum can be any value. The converse is true for a PoDF;

– the interval between two points with a possibility α corresponds to an α-cut and represents the subset with a possibility at least equal to α. There is no equivalent meaning in probability theory.

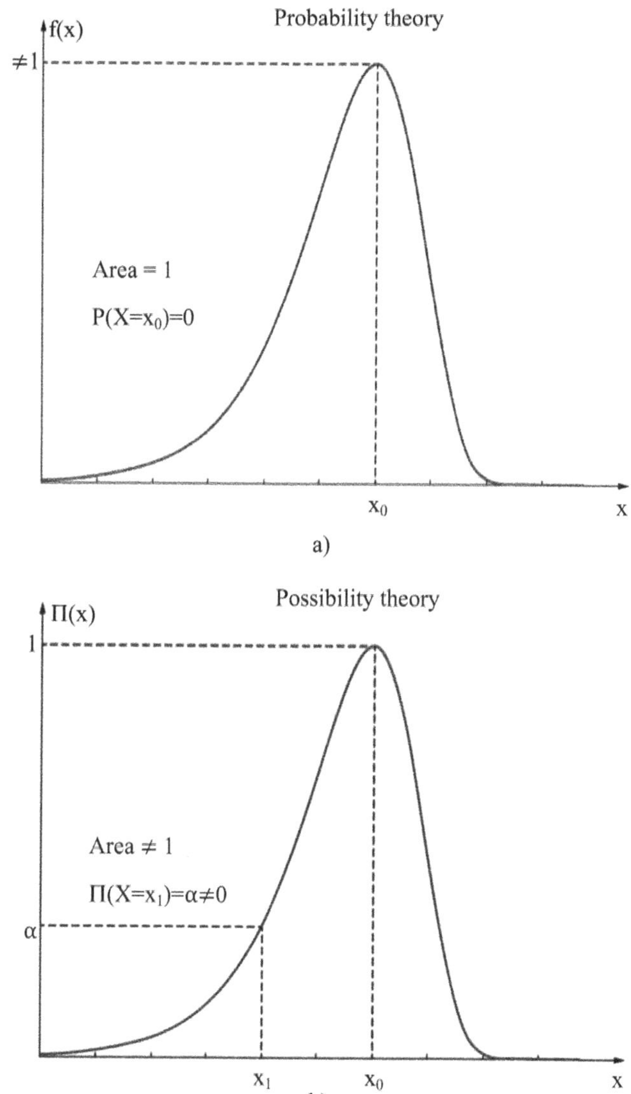

Figure 1.8. *Comparison between a probability density function f(x) and a possibility distribution function* $\Pi(x)$

In order to ensure consistency between variables modeled in a probabilistic manner and variables modeled in a possibilistic manner, the following condition is typically imposed: the possibility of an event must be greater than or equal to its probability. This is consistent with the usual meaning of the words probable and possible: when we say "it is probable that it will snow tomorrow", this shows a stronger conviction than when we say "it is possible that it will snow tomorrow". This condition of consistency implies that reasoning in terms of possibilities is more conservative than reasoning in terms of probabilities.

The differences in the probabilistic and possibilistic axioms, as well as the consistency requirement specified above, imply a very different interpretation of distribution functions (CDF or CPoF) in these two frameworks. Let us examine how the two theories model the information X in the interval [8,12]. If we want to model this in a purely probabilistic framework (and not in a context of probability boxes), we need to make an additional assumption, such as the principle of indifference between values in this interval, which leads to modeling by a uniform distribution between 8 and 12. The corresponding probability density is shown in Figure 1.9. The figure also represents a PoDF modeling the same information (X is in the interval [8,12]). Note that multiple possibility distributions could have been plotted. In particular, a uniform distribution with a possibility value of 1 might have been suitable. Such a distribution would have been very conservative, being the most uninformative distribution that could have been constructed. On the other hand, the triangular distribution plotted in Figure 1.9 is the most informative distribution that can be plotted and at the same time meets the consistency condition, which requires that the possibility of any event X be greater than or equal to its probability.

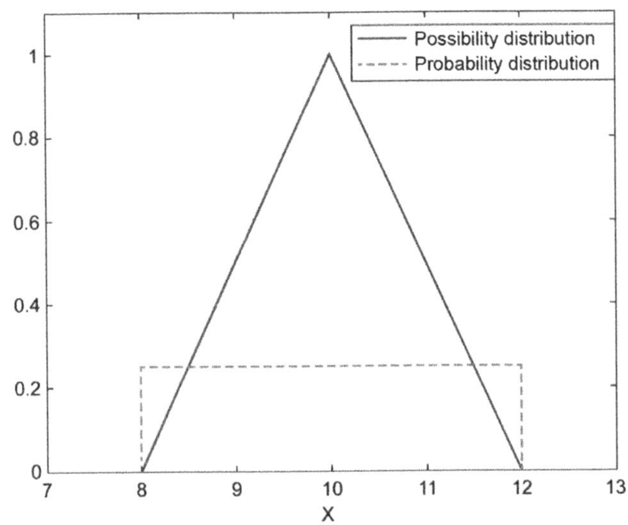

Figure 1.9. *Probability density function and possibility distribution function satisfying the consistency condition*

To conclude this comparison, it should be noted that possibility and necessity can be considered as upper and lower bounds of true probability in the presence of epistemic uncertainty. This implies that the cumulative functions of possibility (CPoF) and necessity (CNeF) will bound the cumulative probability distribution function (CDF).

This is illustrated in Figure 1.10 for the previous example of the variable *X* within the range [8,12]. The CPF of the uniform probability distribution is effectively bounded by the CPoF and the CNeF. This bounding models the presence of epistemic uncertainty. Indeed, the statement *X* in the interval [8,12] is very vague and many probability distributions could have been suitable. We had to introduce the principle of indifference, which may be debatable, to obtain the uniform distribution and its associated CDF. In the absence of this assumption, any PDF whose CDF would have been between the CPoF and CNeF curves could have been suitable. The CPoF and CNeF bounding is thus similar in spirit to that obtained in the context of probability boxes (*p-boxes*), but obtained within the context of possibility theory, based on axioms that are quite different from those of probability theory.

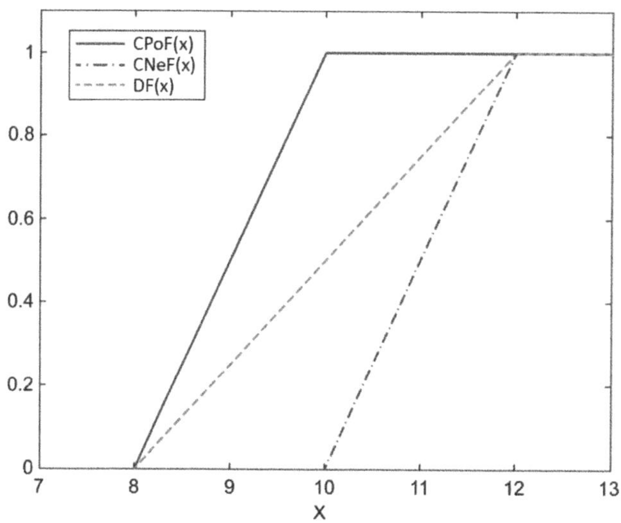

Figure 1.10. *Distribution function (DF), cumulative possibility function (CPoF) and cumulative necessity function (CNeF). For a color version of this figure, see www.iste.co.uk/gogu/uncertainties.zip*

1.7.3. *Rules for combining possibility distributions*

Possibility theory addresses the problem of quantifying uncertainties when solely based on expert opinion, which will assign likelihood levels to different values of the quantity of interest via possibility distributions $\Pi(x)$ (typically via triangular or trapezoidal distributions). One of the fundamental questions that then arises is how to deal with divergent expert opinions. For this purpose, different rules for combining possibility distributions have been established.

Let Π_1 and Π_2 be two possibility distributions. These distributions can be aggregated according to the following rules:

– the conjunctive mode: this is the equivalent of the intersection of events. It corresponds to retaining only the consensus (the common area under the two distributions). This consensus is typically renormalized in order to satisfy the possibilistic axioms;

– the disjunctive mode: this is the equivalent of the union of events. It corresponds to the union of the two distributions;

– the intermediate mode (proposed by Dubois and Prade 1992): as its name indicates, this is an intermediate mode between the two previous ones. By defining the consensus, that is, the upper possibility bound following the intersection of the two distributions, by h, the distribution of the intermediate mode is defined by:

$$\Pi(x) = Max\left(\frac{Min(\Pi_1(x), \Pi_1(x))}{h}, Min(Max(\Pi_1(x), \Pi_1(x)), 1 - h)\right) \quad [1.19]$$

These three modes of combination usually make it possible to combine distributions of possibility from different sources (experts, for example). If there is good agreement between the sources (for example, trapezoidal distributions with overlapping cores), then either the connective or disjunctive modes are both well suited. The choice between the two depends on whether one wishes to consider only consensus or whether one wishes to integrate divergent views as well.

In the absence of good agreement between sources (for example, trapezoidal distributions with non-overlapping kernels), the conjunctive and disjunctive modes are less well suited. The disjunctive mode introduces a multimodal distribution (two peaks), which is generally not desirable in terms of quantifying and propagating uncertainties. The conjunctive mode by itself is too limiting, as it only considers consensus, which can be very limited or even null. The intermediate mode is a proposed solution to resolve these two problems associated with the conjunctive and disjunctive modes. It should be noted that variations on the intermediate mode have also been proposed in order to give more confidence to one distribution (in other words, an expert) than to another. The reader may refer to Dubois and Prade (1992) for more information.

1.8. Evidence theory

The theory of belief functions (or evidence theory) is another theory for modeling uncertainties of an epistemic nature. It was developed by Dempster (1967) and Shafer (1976) and is consequently sometimes known as the Demspter–Shafer theory. This approach is similar in spirit to the probability box theory and possibility theory, in that it seeks to obtain a bounding for a CDF.

1.8.1. *Theoretical context*

Formally, evidence theory is defined on a belief space *EE* defined as follows:

DEFINITION 1.20.– *Let* Ω *be a set and E the set of the subsets of* Ω. *A function* $m: E \rightarrow \mathbb{R}$ *is called belief mass function if it satisfies the following axioms:*

– *the function has values between 0 and 1*: $\forall A \in E, 0 \leq m(A) \leq 1$;

– *the image of the empty set is 0*: $m(\varnothing) = 0$;

– *the image of all the events in the universe is 1*: $m(\Omega) = 1$;

– *the sum of the masses of the events of E is 1*: $\forall A_n \in E, \sum_n m(A_n) = 1$.

DEFINITION 1.21.– *Let* Ω *be a set, called universe, E the set of subsets of* Ω *and m a belief mass function. The triplet* (Ω, *E, m*) *is called the belief space EE.*

Note that the belief mass function *m* provides, as was the case for the possibility distribution, a measure of the relative likelihood of each element of *E*. In the terminology of the theory of belief functions, an element of *E* that has a non-zero mass is called a focal element, denoted Ω_i. Note that these focal elements are, most of the time, defined such as to form a set of disjoint elements. The quantity $m_i = m(\Omega_i)$ is then called mass (or sometimes basic probability assignment [BPA]) associated with this focal element.

In order to be able to compare the likelihood of different events, two quantities, plausibility and belief of the event, are introduced.

DEFINITION 1.22.– *Let (Ω, E, m) be a belief space. We call belief (noted Bel) and plausibility (denoted Pl) of an event $e \in E$, respectively:*

$$Bel(e) = \sum_{\Omega_i \subset e} m(\Omega_i) \quad [1.20]$$

$$Pl(e) = \sum_{\Omega_i \cap e \neq \varnothing} m(\Omega_i) \quad [1.21]$$

From a conceptual point of view, the mass of a focal element $m(\Omega_i)$ provides the true likelihood that is attributed to that focal element, but without specifying how this likelihood is distributed among the subsets of Ω_i. Therefore, the belief *Bel(e)* of an event *e* represents the minimal likelihood that can be associated with *e*, while the plausibility *Pl(e)* represents the maximum likelihood that can be associated with *e*.

In the vast majority of applications of evidence theory, focal elements are defined such as to form a set of adjacent polytopes (or even orthotopes). To illustrate this, consider a universe with nine focal elements Ω_1 to Ω_9 of mass m_1 to m_9 (see Figure 1.11). Let us now consider the belief and plausibility of the event *e*, shaded in the figure. According to the definition, the plausibility *Pl*(*e*) is obtained by adding the masses of the focal elements that have a non-zero intersection with *e* (in our case Ω_2, Ω_3, Ω_5, Ω_6, Ω_8 and Ω_9). This means that each focal point included even partially in *e* contributes to the plausibility of *e*. The belief *Bel*(*e*) is calculated by summing only the masses of the focal elements entirely included in *e* (in our case, Ω_3 and Ω_6). Belief and plausibility can thus be seen as a bounding for the likelihood of an event, reflecting the fact that the distribution of likelihood within a focal element is not known, due to epistemic uncertainty. By considering Ω_8 we can, for example, consider that the likelihood of the small gray triangle included in Ω_8 is very low and that the mass m_8 thus contributes almost nothing to the likelihood of *e*. This hypothesis comes closer to the calculation of belief. On the other hand, we can also assume that all the likelihood of Ω_8 is concentrated in this small gray triangle and all the rest of Ω_8 is not at all likely. In this case, m_8 must actually contribute in full to the likelihood of *e*. This assumption is thus close to the plausibility calculation. Both cases are thus extreme cases, bounding the true likelihood.

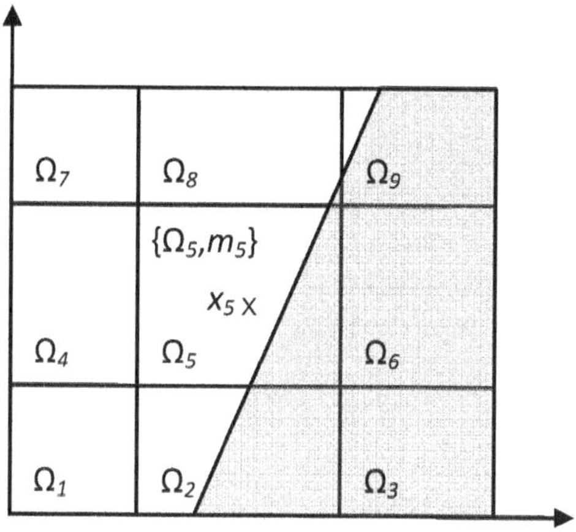

Figure 1.11. *Example of an uncertainty structure (focal elements Ω_i of mass m_i) in the theory of belief functions. An event e is grayed out here*

Note that, in a similar way to what has been done in probability and possibility theory, a cumulative plausibility function (CPlF), a cumulative belief function (CBF), a complementary cumulative plausibility function (CCPlF) and a complementary cumulative belief function (CCBF) can be associated with a BPA function. Figure 1.12 illustrates these different functions in an example where a BPA function assigns a mass of 0.1 to the following intervals: [1,3], [1,4], [1,10], [2,4], [2,6], [5,8], [5,10], [7,8], [7,10], [9,10].

Note that, while the curves in Figures 1.7 and 1.12 are similar in terms of their stepwise nature, there are some notable differences. The major difference is that in possibility theory, the necessity must be zero for the possibility to be different from 1 and vice versa. Such a constraint does not exist in evidence theory, which allows plausibility and belief to be simultaneously different from 0 and 1. This thus allows for a much narrower bounding of an unknown cumulative probability function than could be achieved in possibility theory.

In terms of uncertainty propagation, it is typically achieved for an orthorectangular uncertainty structure using interval propagation for each focal element, using the methods mentioned in section 1.5 for interval analysis.

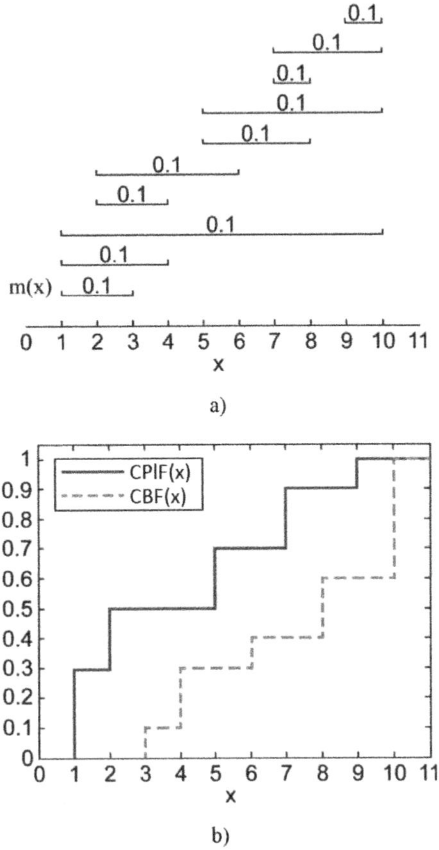

Figure 1.12. *Illustrations of (a) a BPA function m(x); (b) the corresponding cumulative plausibility function (CPIF) and cumulative belief function (CBF)*

1.8.2. *Rules for combining belief mass functions*

Evidence theory is based on the definition of a structure of uncertainties in the form of masses attributed to different focal elements. The definition of this uncertainty structure comes from experts. Analogously, as in the framework of possibility theory, rules have been defined in order to be able to aggregate different and sometimes divergent sources of uncertainty, for example, from different experts.

Let m_1 and m_2 denote two BPA functions with the respective sets C_i and C_j as focal elements. These functions can be aggregated according to the following rules:

– the Dempster rule, defined by:

$$m(A) = \frac{\sum_{C_i \cap C_j = A} m_1(C_i) m_2(C_j)}{1 - \sum_{C_i \cap C_j = \varnothing} m_1(C_i) m_2(C_j)}, \quad \forall A \neq \varnothing \quad [1.22]$$

The denominator reflects the conflict between the two sources *through* normalization. Note that therefore only the agreement between sources is retained in the aggregate uncertainty structure. A disadvantage of this rule is in the way it deals with divergent opinions and it is preferable to apply this rule only in cases where there is reasonable agreement between sources;

– the Yager rule, defined by:

$$q(A) = \sum_{C_i \cap C_j = A} m_1(C_i) m_2(C_j) \text{ for } A \neq \varnothing, \Omega \quad [1.23]$$

$$q(\varnothing) = \sum_{C_i \cap C_j = \varnothing} m_1(C_i) m_2(C_j) \quad [1.24]$$

$$m(A) = q(A) \text{ for } A \neq \varnothing, \Omega \quad [1.25]$$

$$m(\Omega) = q(\Omega) + q(\varnothing) \quad [1.26]$$

The Yager rule attributes the conflict to the entire universe **Ω**, thus increasing the degree of ignorance associated with it;

– the weighted mean rule:

$$m(A) = \frac{1}{2} \sum_{i=1}^{2} w_i m_i(A) \quad [1.27]$$

In this rule, the conflict is simply averaged with a weighting based on the confidence attributed to each expert. Other rules, as well as variations around the three presented, have been proposed and the reader can refer to Sentz and Ferson (2002) for more information.

1.9. Evaluation of epistemic uncertainty modeling

The quantification of epistemic uncertainties is a major challenge for achieving an integral framework for the quantification and propagation of uncertainties in industrial problems. Depending on the nature of the lack of knowledge, these uncertainties can be modeled in different ways. In some situations, epistemic uncertainties may be due to insufficient sampling, in which case the probabilistic framework, through the use of probability boxes (*p-boxes*), is probably the most relevant framework.

In other situations, no samples may be available and obtaining them would be too time consuming or costly, so only expert opinion can provide information on the uncertainties involved. Some consider that such advice cannot be objective and does not need to be mixed with rigorous, sample-based uncertainty quantification within a probabilistic framework. Others consider that, although imperfect, because it is subjective, expert opinions can contain useful information for uncertainty quantification and, rather than ignoring them, have focused on developing theories in order to be able to model the uncertainties expressed in this way. This is particularly the case with the theories of possibility and evidence that we have reviewed. Basically, both theories have the same objective, which is to bound an unknown probability distribution function by the construction of an uncertainty structure based on expert opinion. It is worth noting, however, that possibility theory is based on axioms which are quite different from those of probability theory, which can sometimes lead to counterintuitive results. These axioms also imply a relatively wide bounding for the unknown probability distribution function. Evidence theory, on the other hand, allows a narrower bounding for the distribution function, provided that it is possible to construct BPA functions with a very fine granularity. It also does not involve the more constraining axioms of possibility theory. To conclude, let us also note that in most real problems, epistemic and aleatory uncertainty modeling must be simultaneously considered. This combination of different approaches remains a challenge, especially from the point of view of the associated numerical costs.

1.10. References

Apostolakis, G. (1999). The distinction between aleatory and epistemic uncertainties is important: An example from the inclusion of aging effects into PSA. *Proceedings of PSA '99, International Topical Meeting on Probabilistic Safety Assessment*, 135–142.

Dempster, A.P. (1967). Upper and lower probabilities induced by a multivalued mapping. *Ann. Math. Stat.*, 38, 325–339.

Der Kiureghian, A. and Ditlevsen, O. (2009). Aleatory or epistemic? Does it matter? *Structural Safety*, 31(2), 105–112.

Dubois, D. and Prade, H. (1992). Combination of fuzzy information in the framework of possibility theory. In *Data Fusion in Robotics and Machine Intelligence*, Della Riccia, G., Lenz, H.J. and Fruse, R. (eds). Academic Press, New York.

Dvoretzky, A., Kiefer, J. and Wolfowitz, J. (1956). Asymptotic minimax character of the sample distribution function and of the classical multinomial estimator. *The Annals of Mathematical Statistics*, 27(3), 642–669.

Ferson, S., Joslyn, C.A., Helton, J.C., Oberkampf, W.L. and Sentz, K. (2004). Summary from the epistemic uncertainty workshop: Consensus amid diversity. *Reliability Engineering and System Safety*, 85, 355–369.

Ferson, S., Kreinovich, V., Grinzburg, L., Myers, D. and Sentz, K. (2015). Constructing probability boxes and Dempster–Shafer structures. Technical report, SAND-2015-4166J, Sandia National Lab, Albuquerque.

Hansen, E. and Walster, G.W. (2003). *Global Optimization Using Interval Analysis: Revised and Expanded.* CRC Press, Boca Raton.

Hoffman, F.O. and Hammonds, J.S. (1994). Propagation of uncertainty in risk assessments: The need to distinguish between uncertainty due to lack of knowledge and uncertainty due to variability. *Risk Analysis*, 14(5), 707–712.

Klir, G.J. (2004). Generalized information theory: Aims, results, and open problems. *Reliability Engineering and System Safety*, 85, 355–369.

Kreinovich, V. and Xiang, G. (2008). Fast algorithms for computing statistics under interval uncertainty: An overview. In *Interval/Probabilistic Uncertainty and Non-classical Logics*, Huynh, V.N., Nakamori, Y., Ono, H., Lawry, J., Kreinovich, V., Nguyen, H.T. (eds). Springer, Berlin.

Lemaire, M. (2014). *Mechanics and Uncertainty.* ISTE Ltd, London and John Wiley & Sons, New York.

National Research Council (2009). *Evaluation of Quantification of Margins and Uncertainties Methodology for Assessing and Certifying the Reliability of the Nuclear Stockpile.* The National Academies Press, Washington, D.C.

Nikolaidis, E., Chen, S., Cudney, H., Hatftka, R.T. and Rosca, R. (2004). Comparison of probability and possibility for design against catastrophic failure under uncertainty. *Journal of Mechanical Design*, 126(3), 386–394.

Oberguggenberger, M., King, J. and Schmelzer, B. (2009). Classical and imprecise probability methods for sensitivity analysis in engineering: A case study. *International Journal of Approximate Reasoning*, 50(4), 680–693.

Roy, C.J. and Balch, M.S. (2012). A holistic approach to uncertainty quantification with application to supersonic nozzle thrust. *International Journal for Uncertainty Quantification*, 2(4), 363–381.

Sentz, K. and Ferson, S. (2002). Combination of evidence in Dempster–Shafer theory. Technical report, SAND2002-0835, Sandia National Lab, Albuquerque.

Shafer, G. (1976). *A Mathematical Theory of Evidence*. Princeton University Press, Princeton.

Silverman, B.W. (2018). *Density Estimation for Statistics and Data Analysis*. Routledge, New York.

Vose, D. (2008). *Risk Analysis: A Quantitative Guide*. John Wiley & Sons, New York.

Wagner, R.L. (2003). Science, uncertainty and risk: The problem of complex phenomena. *APS News*, 1(12), 8.

Walley, P. (1991). *Statistical Reasoning with Imprecise Probabilities*. Chapman and Hall, London.

Walley, P. (2000). Towards a unified theory of imprecise probability. *International Journal of Approximate Reasoning*, 24, 125–148.

Zhang, H., Mullen, R.L. and Muhanna, R.L. (2011). Structural analysis with probability-boxes. *International Journal of Reliability and Safety*, 6(1–3), 110–129.

Zimmermann, H.J. (2011). *Fuzzy Set Theory and Its Applications*. Springer, The Hague.

2

Microstructure Modeling and Characterization

François WILLOT
Mines ParisTech, France

2.1. Introduction

The fine structure of metals, composites and other materials, when examined under a microscope, depends on the scale at which it is viewed. A typical example is concrete, which at smaller scales consists of mortar, aggregates and cement (Stroeven 2000). In most cases, such a microstructure appears as a random arrangement, in space, of "phases", namely of its different constituents at the scale under consideration. As noted by Matheron (1967), behind the multiplicity of shapes and sizes, and the variety of networks of interstices between aggregates and clusters of particles, these random arrangements also have a repetitive character (see Figure 2.1). Random set theory, whose modern developments originate in the works of Choquet (1954), Matheron (1965) and Kendall (1974), aims to quantify and simulate the morphology of heterogeneous media using probabilistic methods. Reconsidering the example of concrete, the forms of aggregates present in such materials are always different from one another. However, they present specific characteristics, such as their granulometry (Escoda *et al.* 2015). It is therefore natural to study whether it is possible, by means of a limited number of "morphological" descriptors, to accurately model a given microstructure, seen as the realization of a random set (Molchanov 2005), with the aim of predicting its mechanical properties, in particular by homogenization (Milton 2002).

Mechanical Engineering under Uncertainties,
coordinated by Christian GOGU.

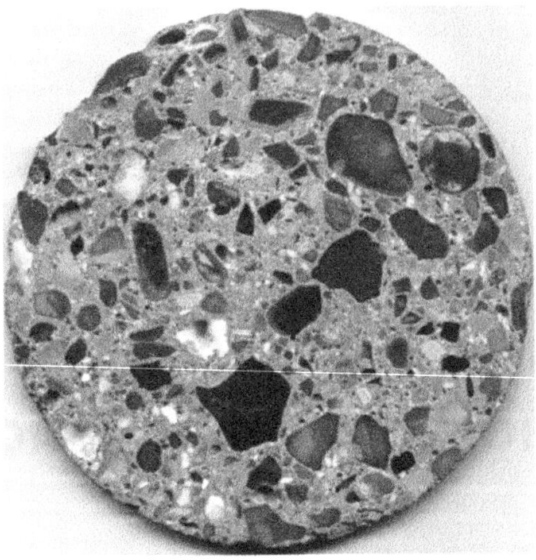

Figure 2.1. *"Biloba" concrete core, according to Escoda (2012) (source: EDF). For a color version of this figure, see www.iste.co.uk/gogu/uncertainties.zip*

The first question, addressed in this chapter, is the subject of integral geometry (or stochastic geometry) developed by Matheron (1975) and Kendall (1974), among others. Two main problems arise. First, the development of morphological criteria to quantify the shape of objects and their distribution in space. These criteria make use of image transformations, which are most often non-linear operators derived from mathematical morphology (Matheron and Serra 1982). These methods apply to random sets or functions that can be very general, for example tensor-valued functions (Angulo 2012), such as a strain or stress field. The second problem is the simulation of random microstructures, representative of real structures, based on morphological criteria. Probabilistic models of structures, some of which rely on analytical tools, provide a means to define and simulate the representative random realizations of real microstructures. Among numerous examples, the following works can be cited: Greco *et al.* (1979), Jeulin *et al.* (1995), Redenbach *et al.* (2011) and Bortolussi *et al.* (2018).

This chapter is organized as follows. Section 2.2 is dedicated to the main theoretical and numerical tools for characterizing microstructures from two- or three-dimensional images. Random point processes are introduced in section 2.3, Boolean models in section 2.4 and repulsion models in section 2.5. The main random partition models are introduced in section 2.6 and random structures from Gaussian fields are briefly discussed in section 2.7. The chapter is concluded by section 2.8.

2.2. Probabilistic characterization of microstructures

2.2.1. *Random sets*

A random set is a stochastic model whose realizations are sets of $\mathbb{R}^d$ (where d = 2 or 3 is the dimension). For reasons that will be explained later, we are interested in closed sets comprised in $\mathbb{R}^d$. The distribution of a random set is completely specified by a probability measure defined on a σ-algebra, that is, a space containing $\mathbb{R}^d$, the empty set, and which is stable by a countless number of unions and intersections. This algebra is used to define measures on sets of $\mathbb{R}^d$. These definitions are discussed by Lantuéjoul (2002, Chapter 2).

A fundamental theoretical tool which enables random-orthogical sets to be characterized, is the Choquet capacity. It can be written as follows:

$$T(K) = P\{X \cap K \neq \varnothing\} \quad [2.1]$$

where K is a compact set of $\mathbb{R}$ and X is a realization of a random set. It satisfies the following properties:

i) $0 \leq T(K) \leq 1$ for any compact subset K, and $T(\varnothing) = 0$ $T(\mathbb{R}^d) = 1$;

ii) for $T(K) \leq T(K \cup K')$ all compact subsets K and K';

iii) if K_n is a sequence of decreasing compact subsets (as inclusion defines it) in $\mathbb{R}^d$ with the limit $\mathcal{K}$, then:

$$\lim_{n \to \infty} T(K_n) = T(\mathcal{K})$$

The "hitting functional" T can be seen as a generalization of the cumulative distribution function for random sets (Matheron 1975). This interpretation is justified by the following theorem (Choquet 1954; Kendall 1974; Matheron 1975):

THEOREM 2.1.– Let T be a functional defined on the set of compact subsets $\mathbb{K}$ of $\mathbb{R}^d$. Then a single probability measure P defined on the σ-algebra $\mathbb{F}_K$ exists such that:

$$P(\mathbb{F}_K) = T(K)$$

if, and only if, T is a Choquet capacity verifying (i), (ii) and (iii) above.

The σ-algebra $\mathbb{F}_K$ is the smallest σ-algebra containing the closed sets that meet the compact subsets of $\mathbb{R}^d$

$$\mathbb{F}_K = \{F \in \mathbb{F} : F \cap K \neq \varnothing\}, \qquad K \in \mathbb{K} \qquad [2.2]$$

where $\mathbb{F}$ is the set of the closed subsets of $\mathbb{R}^d$ and $\mathbb{K}$ is the set of compact subsets.

The realizations of the random set are in $\mathbb{F}$, that is, in the closed set. This restriction comes from the fact that the functional $T(K)$ does not allow the distinguishing of a set from its closure. Note that a dual theory involving the open sets of $\mathbb{R}^d$ exists (Lantuéjoul 2002). However, it is generally preferable to work on closures, because they include random point or line processes that are very important for a large number of models.

When restricted to a finite set of points, the functional $T(K)$ defines the *spatial distribution* of the random set X. If K is a finite set of n points:

$$T_n(x_1,...,x_n) = P\{x_i \in X, 1 \le i \le n\} \qquad [2.3]$$

The spatial distribution does not allow us to fully characterize X. For example, if X is itself a finite set of points, it is clear that $T_n \equiv 0$ for every $n \ge 0$. On the other hand, the spatial distribution makes it possible to define a single random set, which is a minimizer of $T(K)$; see the works of Matheron (1975) on this subject.

DEFINITION 2.1.– *A random set is said to be stationary if its Choquet capacity* $T(K)$ *is translation invariant of K, that is*:

$$T(K) = T(K_x), \qquad K_x = \{k + x; \quad k \in K\} \qquad [2.4]$$

Moreover, if the values $T(K)$ *taken by the functional remain unchanged after a rigid body movement of K, including rotations, the random pattern is said to be isotropic.*

It is assumed in the following that X is a stationary, ergodic random set. There are several definitions of ergodicity in the literature, one of the most general being given by Heinrich (1992). This definition, which is somewhat technical, is not detailed here.

Intuitively, ergodicity implies that the Choquet capacity $T(K)$ can be calculated from a single realization X on $\mathbb{R}^d$ and more precisely, from the average density of the set:

$$\{x \in \mathbb{R}^d : K_x \cap X \neq 0\} \qquad [2.5]$$

In such a case, for example, the density of X (volume or surface area fraction), calculated over a sufficiently large domain, is no longer, asymptotically, a random variable. This property is in line with a more operational definition of ergodicity proposed by other authors (Lantuéjoul 1991). In practice, the Choquet capacity is calculated based on a large-sized realization of X using:

$$T(K_x) = P\{K_x \cap X \neq \varnothing\} = P\{x \in X \oplus \breve{K}\} \qquad [2.6a]$$

$$X \otimes \breve{K} = \{x - k, x \in X, k \in K\} \qquad [2.6b]$$

operators $\oplus$ and $\check{}$ defining morphological dilation, in this case applied to X by a structuring element K (Serra 1980). The operator $\oplus$ is the Minkowski addition:

$$K \oplus K' = \{x + x'; x \in K, x' \in K'\} \qquad [2.7]$$

Random set theory has been extended by Matheron (1969) to upper or lower semi-continuous random functions $Z(x)$ $(x \in \mathbb{R}^d)$. Indeed, if, for example, $Z(x)$ is upper semi-continuous, the supergraph:

$$\{(x, z) \in \mathbb{R}^d \times \mathbb{R} : \quad Z(x) \geq z\} \qquad [2.8]$$

is a random set in $\mathbb{R}^{d+1}$ (Lantuéjoul 2002).

Tensor (or "spectral") valued functions require more sophisticated mathematical morphology tools (Angulo 2012). Random sets can also be defined on non-ordinary topological sets such as point clouds and random trees. These theories are not discussed in this chapter.

2.2.2. *Covariance*

In the event that K is a bipoint, the functional $T(K)$ is the covariance of the set X.

DEFINITION 2.2.– *The covariance of the random set* X *is the function defined on* $\mathbb{R}^d \times \mathbb{R}^d$

$$C(x, x+h) = P\{x \in X, x+h \in X\}, \quad x, h \in \mathbb{R}^d \qquad [2.9]$$

When X *is stationary, we have* $C(x, x+h) = C(h)$. *The correlation function is denoted as* $\rho(h) = C(h) / C(0)$.

Moreover, it is clear that $C(h)$ only depends on the norm of h in the case of an isotropic random set. The covariance is estimated from a realization of an ergodic stationary set by:

$$C(h) = P\{x \in X \cap X_{-h}\} = V(X \cap X_{-h}) \qquad [2.10]$$

where $V(\cdot)$ designates the mean density (area or volume fraction) of the set being considered.

NOTE.– The covariance provides access to statistical information on the set X, in particular (Matheron 1975; Lantuéjoul 2002):

i) $C(0) = V(X)$;

ii) $\lim_{|h| \to \infty} C(h) = C(0)^2$;

iii) $C_{X^c}(h) = 1 - 2C(0) + C(h)$, where C_{X^c} is the covariance of the complement of X.

The quantity in (i) is the volume or surface fraction of the set X, property (ii) is obtained if the two events $x \in X$ and $x + h \in X$ are asymptotically independent, when $h \to \infty$. This property is valid for an ergodic stationary ensemble. It does not hold if, for example, the set X is periodic (in this case, the covariance is itself a periodic function). Finally, property (iii) shows that the covariance of a set and its complement can be inferred from one another.

In addition, the specific area ($d = 3$) or perimeter ($d = 2$) $|\partial X|$ of X are given by:

$$|\partial X| = -\frac{1}{\omega_{d-1}} \int_{S_d} \mathrm{d}\alpha \left. \frac{\partial C(h)}{\partial h} \right|_{h=0} \qquad [2.11]$$

where the sum is carried out on all of the angular sectors of the unit sphere S_d in dimension d, and ω_{d-1} designates the volume of the unit ball of dimension $d-1$. Therefore, in dimension $d=2$ and 3:

$$S_V(X) = -\frac{1}{\pi}\int_{S_3} \mathrm{d}\alpha \left.\frac{\partial C(h)}{\partial h}\right|_{h=0} \qquad [2.12a]$$

$$P_S(X) = -\frac{1}{2}\int_{S_2} \mathrm{d}\alpha \left.\frac{\partial C(h)}{\partial h}\right|_{h=0} \qquad [2.12b]$$

S_V designates the specific surface area (surface density of the boundary X), and P_S the specific perimeter. Both are expressed in an inverse length unit. It should be noted that in an isotropic model with $\partial C(h)/\partial h = +\infty$ in $h=0$, the specific surface area is then infinite (for a fractal object). More precisely, a power law behavior $C(h) \sim h^{\beta}$ can be related to the origin when $h \to 0$ ($0 < \beta < 1$), with the fractal dimension (Hausdorff dimension) of the surface $d_s = 3 - \beta$ (Matheron 1989b).

The covariance function provides access to other types of information. For example, inflection points can reflect the presence of nested structures or clusters, the values $C(h) < C(0)^2$ of anti-correlation phenomena. The latter appear, in particular, in inclusion matrix-based models in which inter-penetration between particles is prohibited.

Other properties relate to the second derivative of the original variogram. It is infinite in absolute value when the surface of the random set presents a cusp, with a finite probability within a bounded domain. In the case of angular points, the second derivative is finite. It is finally zero if the surface is regular (Emery and Lantuéjoul 2011).

NOTE.– Covariance functions are not arbitrary. In particular, they are a definite positive or more precisely:

$$\sum_{\alpha,\beta=1}^{n} \lambda_\alpha \lambda_\beta C(x_\alpha - x_\beta) \geq 0 \qquad [2.13]$$

for any sequence of points x_α and any finite sequence of real numbers λ_α ($\alpha = 1$, ..., n).

The Bochner theorem shows that the above property is verified if, and only if, the covariance is the Fourier transform of a positive measure (Bochner 1959; Christakos 1984). In the case of a homogeneous random function $Z(x)$, whose auto-covariance function is written as:

$$\chi(h) = \langle [Z(x) - \mu][Z(y) - \mu] \rangle, \qquad \mu = \langle Z(x) \rangle \tag{2.14}$$

It is then showed that (Torquato 2013):

$$\tilde{\chi}(k) = \int_{\mathbb{R}^d} \mathrm{d}r\, \chi(r) e^{-ik \cdot r} \geq 0 \tag{2.15}$$

for every k, as soon as χ is integrable, that is:

$$\int_{\mathbb{R}^d} \mathrm{d}r |\chi(r)| \leq \infty$$

NOTE.– The above property applies to random functions. However, it does not make it possible to fully characterize functions that are covariances of random *sets*. A counter-example is given by Torquato (2013).

2.2.3. *Granulometry*

Granulometries are defined using a family of operators Φ_λ depending on a length parameter. The properties of this increasing, anti-extensive and idempotent operator were introduced by Matheron (1975). In practice, a convex set K and a family of sets λK ($\lambda > 0$), obtained by homothety, are considered, which are used as parameters for the operator Φ.

In the case of opening granulometry, the operator Φ_λ is written as:

$$\Phi_\lambda(X) = (X \ominus K) \oplus \lambda K \tag{2.16}$$

where $\ominus$ is the Minkowski subtraction:

$$K \ominus K' = \{x - x'; x \in K, x' \in K'\} \tag{2.17}$$

such that Φ_λ is an opening through the structuring element K. Similarly, a closing can be defined $\Phi_\lambda(X) = (X \oplus K) \ominus K\lambda$.

Erosions or dilations by λK also define granulometries. In practice, however, they are less sensitive to microstructures than openings and closings (Escoda *et al.* 2015) and are therefore less frequently used.

2.2.4. *Minkowski functionals*

We examine the set $\mathcal{C}^d$ of the convex compact subsets of $\mathbb{R}^d$, and the real-valued functionals $\Phi(K)$ defined in $\mathcal{C}^d$.

The following properties are defined:

DEFINITION 2.3.– *The functional* ϕ *defined on* $\mathcal{C}^d$ *is:*

– *isometrically invariant if for any isometry* $\mathcal{G}$ *(translation, rotation, reflection, etc.) and compact convex subset* K, $\Phi(\mathcal{G}K) = \Phi(K)$;

– *monotonic for inclusion if* $K_1 \subseteq K_2 \Phi(K_1) \leq \Phi(K_2)$;

– C^d*-additive if for any pair of convex subsets* K_1, K_2, *such that* $K_1 \cup K_2 \in \mathcal{C}^d$, $\phi(K_1 \cup K_2) = \phi(K_1) + \phi(K_2) - \phi(K_1 \cap K_2)$.

The following theorem is due to Minkowski:

THEOREM 2.2.– The isometrically invariant, monotonic and C^d-additive functions are positive linear combinations of $d+1$ functions, known as homogeneous Minkowki functions of degree W_i, that is verifying $d-i$:

$$\lambda > 0 \Rightarrow W_i(\lambda K) = \lambda^{d-i} W_i(K)$$

By convention, Minkowski's functionals all take the same value in the unit ball in dimension d.

NOTE.– Let us take a closer look at the cases $d = 2$ and $d = 3$. We have:

– $W_0(K) = |K|$ the volume ($d = 3$) or the surface area ($d = 2$) of K;

– $W_1(K) = |\partial K|/d$, where $|\partial K|$ is the length ($d = 2$) or area ($d = 3$) of the boundary of K;

– $W_{d-1}(K) = \omega_d \gamma(K)/2$, where $\gamma(K)$ designates the average extension of K. If $\gamma(K,v)$ is the extension of K in the direction v (the length of the projection of K on a straight line in direction v), then:

$$\gamma(K) = \frac{\int_{v \in S_d} \mathrm{d}v \gamma(K,v)}{\int_{v \in S_d} \mathrm{d}v}$$

– $W_d(K) = \omega_d$ if $K \neq \varnothing$, and 0 otherwise.

Thereby, for $d = 2$:

$$b(K) = \frac{1}{\pi} |\partial K|$$

When $d = 3$, W_2 is proportional to the integral of the mean curvature:

$$M(K) = \int_{\partial K} \mathrm{d}S m(x)$$

$$m(x) = \frac{1}{2}\left(\frac{1}{R_1} + \frac{1}{R_2}\right) \tag{2.18}$$

where $R_1(x)$, $R_2(x)$ are the radii of curvature in x on the surface of the convex.

Steiner's formula can be used to connect the volume of the dilation of a convex set K by a ball $B(0,r)$, of radius r, knowing the Minkowski functionalities of K.

THEOREM 2.3.– Let $K \in C^d$ and $r > 0$. The volume of the dilated set of K by the ball of radius r is:

$$|K \oplus B(0,r)| = \sum_{j=0}^{d} \binom{d}{j} W_j(K) r^j \tag{2.19}$$

where $\binom{d}{j}$ designates the binomial coefficient in d, j.

2.2.5. *Stereology*

An affine space A_i of dimension $i < d$ cuts the convex set K into a convex subset $A_i \cap K$. An affine space of dimension i is denoted by $\mathcal{A}_i$, taken uniformly among all those that cut K. The Crofton formula (Hadwiger 1957) gives the average of the Minkowski functionals of $A_i \cap K$ (where $A_i \in \mathcal{A}_i$), as a function of those of K.

THEOREM 2.4.– The Minkowski functionals $W_j^{(i)}$ of the intersection of $K \in C^d$, with an affine space of dimension i < d, are written as:

$$E\left\{W_j^{(i)}(K \cap A_i)\right\}_{A_i \in \mathcal{A}_i} = \frac{\omega_i \omega_{d-j}}{\omega_{d-i}\omega_{i-j}} \frac{W_j(K)}{W_i(K)}, \qquad 0 \leq j < i \qquad [2.20]$$

Given that isotropy and stationarity are assumed, these properties allow the calculation of the volume of 3D objects based on a two-dimensional slice assumed to be representative, or to infer the surface area of these objects from an estimate of the perimeter on a slice (Schneider and Weil 2008).

2.2.6. *Linear erosion*

Instead of calculating the Choquet capacity for a compact subset reduced to a bi-point (or n points), it is now applied to a segment L of length h and orientation α. This is obtained by linear erosion:

$$X \ominus L = \bigcap \{X - u : u \in [0;h]\} \qquad [2.21]$$

For a stationary set, the moment is defined by linear erosion $P(h)$ of X using the Lebesgues measure of $X \ominus L$:

$$P(h) = |X \ominus L| = P\{L \subset X\} \qquad [2.22]$$

The following properties are easily verified:

– $P(0) = C(0)$ is the volume (or surface) fraction of the set X;

– the linear erosion moment and the covariance have the same slope at the origin, namely $P'(0) = C'(0)$;

– for a fixed angle α, $C(h) \geq P(h)$;

– at a fixed angle α, $P(h)$ is a decreasing function of h;

– $P(\infty) = 0$.

In addition, the opening granulometry of a segment L can be characterized by:

$$P\{x \in X \circ L\} = P(h) - hP'(h), \qquad X \circ L = (X \ominus L) \oplus L \tag{2.23}$$

where $P'(h)$ is the derivative of $P(h)$.

2.2.7. *Representative volume element*

We consider a measure of the volume fraction of an ergodic random set X, estimated from a realization in a bounded domain:

$$f_V = \frac{1}{|V|}\int_V 1_X(x)\mathrm{d}x \tag{2.24}$$

DEFINITION 2.4.– *The integral range of the ergodic random set X is the d-volume quantity:*

$$A_d = \int_{\mathbb{R}^d} \mathrm{d}h \frac{C(h) - C(0)^2}{C(0) - C(0)^2} \tag{2.25}$$

It is shown that if X is of the finite integral range A_d, the variance of the estimates f_v of f varies asymptotically as (Matheron 1989a):

$$\mathbb{E}\{(f_V - f)^2\} = \frac{\sigma^2 A_d}{|V|} + O\left(\frac{1}{|V|}\right), \qquad |V| \to \infty \tag{2.26}$$

where the volumes V are increasing compacts that asymptotically overlap $\mathbb{R}^d$. Furthermore, the above equation provides a good estimate of the variance when $|V| \gg A_d$. This relation shows that the volume V behaves as $n = |V|/A_d$ volume-independent domains of integration range. In the case where $A_d = +\infty$, that is when the quantity is $C(h)-f^2$ not infinitely integrable, the asymptotic relation [2.26] is no longer valid. This is particularly the case for models containing Poissonian varieties of an infinite size, in which correlations exist at scales as large as desirable (Lantuéjoul 1991; Jeulin 2016). Exact results for the asymptotic behavior of the variance of f_V of some cylinder Boolean structures are given by Willot (2017). For three-dimensional Boolean fibrous materials containing large-sized elongated particles, for example, two regimes appear. The variance of the estimate f_V behaves as $\sim 1/|V|^{2/3}$ in a domain $h \ll \ell$, and as $\sim 1/|V|$ when $h \gg \ell$, the change of regime occurs over the length $h \approx \ell$ of the fibers.

Mechanical fields (for example, the strain tensor) can also be associated with an integral range; by studying the asymptotic behavior of the average field over sub-volumes. This average reflecting the apparent properties is (for example, the apparent elastic modulus) associated with a finite size volume (Altendorf *et al.* 2014; Azzimonti *et al.* 2013), for example, numerically calculating separate exponents for the microstructure and the displacement field (in optics) of a repository model.

2.3. Point processes

A random point process is defined as a random set of points, each realization of which is a closed set that contains a finite or countable number of points. In general, it is assumed that each realization is locally finite, that is, the number of points contained in a compact set is almost surely finite.

Random point processes are a very rich class of stochastic models. They are useful for defining models of continuous random sets. For example, a straight line in the plane can be represented by an angle, measured with respect to a fixed axis, and a positive real, its distance to a fixed point. A process of random points in a semi-infinite band $[0;2\pi]\times\mathbb{R}_+$ consequently corresponds to a random process of straight lines.

In the case of a Poisson point process (that will be defined below), a pattern of Poissonian lines is obtained (see Figure 2.2(f)). Similarly, a random point process can be used as germs for Boolean schemes (see Figures 2.2(a) and (b)), or for a dead-leaves models (see Figure 2.2(c)) or random partitions (see Figures 2.2(d) and (e)).

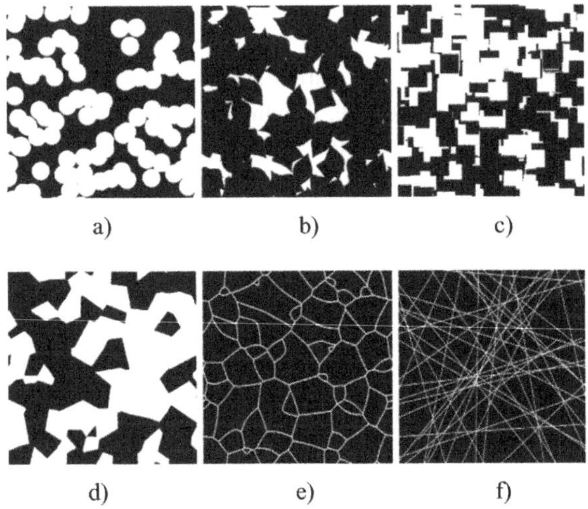

Figure 2.2. *Examples of stochastic models derived from random point processes: (a and b) Boolean models; (c) dead-leaves model; (d) labeling of a Poisson–Voronoi random partition; (e) Johnson–Mehl random partition; (f) Poisson line processes. For a color version of this figure, see www.iste.co.uk/gogu/uncertainties.zip*

The remainder of this section focuses exclusively on Poisson point processes. The reader can refer to Van Lieshout (2000) and Daley and Vere-Jones (1988) for a description of other random point processes.

2.3.1. *Homogeneous Poisson point processes*

Let X be a realization of a random point model and $T(K)$ its Choquet capacity. We consider the spatial distribution:

$$P_K(n) = P\{N(K) = n\}, \qquad n \geq 0$$

defined for any compact set K and integer n, with $N(K)$ as the “count function”, namely the function that assigns with K, the number of points of X contained in K. The Choquet capacity is written as $T(K) = 1 - P_K(0)$.

It is natural to assume that in a compact set δK of infinitesimal volume, the probability that a point is included in δK is of the order of the volume of δK :

$$\lim_{|\delta K| \to 0} T(\delta K) \propto |\delta K| \qquad [2.27]$$

Subsequently, it is assumed that:

$$\lim_{|\delta K| \to 0} T(\delta K) = \theta |\delta K| \qquad [2.28]$$

for any series of compact sets δK of volume tending toward 0 containing a point x, where θ (with the unit as the inverse of a d-volume) is a constant. The parameter θ is the *intensity* of the Poisson point process and designates a point density (number of points per unit of d-volume).

DEFINITION 2.5.– *A homogeneous Poisson point process of intensity* $0 < \theta < \infty$ *is a point process verifying [2.28], and such that the events* $K \cap X \neq \varnothing$ *and* $K' \cap X \neq \varnothing$ *for* K *and* K' *included in* $\mathbb{R}^d$, *are independent as soon as* K *and* K' *are disjoint.*

Let us calculate the spatial distribution of a homogeneous Poisson point process of intensity θ. Consider a compact set K that we partition into N sub-domains of infinitesimal volume $|\delta K| = |K|/N$:

$$P_K(n) = (\theta |\delta K|)^n (1 - \theta |\delta K|)^{N-n} \frac{N!}{n!(N-n)!} = \frac{\theta^n |K|^n e^{-\theta |K|}}{n!} \qquad [2.29]$$

after using the Stirling $z! = z^z e^{-z} \sqrt{2\pi z}$ formula. According to [2.29], the spatial distribution of the point process is therefore, of fixed K, a Poisson distribution of mean $\theta |K|$, hence the name *Poisson point process*. Moreover, it is clear that:

$$T(K) = 1 - e^{-\theta |K|} \qquad [2.30]$$

Using [2.29], the probability of simultaneous events involving N can be calculated such that:

$$P\{N(K) = n, N(K') = n'\}$$

$$= \sum_{i=0}^{n'} P\{N(K \cap K') = i\} P\{N(K \setminus K') = n - i\} P\{N(K' \setminus K) = n' - i\}$$

for any pair of compact set K, K' with $n' \leq n$. A similar general formula can be used to calculate the probability of the event $(N(K_1)=n_1)\cap \ \dots \ \cap(N(K_N)=n_N)$ for $K_i \in \mathbb{K}$, $n_i \geq 0$, $N \geq 1$.

NOTE.– Most authors define a Poisson point process as a spatial distribution process [2.29], such that $N(K)$ and $N(K')$ are independent random variables for disjoint K and K' (see, for example, Lantuéjoul 2002).

THEOREM 2.5.– Let us assume that $N(K)=n$ in a certain domain K. Based on property [2.28], the n points are uniformly distributed in K and it can be shown that:

$$P\{N(K')=n' \mid N(K)=n\} = \binom{n}{n'}\left(\frac{|K'|}{|K|}\right)^{n'}\left(1-\frac{|K'|}{|K|}\right)^{n-n'} \qquad [2.31]$$

for $K' \subset K$, $n' \leq n$. Therefore, to simulate a Poisson process in a finite volume domain K, one simply has to randomly draw an integer n according to the probability distribution [2.29], then implant n points uniformly into K.

2.3.2. *Inhomogeneous Poisson point processes*

In an inhomogeneous Poisson point process, the density θ is now a function of x:

$$\lim_{|\delta K(x)|\to 0} T(\delta K(x)) = \theta(x)\,|\,\delta K(x)\,| \qquad [2.32]$$

where the $\delta K(x)$ are a sequence of volume compact sets tending to 0 and containing the point x. One further defines:

$$\theta(K) = \int_K \mathrm{d}x\,\theta(x) \qquad [2.33]$$

As with a homogeneous Poisson point process, the random variables $N(K)$ and $N(K')$ are independent if the compact sets K and K' are disjoint.

Equation [2.29] then remains valid, provided $\theta\,|\,K\,|$ is replaced by $\theta(K)$:

$$P_K(n) = \frac{\theta^n(K)e^{-\theta(K)}}{n!} \qquad [2.34]$$

which is still a Poisson distribution. Similarly, we have:

$$T(K) = 1 - e^{-\theta(K)} \quad [2.35]$$

NOTE.– In the case where the intensity $\theta(x)$ is a random, locally integrable function, a Cox process is obtained (Lantuéjoul 2002). This random point process is not a Poisson point process, since $N(K)$ is not generally a random Poisson variable (see Lantuéjoul (2002) for examples of non-Poisson distributions). A relevant example of a Cox process is obtained when $\theta(x)$ is the indicator of a random set, which allows the construction of multiscale sets (Jeulin and Ostoja-Starzewski 2001; Willot and Jeulin 2011).

2.4. Boolean models

2.4.1. *Definition and Choquet capacity*

The Boolean model is based on two fundamental components:

– a random Poisson point process of intensity $\theta(x)$ (see section 2.3);

– a family $A(x)$ ($x \in \mathbb{R}^d$) of random non-empty compact sets included in $\mathbb{R}^d$.

They are used in the following definition:

DEFINITION 2.6.– *The Boolean model of intensity θ, associated with the family $(A(x))$, is the union of all the compact sets $A(x)$ implanted on the germs of a Poisson point realization $\mathcal{P}$ of intensity θ:*

$$X = \bigcup_{x \in \mathcal{P}} A(x) \quad [2.36]$$

THEOREM 2.6.– Now let K be a compact set of $\mathbb{R}^d$ and X a Boolean model of primary grain A and intensity θ. The number $N(K)$ of primary grains intersected by K follows a Poisson distribution of parameter $\mathbb{E}\{\theta(A \oplus K)\}$:

$$P\{N(K) = n\} = \frac{\mathbb{E}\{\theta(A \oplus K)\}^n}{n!} e^{-\mathbb{E}\{\theta(A \oplus K)\}} \quad [2.37]$$

This property leads to the following theorem:

THEOREM 2.7.– If the average number of primary grains encountered by K is finite $(\mathbb{E}\{\theta(A \oplus K)\} < \infty)$, the Choquet capacity of the Boolean model X is written as:

$$T(K) = 1 - e^{-\mathbb{E}\{\theta(A \oplus K)\}} \quad [2.38]$$

for any compact set K included in $\mathbb{R}^d$, where we set:

$$\theta(B) = \int_B \mathrm{d}x \theta(x) \quad [2.39]$$

For a stationary model:

$$T(K) = 1 - e^{-|\theta(A \oplus K)|}. \quad [2.40]$$

Finally, for a stationary model of homogeneous intensity:

$$T(K) = 1 - e^{-\theta|A \oplus K|} \quad [2.41]$$

where $|A \oplus K|$ is the mean of the Lebesgue measure of the primary grain A dilated by the compact set K.

The above theorem gives, in particular, an expression of probability $P\{x \in X\}$.

THEOREM 2.8.– The volume fraction of the stationary Boolean model X of primary grain A and homogeneous intensity θ is written as:

$$T(\{x\}) = P\{x \in X\} = 1 - e^{-\theta|A|} \quad [2.42]$$

The above formula can be demonstrated by elementary means. Let f be the volume fraction of the random set X, a Boolean model of primary grain A and intensity θ. Assume that X is ergodic and denote $\theta = N/V$, where N is the number of grains included in a large size domain V. After adding dN grains, the volume fraction f is changed into $f + \mathrm{d}f$, and the intensity into $\theta + \mathrm{d}\theta$ with $\mathrm{d}\theta = \mathrm{d}N/V$. Now we have on average:

$$\mathrm{d}f = \frac{\mathrm{d}N|A|}{V}(1-f)$$

Indeed, it is clear that $df = dN|A|/V$ if $f = 0$. On the other hand, if $f > 0$, this value must be weighted by the volume intersected by X^c only, from which the term $1 - f$ is derived. Hence, one has $df/(1-f) = |A|\,d\theta$ by integration $1 - f = const + e^{-|A|\theta}$, with the constant being 0.

NOTE.– Theorem [2.42] is also known in crystallization as the Avrami equation (Avrami 1939).

We can demonstrate [2.41] from [2.42]. In effect, one has $1 - T(K) = P\{K \subset X^c\}$. Let K_x be a compact set centered at $x \in \mathbb{R}^d$. Then, we have:

$$K_x \subset X^c \Leftarrow x \in (X^c \ominus K) \Leftarrow x \in (X \oplus K)^c = \bigcup_{y \in \mathcal{P}} [A(y) \oplus K]^c$$

Therefore, K_x is included in the complement of X if, and only if, x belongs to the complement of the Boolean of germs $\mathcal{P}$ and primary grain $A \oplus K$. We can thus exactly obtain formula [2.41].

The volume fraction of X is denoted by $f = P\{x \in X\}$. We can rewrite [2.41] in the form:

$$T(K) = 1 - (1 - f)^{|A \oplus K|/|A|} \qquad [2.43]$$

2.4.2. *Properties*

Boolean random sets verify the following fundamental properties:

– Matheron (1975) showed that Boolean models are "infinitely divisible". A random set X is infinitely divisible if, for any integer $n > 0$, X is the union n of Y_i independent random sets $\cup_i Y_i$. This property implies that the union of two realizations of a Boolean model is still Boolean (Serra 1981);

– a section (intersection with a space smaller than d) of a Boolean set is still Boolean (Serra 1983). It should be noted that one cannot generally infer all the characteristics of the model in space $\mathbb{R}^d$ from those obtained in a section. A counter-example is given by a Boolean random set, in which some primary grains are reduced to points. Nevertheless, all the information provided by the functional $T(K)$ can be found when K is itself included in the space of the smaller dimension. This includes the volume fraction. For a 3D isotropic model, the covariance is recovered from a section according to a straight line in the model, thus providing access to, for example, the specific surface;

– Boolean models are stable by dilation. This property is derived from the Choquet capacity formula, where the dilation of a Boolean model of primary grain A, by a compact set K, is a Boolean model of primary grain $A \oplus K$ (Serra 1981).

Let us examine the "domain of attraction" of the Boolean model. Some limits of the unions of random partitions are Boolean models (Serra 1981, 1983; Chiles and Delfiner 2009), in the following sense. Let X be a random partition of $\mathbb{R}^d$ and $\{X_i'\}$ ($i \in \mathcal{I}$) be a family of random sets obtained by choosing each cell of X independently with a probability p, and taking the union closure of the chosen cells.

– We set:

$$Y_n = \bigcup_{i \leq n} X_i' \qquad [2.44]$$

We leave $n \to \infty$, $p \to 0$ such that $pn = \theta$ with fixed $0 < \theta < \infty$. Then we show that, for any compact set K:

$$\lim_{n \to \infty} T_n(K) = 1 - e^{-\theta |X'|^{-1} \mathbb{E}\{|X' \oplus K|\}} \qquad [2.45]$$

where X' is the class of the cells of the random partition X, and T_n is the Choquet capacity associated with Y_n. According to the Choquet–Matheron–Kendall theorem, the limit of Y_∞ of the $(Y_n)_n$ exists and is unique, and in addition, Y_∞ is Boolean.

NOTE.– The above property shows that Boolean models are obtained as the limit of the non-Boolean random partition unions. This property can be seen as a central limit theorem for random sets (Serra 1981; Cressie and Laslett 1987), the union [2.44] of random sets acting as the sum for random variables. This property further suggests that the Boolean model, for random variables, plays an analogous Gaussian-like role for random sets.

2.4.3. *Covariance*

The covariance $C(h)$ of X is calculated using the Choquet capacity. It is written as:

$$C(h) = 2f - 1 + (1-f)^2 e^{\theta|A \cap A_h|} = 2f - 1 + (1-f)^{2-\gamma(h)/\gamma(0)} \qquad [2.46]$$

where $\gamma(h) = | A \cap A_h |$ is the geometric covariogram of A.

Simple analytical formulae for the geometric covariogram are known for certain primary grain forms only. For a disc of radius R:

$$\gamma(h) = 2R^2 \left[\cos^{-1}\left(\frac{h}{2R}\right) - \frac{h}{2R}\sqrt{1-\left(\frac{h}{2R}\right)^2} \right] H(2R-h) \qquad [2.47]$$

where $H(\cdot) = 1_{\mathbb{R}_+}(\cdot)$ is the Heaviside function. For a sphere of radius R:

$$\gamma(h) = \frac{4\pi R^3}{3}\left(1 - \frac{3h}{4R} + \frac{h^3}{16R^3}\right) H(2R-h) \qquad [2.48]$$

Although its definition is very simple, the geometrical covariogram is known only for a few elementary forms, including cylinders of revolution (Gille 1987; Willot 2017), ellipsoids, half-spheres and a few other forms (Gille 2016). It is also known in random forms derived from specific partitions such as Poisson polyhedra (Matheron 1972), cells of a Voronoi partition (Brumberger and Goodisman 1983), or from the typical grain of a dead-leaves pattern (Gille 2002).

2.4.4. *Other characteristics*

2.4.4.1. *Three-point function*

Let $K = \{x, x+h_1, x+h_2\}$, according to [2.41] we have:

$$C(h_1, h_2) = P\{x \in X^c, x+h_1 \in X^c, x+h_2 \in X^c\} = e^{-\theta|A \cup A_{-h1} \cup A_{-h2}|}$$

$$= (1-f)^{3-r(h_1)-r(h_2)+s(h_1,h_2)} \qquad [2.49]$$

with:

$$r(h) = \frac{\gamma(h)}{\gamma(0)}, \qquad s(h_1, h_2) = \frac{|A \cup A_{-h1} \cup A_{-h2}|}{\gamma(0)} \qquad [2.50]$$

Here again, the main difficulty lies in evaluating the functional s for non-trivial primary grain forms.

2.4.4.2. *Contact distribution*

Assume a point x is in the complementary of X and let ℓ be its distance from X. The distribution function of ℓ is given by (Serra 1980):

$$P\{\ell \leq L\} = 1 - \frac{1 - T(B_L)}{1 - f} \qquad [2.51]$$

where B_L is the ball of radius L.

2.4.4.3. *Specific surface area*

The specific surface area of a three-dimensional Boolean model X is obtained using [2.12a]:

$$S_V(X) = 4\frac{\gamma'(0)}{\gamma(0)}(1-f)\log(1-f) \qquad [2.52]$$

where $\gamma(h)$ is the geometric covariogram of the primary grain of the model X. We note that $S_V / f \sim const$ when $f \rightarrow 0$, which is expected for grains almost isolated from each other, without interpenetration. On the other hand, $S_v / (1-f) \sim \log(1-f) \rightarrow \infty$ in the dual case $f \rightarrow 1^-$. Indeed, when $f \rightarrow 0$, the Boolean model is composed of a set of very elongated objects that are the (uncovered) interstices between primary grains, and these interstices have a surface/volume ratio that becomes infinite.

2.4.4.4. *Linear erosion curves for convex primary grains*

Assume that A is a convex random compact set. Steiner's formula [2.19] allows us to explicitly calculate $T(K)$. The measure of $A \oplus \lambda K$ is, in particular, a polynomial of degrees d in λ, whose coefficients depend on the Minkowski functionals of A. If, for example, $K = [x; x+h]$ is a segment of length h and orientation, $|A \oplus K| = |A| + h|A_\alpha|$, where A_α represents the length of the boundary of A in the direction of dilation α. The probability P_X^c that a segment L of length h is included in the complement of the Boolean model is then (see equation [2.22]):

$$\begin{aligned} P_X^c(h) &= P\left\{L \subset X^c\right\} = e^{-\theta|A\oplus K|} = e^{-\theta(|A|+h|A_\alpha|)} \\ &= e^{-\theta\left[\gamma(0)-h\gamma'(0)\right]} = (1-f)^{1-h\gamma'(0)/\gamma(0)} \end{aligned} \quad [2.53]$$

This relation is particularly useful because it allows the testing of the validity of the Boolean assumption: *for any convex primary grain, the linear erosion curve of a Boolean model varies exponentially with the length.*

NOTE.– By taking a Cox process as a set of germs, itself based on a Boolean model, multiscale structures can be generated (see, for example, Jean *et al.* 2011).

An upper bound on the length of the minimum path length, going through a Boolean medium of uniform intensity, having disks, squares ($d = 2$) or hyperspheres ($d \geq 2$) as the primary grain, was calculated by Willot (2015) for a diluted volume fraction ($f \ll 1$).

This bound shows non-trivial exponents ($\sim f^{2/3}$ for $d = 2$, $\sim f^{1/2}$ for $d = 3$), and a diameter-dependent prefactor in the direction of propagation of the primary grain.

Doi (1976) calculated the surface correlation functions of a three-dimensional Boolean model of spheres. For a random set X, these correlation functions are defined by the quantities:

$$F_{sv}(x_1, x_2) = \langle \chi(x_1) s(x_2) \rangle \quad [2.54a]$$

$$F_{ss}(x_1, x_2) = \langle s(x_1) s(x_2) \rangle \quad [2.54b]$$

with $\chi(\cdot)$ as the indicator of X, $1-\chi(\cdot)$ that of X^c and:

$$s(x) = |\nabla \chi(x)| \quad [2.55]$$

where ∇ is the nabla vector-operator of components $\partial / \partial x_i$ ($i = 1, ..., d$) and x_i being the ith component of x in a Cartesian coordinate system. For a stationary random set, the correlation functions F_{ss} and F_{sv} only depend on $x_1 - x_2$, and on $h = |x_1 - x_2|$ for an isotropic model.

When $h \to \infty$ (Torquato 2013), we have:

$$F_{sv}(h) \sim fs, \qquad F_{ss}(h) \sim s^2, \quad r \to \infty \qquad [2.56]$$

where s is the specific surface area of ∂X.

For the Boolean sphere model, in three dimensions, Torquato (2013) gives:

$$F_{sv}(r) = \frac{-3\log(1-f)}{R}\left[2 - \frac{\gamma(h)}{\gamma(0)}\right]\left[1 - \left(\frac{1}{2} - \frac{h}{4R}\right)H(2R-h)\right] \qquad [2.57a]$$

$$F_{ss}(r) = \left[2 - \frac{\gamma(h)}{\gamma(0)}\right]\left\{\frac{9\log^2(1-f)}{R^2}\left[1 - \left(\frac{1}{2} - \frac{h}{4R}\right)H(2R-h)\right]^2 - \frac{3\log(1-f)}{2hR}H(2R-h)\right\} \qquad [2.57b]$$

The functions F_{ss} and F_{sv} appear, in particular, in the expressions of certain bounds of the permeability of porous media (Doi 1976; Berryman and Milton 1985; Willot *et al.* 2016; Bignonnet 2018).

2.5. RSA models

The random sequential adsorption (RSA) model (Talbot *et al.* 2000) is a model in which, unlike the Boolean model, grains are added sequentially by prohibiting interpenetration. In the simplest case, the intensity is taken uniformly in $\mathbb{R}^d$. When a grain partially overlaps a grain previously positioned, it is rejected (see Månsson and Rudemo (2002) for other disjoint grain models). Few exact results are known for the RSA model, except in the one-dimensional case (Torquato 1995).

For a small volume fraction of the random set, the RSA model is asymptotically equal to a Boolean medium and the grain positions are those of a Poisson point process (Månsson and Rudemo 2002). The RSA model does not allow high-density realizations to be generated. In this case, special techniques must be implemented, in which objects are placed and restoring forces are imposed between interpenetrating objects. This approach was proposed, in particular, for the modeling of fibrous media (Altendorf and Jeulin 2011).

The images in Figure 2.3(a) show two-dimensional, cross-sections following two orthogonal directions, of a composite material, and Figure 2.3(b) shows a random fiber model reproducing the microstructure characteristics, with, in particular, the radius of curvature of the fibers and their tortuosity.

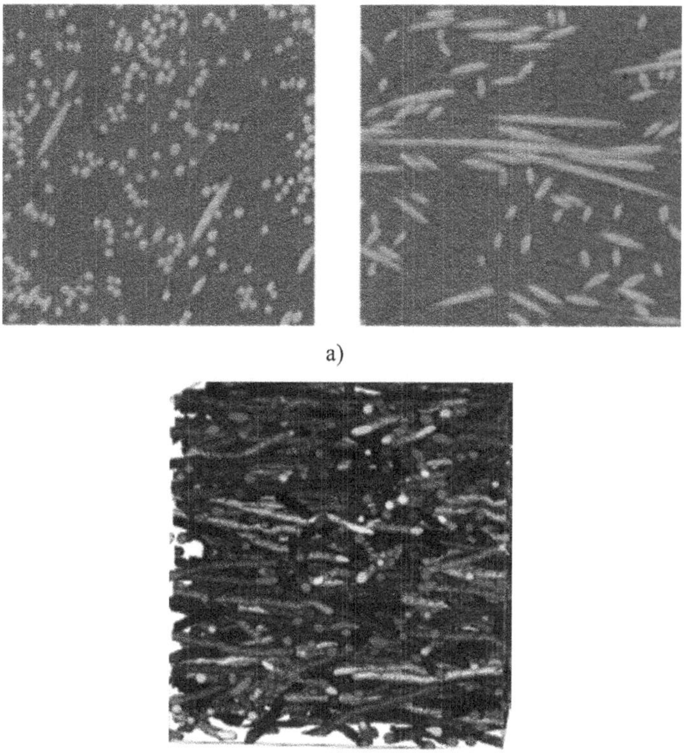

Figure 2.3. *(a) Composite fibrous medium; (b) model (source: Altendorf* et al. *2014). For a color version of this figure, see www.iste.co.uk/gogu/uncertainties.zip*

2.6. Random tessellations

A random tessellation or partition of space is by definition a set of pairwise disjoint bounded open subsets (cells) of $\mathbb{R}^d$, whose union closure covers the whole space. Matheron (1969) showed that probabilities could be defined on a σ-algebra, based on tessellations of $\mathbb{R}^d$, which the set of points located at the edges of the cells is added to (see Lantuéjoul 2002).

Two approaches are possible in order to calculate properties relating to grain characteristics (number of neighbors, average surface area, etc.). In the first case, the focus is on the cell containing the origin (there is one with a probability of 1), called the *fundamental cell*. In the second approach, the distribution of a characteristic is calculated from all the cells of all the realizations, or from all the cells in a realization of an ergodic model. Although no particular cell has been chosen, it is referred to as *typical cell* in this case. In the first case, which is a volume distribution, large volume cells will be represented more than in the second case, which consists of calculating a number distribution.

In all the following examples, it is assumed that the number of cells intersected by a compact set is finite with a probability of 1.

2.6.1. *Voronoi tessellation*

The Poisson–Voronoi tessellation is defined by the influence zones $C(g,X)$ of the germs g of a Poisson point model $\mathcal{P}$. The cell $C(g,X)$ associated with the germ $g \in \mathcal{P}$ is:

$$C(g,X) = \{x \in \mathbb{R}^d : \ |x-g| < |x-g'| \forall g' \in \mathcal{P}, g' \neq g\} \qquad [2.58]$$

This cell is a convex polytope delimited by parts of hyperplanes. Despite the simplicity of the model, the analytical results from the Voronoi model are limited. In three dimensions, the volume distribution of the surface of the typical cell C_t is known, whose mean is written as (Miles 1974; Møller 1989):

$$\mathbb{E}\{|\partial C_t|\} = \frac{2\sqrt{\pi}(d-1)!\Gamma(1+d/2)^{1-1/d}\,\Gamma(2-1/d)}{\theta^{1-1/d}\,\Gamma((d+1)/2)\Gamma(d-1/2)} \qquad [2.59]$$

where $\Gamma(\cdot)$ is the extension of the factorial function to $\mathbb{R}$ $\Gamma(n) = (n-1)!$ (n being an integer). Similar results are known for the mean lengths of the triple lines of the typical grain, and more generally for the d'-volumes of dimension $d' < d$, of the d'-facets of the typical cell of the Voronoi tessellation of dimension d.

The *number* distribution of the area or volume of Voronoi cells is often modeled, in an approximate way, by gamma distributions (Ferenc and Néda 2007; Farjas and Roura 2008); this result being exact in one dimension. It should be noted that the 2D section of a 3D Voronoi tessellation is not a two-dimensional Voronoi tessellation. The latter presents, in particular, distributions of distinct cell sizes.

The distribution of the number of sides, in the typical cell of a two-dimensional Voronoi partition, was calculated by Calka (2003a). The same author studied the distribution of the area and perimeter of the same cell (Calka 2003b) and other quantities related, in particular, to the size of the largest disk included in the typical cell, which has a point in the cell as its center (Calka 2002).

2.6.2. *Johnson–Mehl tessellation*

A Johnson–Mehl tessellation is defined by a Poisson points process in $\mathbb{R}^d \times \mathbb{R}_+$, the last dimension being a time dimension. Germs g_i $(i = 1, ..., N)$ are implanted sequentially at times t_i $(i = 1, ..., N)$. The associated classes $C(g_i, X)$ are given by:

$$C(g_i, X) = \{x \in \mathbb{R}^d : \ t_i + \frac{|x - g_i|}{v} < t_j + \frac{|x - g_j|}{v} \forall j \neq i,\} \qquad [2.60]$$

where v is a velocity of propagation of the edge cells. A different choice of v is equivalent to changing the spatial scale and does not fundamentally alter the model. The Johnson–Mehl model is perhaps the simplest random partition with non-convex grains.

2.6.3. *Laguerre tessellation*

Poisson–Voronoi and Poisson–Johnson–Mehl random tessellations essentially depend on one parameter, the intensity of the Poisson point process, which is a scaling factor and only trivially modifies the resulting structure. In many applications, it is desirable to generate random partitions with a prescribed distribution of a geometric characteristic in the cells. For example, consider the number-weighted distribution of the volume of cells. Each grain must be assigned a weight representative of the characteristics of the grain.

Given a process of random Poisson points g_i, the Laguerre tessellation consists of dividing $\mathbb{R}^d$ into cells defined as:

$$C(g_i, X) = \{x \in \mathbb{R}^d : \ |x - g_i|^2 - r_i^2 < |x - g_j|^2 - r_j^2 \ \forall j \neq i,\} \qquad [2.61]$$

where r_i are positive weights assigned to each germ, which follow a certain distribution $\mathcal{R}$. Similarly to the Voronoi tessellation, the cells of a Laguerre tessellation are convex polytopes; however, unlike the Voronoi model, some cells may be empty, and some may not contain their germ. The Laguerre model offers extra degrees of freedom compared to the Voronoi tessellation. The distribution of size $\mathcal{R}$ can be adjusted to simulate cells with volume distributions more widely spread out than the random Voronoi tessellation (Lautensack and Zuyev 2008), which makes it possible to model cellular media or foams (Redenbach 2009), or even polycrystals.

2.6.4. *Random Poisson tessellation*

A Poisson tessellation is defined by a set of hyperplanes in $\mathbb{R}^d$. Each hyperplane is parameterized by its distance from the origin and a direction in the half-hypersphere S_d^+. The model can then be seen as a Poisson process in $\mathbb{R}^d \times S_d^+$. Thereby, it is completely determined by a single parameter, the intensity θ of the Poisson point process. The cells of the Poisson random tessellation are the subsets of $\mathbb{R}^d$, delimited by the boundaries of the Poisson hyperplanes.

A large number of exact results are available for the typical polytope of the Poisson partition. Miles (1964, 1974) calculated its average perimeter ($d = 2$), the average number of vertices, the average total length of its edges, and its average area ($d = 3$). In general, for the volume $F_{d'}$ of the $d' \leq d$ -dimension variety of the typical polytope C_t:

$$\mathbb{E}\{F_{d'}\} = \binom{d}{d'} \frac{2^d}{\omega_{d'}(\omega_{d-1}\theta)^{d'}} \qquad [2.62]$$

Matheron (1975) gives the mean value of the Minkowski function of C_t:

$$\mathbb{E}\{W_i(C_t)\} = \frac{\omega_i}{\omega_{d-i}} \left(\frac{2}{\omega_{d-1}\theta} \right)^{d-i}, \qquad 0 \leq i \leq d \qquad [2.63]$$

Miles (1974) gives the average number $N_{d'}$ of the d'-face ($d' \leq d$ -dimensional variety) of the typical cell C_t:

$$\mathbb{E}\{N_{d'}\} = 2^{d-d'} \binom{d}{d'} \qquad [2.64]$$

In addition, it is shown that the random partition obtained by the intersection of a Poisson partition in $\mathbb{R}^d$, by a dimensional hyperplane d', is still a Poisson partition in $\mathbb{R}^{d'}$. Its intensity is written as (Miles 1964):

$$\theta' = \frac{\omega_{d-1}}{\omega_{d'-1}}\theta \qquad [2.65]$$

The first three moments of the volume distribution of the fundamental cell C_f of the Poisson partition, were obtained by Matheron (1975). Its mean is more specifically written as:

$$\mathbb{E}\{|C_f|\} = \frac{d!\,\omega_d}{(\omega_{d-1}\theta)^d} \qquad [2.66]$$

Poisson random partition models have been used, in particular, to model the granulometries obtained by screening concrete aggregates (Escoda *et al.* 2015), or to simulate cementitious materials (Heinemann *et al.* 1999). On the other hand, these cells can be used as random primary grains of Boolean models, for example to represent tungsten carbides (Quénec'h *et al.* 1996).

2.6.5. *The dead-leaves model*

The dead-leaves model was originally introduced by Matheron (1968) and extended by Jeulin (1979) (see also Serra 1983). It is a sequential Boolean model that asymptotically fills up the whole space and defines a random partition. It makes use of n "colors" indexed by an integer $i \in \mathcal{I}$. The "leaves" are random compact sets of $\mathbb{R}^d$, which appear at times $-\infty < t < 0$, according to a Poisson point process of intensity θ in $\mathbb{R}^d \times \mathbb{R}_-$. Each leaf is assigned a color in $\mathcal{I}$ independent of the color of the other leaves. The colored leaves i appear in the time interval $[t;t+\mathrm{d}t]$, with a probability p_i. Those that appear in the time interval $[t;t+\mathrm{d}t]$ form a Boolean model, where the random compact set A_i is the primary grain. The grain that appears in x at time t is denoted by $A(x,t)$. The grains cover those that appeared previously.

At time $t = 0$, the space $\mathbb{R}^d$ is covered at every point by a grain with a probability of 1. The random partition is formed by uncovered cells $A(x,t)$. These are additionally identified by a color in $\mathcal{I}$. The cells are not connected in general.

For this model, we know how to calculate the probability that each point y_i, in a finite set of points $Y \subset \mathbb{R}^d$, is of a given color, depending on y_i (Lantuéjoul 2002). In particular, we have access to the two-point spatial distribution, which is written using the Lantuéjoul (2002) notations:

$$P\{Z(x)=i, Z(y)=j\} = \frac{a_i(a_j - b_j) + a_j(a_i - b_i) + ab_i\delta_{ij}}{a(2a-b)} \quad [2.67]$$

where δ_{ij} is the Kronecker symbol, $Z(x)$ is the color of the partition at point x, at the time $t = 0$ and:

$$a_i = \theta p_i \mathbb{E}\{|A_i|\}, \quad b_i = \theta p_i K_i(x-y), \quad a = \sum_i a_i, \quad b = \sum_i b_i \quad [2.68]$$

with $K_i(\cdot)$ as the geometric covariogram of A_i:

$$\mathbb{E}\{|(A_i \ominus \{x,y\}) \setminus (A_i \oplus \{x\})|\} = K_i(0) - K_i(x-y) \quad [2.69]$$

The dead-leaves model is often used to model occlusion phenomena in natural images (Lee *et al.* 2001), or to analyze two-dimensional images obtained by electron microscopy, powders for example (Jeulin *et al.* 1995). Furthermore, the set of intact grains at $t = 0$ is an interesting random model for generating high-density stacks, if the intensities and grains are properly chosen over time (Jeulin 2019).

2.6.6. *Generalized random partition models*

Random partitions based on Poisson point processes can be generalized using local metrics (not necessarily Euclidean) attached to each germ (Jeulin 2013), and reinterpreted from Boolean function models. Figure 2.4(a) shows an example of a non-convex grain random partition (Gasnier *et al.* 2015a), obtained from a generalized Johnson–Mehl model with an anisotropic distance. The latter models a polycrystal whose grain boundaries, seen by scanning electron microscopy (SEM), have been separated from their backgrounds (see Figure 2.4(b)). The model is three-dimensional, and the SEM image is two-dimensional. The cell shapes in the Johnson–Mehl model with an anisotropic distance are more elongated than in the classical Johnson–Mehl model. The model is macroscopically isotropic.

Figliuzzi (2019) proposes an approach enabling grain interfaces to be generated with controlled roughness and a method for the rapid simulation of random partitions on a voxel grid.

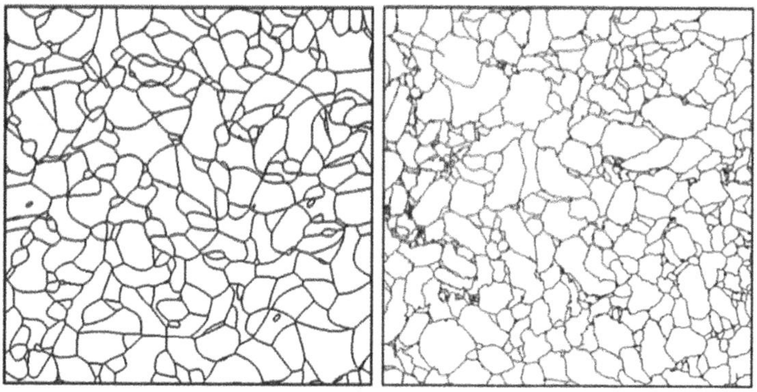

a) b)

Figure 2.4. *(a) Two-dimensional section of a Johson–Mehl random partition modeling a polycrystal; (b) segmented experimental image (source: image (a) Gasnier* et al. *2015b; image (b) Philippe Lambert/CEA le Ripault)*

2.7. Gaussian fields

In this section, we briefly define Gaussian random functions, as well as Gaussian excursions, and describe how they may be simulated. The theory of Gaussian functions was originally studied by Rice (1944) in the one-dimensional case and by Longuet-Higgins (1957) in two-dimensional case. Bardeen *et al.* (1985) define a Gaussian random function as a function whose n-point distribution laws are multi-variate Gaussian.

Gaussian excursion models allow the simulation of microstructures with a fixed covariance function. The covariance, in fact, fully determines the Gaussian random function. The remainder of this section explains how to generate such a model. For a detailed description of Gaussian random functions, see Lantuéjoul (1994, 2002).

Let $U(x)$ be a Gaussian noise, namely such that $U(x)$ is a normal distribution for all x and $U(x)$, $U(x')$ are independent if $x \neq x'$.

We consider the random function $Z(x)$ and the random set X:

$$X=\left\{x\in \mathbb{R}^d ; Z(x)\geq \lambda\right\} \qquad [2.70a]$$

$$Z(x)=(w*U)(x), \qquad U(x)\sim \mathcal{N}(0,1) \qquad [2.70b]$$

where w is a weight function, λ a scalar, designates a convolution product, and $\mathcal{N}(0,1)$ is the normal distribution of zero mean and variance 1. The mean of the indicator function of X is equal to:

$$f = P\{Z(x)\geq \lambda\} = P\{\mathcal{N}(0,1)\geq \lambda\} \qquad [2.71]$$

which determines λ according to f:

$$\lambda = F^{-1}(1-f) \qquad [2.72]$$

where F is the cumulative distribution function of the normal distribution.

The weight function w verifies:

$$\int_{\Omega} w^2(x)\mathrm{d}x = 1, \qquad w(x)=w(-x) \qquad [2.73]$$

This is related to the covariance $C(h)$ by X (Bron and Jeulin 2011):

$$C(h)=\frac{1}{2\pi}\int_0^{\rho(h)} \frac{1}{\sqrt{1-t^2}} e^{\frac{-\lambda^2}{1+t}} \mathrm{d}t \qquad [2.74]$$

where $\rho_X(h)=(w*w)(h)$.

The function w is obtained by means of:

$$w = FT^{-1}\left\{\sqrt{FT\{\rho\}}\right\} \qquad [2.75]$$

where FT and FT^{-1} are the direct and inverse Fourier transforms in $\mathbb{R}^d$. Once λ is calculated, the function ρ is obtained by numerically inverting [2.74] and w is given by [2.75].

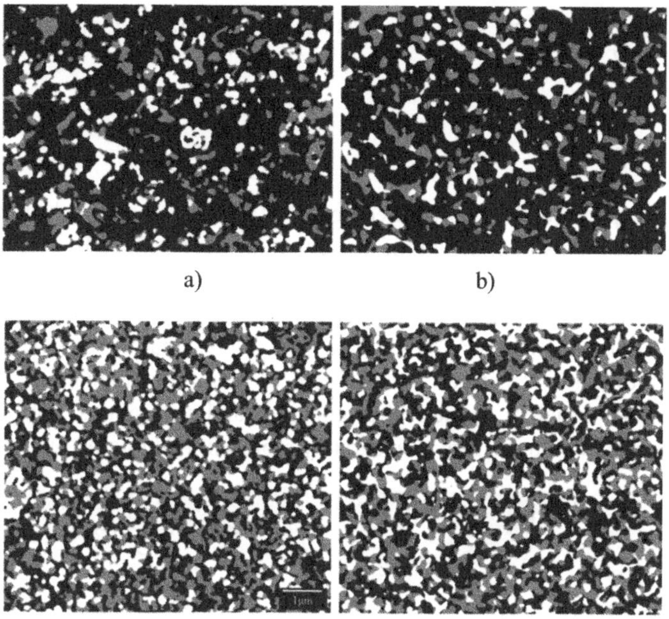

a) b) c) d)

Figure 2.5. *(a and c) SEM image (after segmentation) of fuel cell anode layers; (b and d) plurigaussian model (source: Abdallah* et al. *2015)*

A model based on Gaussian excursions was developed by Abdallah *et al.* (2015) (see Figure 2.5). To simulate a three-phase random medium, it is assumed that there are two independent random sets X and Y, such that one of the phases is given by X, another by $Y \setminus X$ and the third by the complementary of the union of the other two. Due to the independence of the random sets X and Y, the cross-covariance C_{12} of the phases 1 and 2 is given by the covariances C_1 and C_2 of these same phases. In effect:

$$C_{12}(h) = P\{x \in phase(1); x+h \in phase(2)\} = \frac{C_2(0)[C_1(0)-C_1(h)]}{1-C_1(0)} \quad [2.76]$$

This property allows the testing of whether modeling using independent sets X and Y is possible, regardless of the choice of X and Y. Moreover, the covariances of X and Y can be calculated from those of phases 1 and 2. Figure 2.5 represents the results obtained in the case of Gaussian excursions for X and Y. This model is used to represent the microstructure of anode layers consisting of two solid phases and a porous phase, existing in some fuel cells and for which only two-dimensional representations are known. The model has been compared to a graph-based approach (Neumann *et al.* 2019).

2.8. Conclusion

In this chapter, we have presented some of the most commonly used random models. They are parsimonious in the sense that they are defined by few parameters. In general, they can be easily simulated and, in some cases, analytical results are available. They can be directly used to model a particular microstructure, or combined with one another to generate sets representative of microstructures, especially multiscale structures (an example of such a combination is shown in Figure 2.6). Nevertheless, numerical calculations and simulations, not discussed in this chapter, are often necessary to accurately represent real microstructures.

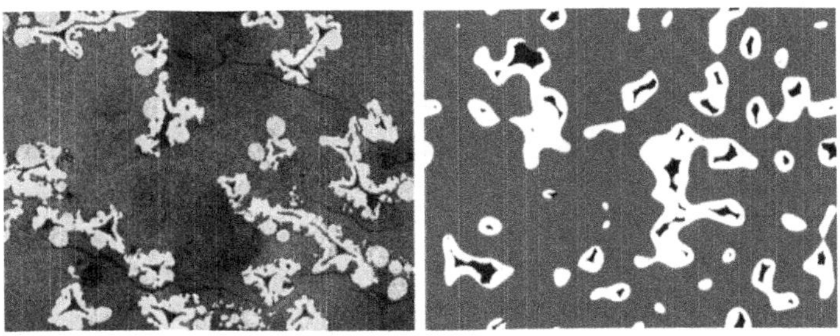

a) b)

Figure 2.6. *(a) SEM image of a fuel cell foam; (b) realization of a "Boolean random shell model" (source: Abdallah (2015), image (a) A. Chesnaud/Mines ParisTech)*

The main difficulty with these approaches lies in the determination of the relevant morphological criteria and the choice of model. To illustrate this problem, refer to Figure 2.7, which represents three isotropic microstructures with the same covariance and same three-point functions (Chilès and Lantuéjoul 2005). This example illustrates the extent to which the covariances and even the trivariate functions remain "blind" and do not allow us to distinguish geometries that are very different from one another. Within the context of random set approaches, the issue surrounding the choice of model remains an open problem.

a) b) c)

Figure 2.7. *Random realizations of three probabilistic models with the same trivariate function: (a) Gaussian excursion; (b) random Poisson partition; (c) dead-leaves model (source: Chilès and Lantuéjoul 2005)*

2.9. Acknowledgments

The author thanks C. Lantuéjoul for the images in Figure 2.7, H. Trumel for the image in Figure 2.4(b), C. Gogu and Delphine Livet for their proofreading and G. Moreira for translating this text into English.

2.10. References

Abdallah, B. (2015). Analyse morphologique et modélisation pour l'optimisation structurelle d'électrodes. PhD Thesis, Paris School of Mines, Paris.

Abdallah, B., Willot, F., Jeulin, D. (2015). Stokes flow through a Boolean model of spheres: Representative volume element. *Transport in Porous Media*, 109(3), 711–726.

Altendorf, H. and Jeulin, D. (2011). Random-walk-based stochastic modeling of three-dimensional fiber systems. *Physical Review E*, 83(4), 041804.

Altendorf, H., Jeulin, D., Willot, F. (2014). Influence of the fiber geometry on the macroscopic elastic and thermal properties. *International Journal of Solids and Structures*, 51(23), 3807–3822.

Angulo, J. (2012). Morphologie mathématique pour des images multi-variées. PhD Thesis, University of Paris-Est, Marne-la-Vallée.

Avrami, M. (1939). Kinetics of phase change. I general theory. *The Journal of Chemical Physics*, 7(12), 1103–1112.

Azzimonti, D., Willot, F., Jeulin, D. (2013). Optical properties of deposit models for paint: Full-fields FFT computations and representative volume element. *Journal of Modern Optics*, 60(7), 519–528.

Bardeen, J., Szalay, A., Kaiser, N., Bond, J. (1985). The statistics of peaks of Gaussian random fields. *The Astrophysical Journal*, 304, 15–61.

Berryman, J. and Milton, G. (1985). Normalization constraint for variational bounds on fluid permeability. *The Journal of Chemical Physics*, 83(2), 754–760.

Bignonnet, F. (2018). Upper bounds on the permeability of random porous media. *Transport in Porous Media*, 122(1), 57–76.

Bochner, S. (1959). *Lectures on Fourier Integrals*. Princeton University Press, Princeton.

Bortolussi, V., Figliuzzi, B., Willot, F., Faessel, M., Jeandin, M. (2018). Morphological modeling of cold spray coatings. *Image Analysis and Stereology*, 37(2), 145–158.

Bron, F. and Jeulin, D. (2011). Modelling a food microstructure by random sets. *Image Analysis and Stereology*, 23(1), 33–44.

Brumberger, H. and Goodisman, J. (1983). Voronoï cells: An interesting and potentially useful cell model for interpreting the small-angle scattering of catalysts. *Journal of Applied Crystallography*, 16(1), 83–88.

Calka, P. (2002). The distributions of the smallest disks containing the Poisson-Voronoï typical cell and the Crofton cell in the plane. *Advances in Applied Probability*, 34(4), 702–717.

Calka, P. (2003a). An explicit expression for the distribution of the number of sides of the typical Poisson-Voronoï cell. *Advances in Applied Probability*, 35(4), 863–870.

Calka, P. (2003b). Precise formulae for the distributions of the principal geometric characteristics of the typical cells of a two-dimensional Poisson-Voronoï tessellation and a Poisson line process. *Advances in Applied Probability*, 35(3), 551–562.

Chilès, J.-P. and Delfiner, P. (2009). *Geostatistics: Modeling Spatial Uncertainty*. John Wiley & Sons, New York.

Chilès, J.-P. and Lantuéjoul, C. (2005). Prediction by conditional simulation. In *Space, Structure and Randomness. Contributions in Honor of Georges Matheron in the Fields of Geostatistics, Random Sets and Mathematical Morphology*, Bilodeau, M., Meyer, F., Schmitt, M. (eds). Springer, New York.

Choquet, G. (1954). Theory of capacities. *Annales de l'Institut Fourier*, 5, 131–295.

Christakos, G. (1984). On the problem of permissible covariance and variogram models. *Water Resources Research*, 20(2), 251–265.

Cressie, N. and Laslett, G.M. (1987). Random set theory and problems of modeling. *SIAM Review*, 29(4), 557–574.

Daley, D.J. and Vere-Jones, D. (1988). *An Introduction to the Theory of Point Processes*. Springer-Verlag, New York.

Doi, M. (1976). A new variational approach to the diffusion and the flow problem in porous media. *Journal of the Physical Society of Japan*, 40(2), 567–572.

Emery, X. and Lantuéjoul, C. (2011). Geometric covariograms, indicator variograms and boundaries of planar closed sets. *Mathematical Geosciences*, 43(8), 905–927.

Escoda, J. (2012). Modélisation morphologique et micromécanique 3D de matériaux cimentaires. PhD Thesis, École nationale supérieure des mines, Paris.

Escoda, J., Jeulin, D., Willot, F., Toulemonde, C. (2015). 3D morphological modeling of concrete using multiscale Poisson polyhedra. *Journal of Microscopy*, 258(1), 31–48.

Farjas, J. and Roura, P. (2008). Cell size distribution in a random tessellation of space governed by the Kolmogorov-Johnson-Mehl-Avrami model: Grain size distribution in crystallization. *Physical Review B*, 78(14), 144101.

Ferenc, J.-S. and Néda, Z. (2007). On the size distribution of Poisson Voronoi cells. *Physica A: Statistical Mechanics and its Applications*, 385(2), 518–526.

Figliuzzi, B. (2019). Eikonal-based models of random tesselations. *Image Analysis and Stereology*, 38(1), 15–23.

Gasnier, J., Figliuzzi, B., Faessel, M., Willot, F., Jeulin, D., Trumel, H. (2015a). 3D morphological modeling of a polycrystaline microstructure with non-convex, anisotropic grains. In *Acta Stereologica. Proceedings of the 14th International Congress for Stereology and Image Analysis (ICSIA)*. Liège, 7–10 July [Online]. Available at: https://hal.archives-ouvertes.fr/hal-01184811 [Accessed 2 October 2020].

Gasnier, J., Willot, F., Trumel, H., Figliuzzi, B., Jeulin, D., Biessy, M. (2015b). A Fourier-based numerical homogenization tool for an explosive material. *Matériaux et techniques*, 103(3), 308.

Gille, W. (1987). The intercept length distribution density of a cylinder of revolution. *Experimentelle Technik der Physik*, 35(2), 93–98.

Gille, W. (2002). The set covariance of a dead leaves model. *Advances in Applied Probability*, 34(1), 11–20.

Gille, W. (2016). *Particle and Particle Systems Characterization: Small-Angle Scattering (SAS) Applications*. CRC Press, Boca Raton.

Greco, A., Jeulin, D., Serra, J. (1979). The use of the texture analyser to study sinter structure: Application to the morphology of calcium ferrites encountered in basic sinters of rich iron ores. *Journal of Microscopy*, 116(2), 199–211.

Hadwiger, H. (1957). Über eibereiche mit gemeinsamer treffgeraden. *Portugaliae mathematica*, 16(1), 23–29.

Heinemann, A., Hermann, H., Wetzig, K., Häussler, F., Baumbach, H., Kröning, M. (1999). Fractal microstructures in hydrating cement paste. *Journal of Materials Science Letters*, 18(17), 1413–1416.

Heinrich, L. (1992). On existence and mixing properties of germ-grain models. *Statistics*, 23(3), 271–286.

Jean, A., Jeulin, D., Forest, S., Cantournet, S., N'Guyen, F. (2011). A multiscale microstructure model of carbon black distribution in rubber. *Journal of Microscopy*, 241(3), 243–260.

Jeulin, D. (1979). Morphologie mathématique et propriétés physiques des agglomérés de minerais de fer et du coke métallurgique. PhD Thesis, École nationale supérieure des Mines, Paris.

Jeulin, D. (2013). Random tessellations and Boolean random functions. In *Mathematical Morphology and Its Applications to Signal and Image Processing*, Luengo Hendriks, C.L., Borgefors, G., Strand, R. (eds). Springer, Berlin.

Jeulin, D. (2016). Power laws variance scaling of Boolean random varieties. *Methodology and Computing in Applied Probability*, 18(4), 1065–1079.

Jeulin, D. (2019). Some dense random packings generated by the dead leaves model. *Image Analysis and Stereology*, 38(1), 3–13.

Jeulin, D. and Ostoja-Starzewski, M. (2001). *Mechanics of Random and Multiscale Microstructures*. Springer, Berlin.

Jeulin, D., Villalobos, I.T., Dubus, A. (1995). Morphological analysis of UO, powder using a dead leaves model. *Microscopy Microanalysis Microstructures*, 6(4), 371–384.

Kendall, D. (1974). Foundations of a theory of random sets. In *Stochastic Geometry: A Tribute to the Memory of Rollo Davidson*, Harding, E.F., Kendall, D.G. (eds). Wiley, London.

Lantuéjoul, C. (1991). Ergodicity and integral range. *Journal of Microscopy*, 161(3), 387–403.

Lantuéjoul, C. (1994). Non conditional simulation of stationary isotropic multi-Gaussian random functions. In *Geostatistical Simulations*, Armstrong, M., Dowd, P.A. (eds). Springer, Berlin, 147–177.

Lantuéjoul, C. (2002). *Geostatistical Simulation: Models and Algorithms*. Springer, Berlin.

Lautensack, C. and Zuyev, S. (2008). Random Laguerre tessellations. *Advances in Applied Probability*, 40(3), 630–650.

Lee, A., Mumford, D., Huang, J. (2001). Occlusion models for natural images: A statistical study of a scale-invariant dead leaves model. *International Journal of Computer Vision*, 41(1–2), 35–59.

Månsson, M. and Rudemo, M. (2002). Random patterns of nonoverlapping convex grains. *Advances in Applied Probability*, 34(4), 718–738.

Matheron, G. (1965). *Les variables régionalisées et leur estimation : une application de la théorie des fonctions aléatoires aux sciences de la nature*. Masson, Paris.

Matheron, G. (1967). *Éléments pour une théorie des milieux poreux*. Masson, Paris.

Matheron, G. (1968). Schéma booléen séquentiel de partition aléatoire. Technical report, N-83, Centre de morphologie mathématique, École nationale supérieure des Mines, Paris. [Online]. Available at: http://cg.ensmp.fr/bibliotheque/public/MATHERON_Rapport_00121.pdf [Accessed 2 October 2020].

Matheron, G. (1969). *Théorie des ensembles aléatoires.* École nationale supérieure des Mines, Paris.

Matheron, G. (1972). Ensembles fermés aléatoires, ensembles semi-markoviens et polyèdres poissoniens. *Advances in Applied Probability*, 4(3), 508–541.

Matheron, G. (1975). *Random Sets and Integral Geometry.* Wiley, New York.

Matheron, G. (1989a). *Estimating and Choosing: An Essay on Probability in Practice.* Springer, Berlin.

Matheron, G. (1989b). The internal consistency of models in geostatistics. In *Geostatistics*, Armstrong, M. (ed.). Springer, Dordrecht.

Matheron, G. and Serra, J. (1982). *Image Analysis and Mathematical Morphology.* Academic Press, London.

Miles, R.E. (1964). Random polytopes: The generalisation to n dimensions of the intervals of a Poisson process. PhD Thesis, University of Cambridge, Cambridge.

Miles, R.E. (1974). A synopsis of "Poisson flats in Euclidean spaces". *Stochastic Geometry*, 2, 202–227.

Milton, G.W. (2002). *The Theory of Composites.* Cambridge University Press, Cambridge.

Molchanov, I.S. (2005). *Theory of Random Sets.* Springer, Berlin.

Møller, J. (1989). Random tessellations in R^d. *Advances in Applied Probability*, 21(1), 37–73.

Neumann, M., Abdallah, B., Holzer, L., Willot, F., Schmidt, V. (2019). Stochastic 3D modeling of three-phase microstructures for the prediction of transport properties in solid oxide fuel cells. *Transport in Porous Media*, 128(1), 179–200.

Quénec'h, J.-L., Chermant, J.-L., Coster, M., Jeulin, D. (1996). Example of application of probabilistic models: Determination of the kinetics parameters during liquid phase sintering. *Microscopy Microanalysis Microstructures*, 7(5–6), 573–580.

Redenbach, C. (2009). Microstructure models for cellular materials. *Computational Materials Science*, 44(4), 1397–1407.

Redenbach, C., Wirjadi, O., Rief, S., Wiegmann, A. (2011). Modeling a ceramic foam for filtration simulation. *Advanced Engineering Materials*, 13(3), 171–177.

Rice, S.O. (1944). Mathematical analysis of random noise. *Bell System Technical Journal*, 23(3), 282–332.

Schneider, R. and Weil, W. (2008). *Stochastic and Integral Geometry.* Springer-Verlag, Berlin, Heidelberg.

Serra, J. (1980). The Boolean model and random sets. *Computer Graphics and Image Processing*, 12(2), 99–126.

Serra, J. (1981). The Boolean model and random sets. In *Image Modeling*, Rosenfield, A. (ed.). Academic Press, Cambridge.

Serra, J. (1983). *Image Analysis and Mathematical Morphology*. Academic Press, Cambridge.

Stroeven, P. (2000). A stereological approach to roughness of fracture surfaces and tortuosity of transport paths in concrete. *Cement and Concrete Composites*, 22(5), 331–341.

Talbot, J., Tarjus, G., Van Tassel, P., Viot, P. (2000). From car parking to protein adsorption: An overview of sequential adsorption processes. *Colloids and Surfaces A: Physicochemical and Engineering Aspects*, 165(1–3), 287–324.

Torquato, S. (1995). Nearest-neighbor statistics for packings of hard spheres and disks. *Physical Review E*, 51(4), 3170.

Torquato, S. (2013). *Random Heterogeneous Materials: Microstructure and Macroscopic properties*. Springer, Berlin.

Van Lieshout, M. (2000). *Markov Point Processes and their Applications*. Imperial College Press, London.

Willot, F. (2015). The power laws of geodesics in some random sets with dilute concentration of inclusions. *Lecture Notes in Computer Science*, 9082, 535–546.

Willot, F. (2017). Mean covariogram of cylinders and applications to Boolean random sets. *Journal of Contemporary Mathematical Analysis*, 52(6), 305–315.

Willot, F. and Jeulin, D. (2011). Elastic and electrical behavior of some random multiscale highly-contrasted composites. *International Journal for Multiscale Computational Engineering*, 9(3), 305–326.

Willot, F., Abdallah, B., Jeulin, D. (2016). The permeability of Boolean sets of cylinders. *Oil and Gas Science and Technology*, 71(2), 1.

3

Uncertainty Propagation at the Scale of Aging Civil Engineering Structures

David BOUHJITI, Julien BAROTH and Frédéric DUFOUR
Grenoble Aples University, France

3.1. Introduction

Strategic civil engineering structures, particularly those with a containment role, are mostly made up of reinforced and prestressed concrete. On the one hand, this material provides a good trade-off between cost, resistance, water tightness and durability[1]. On the other hand, the use of reinforced and prestressed concrete creates additional difficulties in controlling damage or cracking and their predictability over time, under normal and accidental service conditions.

These difficulties can be caused by:

– *the complexity of the concrete material*: like all materials, concrete is, by nature, in permanent thermodynamic imbalance (thermal, hydric, mechanical, hydraulic and chemical imbalances). Several mechanisms take place in order to tend, over time, toward a stable and balanced state between the different phases (liquid, vapor, solid skeleton) that compose concrete. However, given the perpetual changes in the environment, this equilibrium state is hard to reach. This induces an "aging" effect

1 It should be remembered that concrete is 15–20 times cheaper than steel and that, in its healthy state, its intrinsic permeability is of the order of 10^{-17} to 10^{-20} m^2 to air and water, respectively.

Mechanical Engineering under Uncertainties,

on concrete properties and on its response to applied loads of a thermal, hydric or mechanical nature. Among the various processes associated with aging, there are, among others (Bouhjiti 2018): concrete hydration, maturation, delayed strains (endogenous shrinkage and drying shrinkage, creep, etc.), under simultaneous stresses (dead weight, residual stresses, prestressing, fluid thrust, etc.), chemical damage (leaching phenomena, formation of secondary ettringite, etc.), thermal damage (particularly at high temperatures above 80°C) or mechanical damage (by restrained shrinkage or high load level). It is useful to underline that these phenomena are coupled and strongly nonlinear, and that the level of coupling is, to date, only partially explored. However, despite this partial knowledge of the dynamics of concrete aging, several models are proposed in the literature. Provided with the right input parameters, these models are capable of reproducing the behavior observed at the material scale. The blind prediction of the structural behavior remains less obvious;

– *The multitude of sources of uncertainty related to the behavior of concrete*: among the sources of uncertainty, and in accordance with the classification proposed in Baroth *et al.* (2011), Sudret (2007) and Ditlevsen and Madsen (2005), we can distinguish:

- measurable random uncertainties: these are related to the heterogeneous nature of concrete, the variability of its properties in space and time and the variability of the applied structural loads;

- epistemic uncertainties due to simplifying assumptions in the models, the precision of experimental measurements on site, insufficient identification data, etc.;

- ontological uncertainties due to inevitable and unpredictable overflows following a non-compliant implementation procedure or a response outside the scope of the initial assumptions, or regarding the validity of the physical principles under consideration.

At the large structural scale, and given the large number of phenomena involved, these uncertainties, and more specifically the quantifiable ones, induce a heterogeneous response of the concrete volume and deserve to be taken into account during modeling to predict an accurate cracking pattern (number and position of cracks). The latter are a major concern for improving the reliability of the models and their representativeness of the real behavior of concrete structures, especially because defects will play a major role in nonlinear behavior.

Therefore, the foreseen mathematical models and numerical methods would be those verifying (a) an acceptable physical representativeness[2] at material and

2 That is to say, capable of reproducing the observed behavior with errors of the same order of magnitude as the measurement.

structural scales; (b) a guaranteed numerical robustness with respect to spatial and temporal discretizations when solving descriptive equations; and (c) above all, taking into account uncertainties through probabilistic approaches compatible with the large number of physical parameters and with expensive aging calculations at the large structural scale. As such, in this chapter, the objective is to introduce stochastic finite element methods (SFEM) (Matthies *et al.* 1997; Sudret and Der Kiureghian 2000; Soize and Ghanem 2004; Lemaire 2009; Stefanou 2009), and to explore their potential once applied to the problem of aging in concrete structures. At the end of this application, conclusions and questions are drawn regarding the effectiveness and efficiency of such methods to represent realistic behavior of reinforced and prestressed concrete at the structural scale.

3.2. Problem positioning

3.2.1. *Probabilistic formulation*

In the following, Ω, F and P denote a probability space, where Ω is the set of possible outcomes and P is a probability distribution on the set F. The problem to be solved consists of quantifying the effect on the mechanical response **z** of the uncertainty affecting a vector of uncertain parameters **y**. It is assumed that these parameters, variable in space or not, can be modeled by a random vector variable (r.v.) Y, defined by its distribution. The problem is probabilized in the form f(Y), where f is defined by the physical transfer function (analytical or numerical, regardless of the numerical code used). In theory, for any continuous probability distribution, a Gaussian normalization can be performed (Nataf 1962; Der Kiureghian and Liu 1986; Nelson 1999). The use of a Gaussian distribution has the advantage of simplifying the calculations, or even making them explicit. Nevertheless, this distribution has the disadvantage that it is sometimes unrealistic, in particular when a physical parameter takes a negative value.

This normalization consists of defining a transformation, denoted T, which links the input r.v. Y and a standard Gaussian vector X [3.1]:

$$Y = T(X)$$

$$g(X) = f(Y) = f \circ T(X) \qquad [3.1]$$

where X is a standard N-dimensional Gaussian r.v.

The objective is to characterize g by approximating its mean, its covariance matrix, or even its probability density function (PDF) or its cumulative distribution function (CDF).

3.2.2. *Thermo-hydro-mechanical-leakage transfer function*

The transfer function f chosen in this chapter reflects a thermo-hydro-mechanical modeling with posttreatment of leaks at particular times (THM+L).

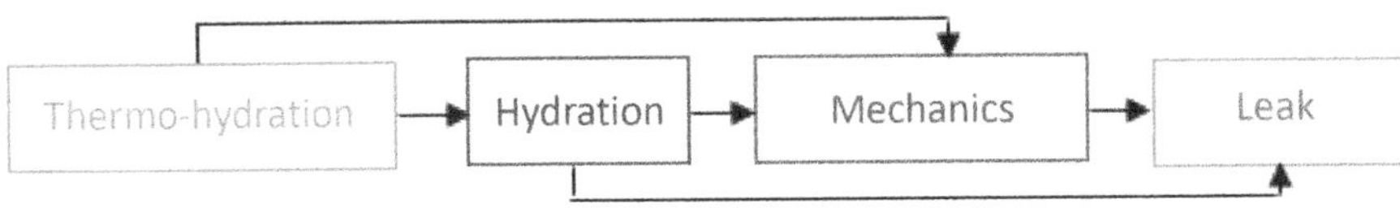

Figure 3.1. *Principle of a weakly chained THM+L modeling. For a color version of this figure, see www.iste.co.uk/gogu/uncertainties.zip*

Under normal service conditions (temperature below 50°C, compressive stress limited to 30% of the strength, hydraulic flows in thickness mainly laminar), weakly coupled (or chained, see Figure 3.1) THM+L modeling is sufficient to describe the aging behavior of concrete (Bouhjiti *et al.* 2018b). This consists of solving, in series (Bouhjiti *et al.* 2018a):

– the thermal problem **T**, with temperature T as the main unknown, described by the heat equation with a source term in [3.2]:

$$\mathbf{T}: \dot{\beta}(T) - \nabla \cdot (\lambda_c \nabla T) = Q \tag{3.2}$$

with $\dot{\beta}(T)$, λ_c the enthalpy and thermal conductivity, respectively, and Q the source descriptive term of heat production in the hydration phase;

– the mass transfer problem **H**, with water content C as the main unknown, described by the diffusion equation with a nonlinear diffusivity term in [3.3]:

$$\mathbf{H}: \dot{C} - \nabla \cdot (D(C, T)\, \nabla C) = 0 \tag{3.3}$$

with $D(C, T)$ the nonlinear term of water diffusivity as a function of water content C and temperature T;

– the mechanical problem **M**, with strain tensor ε and stress tensor σ as the main unknowns[3], described by superimposing the effects of each aging phenomenon in [3.4]:

$$\mathbf{M}: \begin{cases} \varepsilon = \varepsilon_{EL} + \varepsilon_{TH} + \varepsilon_{ES} + \varepsilon_{DS} + \varepsilon_{FL} \\ \sigma = (1-d)\mathbb{C} : \varepsilon_{EL} \end{cases} \quad [3.4]$$

with ε_{EL} the elasto-damageable strain tensor, ε_{TH} the thermal strain tensor associated with the time variation of the temperature, ε_{ES} the endogenous strain tensor associated with the chemical shrinkage induced by the hydration of the cement in the concrete, ε_{DS} the desiccation shrinkage tensor associated with the temporal variation in moisture content, ε_{FL} the creep strain tensor as a function of temperature and water content, $\mathbb{C}$ the elastic stiffness tensor in the absence of damage and d the damage (scalar) variable as a function of a local strain flow rule regulated in cracking energy;

– the hydraulic transfer (or leakage) problem **F**, with permeability k and mass flow q as the main unknowns, described by the diffusion equation [3.5]:

$$\mathbf{F}: \begin{cases} \dot{P} - \nabla \cdot \left(\frac{k}{2\,\mu} \nabla \cdot P^2\right) = 0 \\ q = -\frac{k}{\mu} \nabla \cdot P \end{cases} \quad [3.5]$$

with P the pressure field of the fluid, k = k(C, ε) the fluid permeability field and μ the dynamic viscosity of the fluid.

So the response $\mathcal{R}$ looked for at any point $\vec{x}$ of the space domain of interest $\mathcal{B}$ and at all times t is written by function composition T, H, M, F [3.6] as:

$$\mathcal{R}_{\mathcal{B}} = f(\vec{x}, t) = (\mathbf{F} \circ \mathbf{M} \circ \mathbf{H} \circ \mathbf{T})(\vec{x}, t) \quad [3.6]$$

3.2.3. *Resulting probabilistic THM-F problem*

The problem is therefore probabilized in the form of $\mathcal{R}_{\mathcal{B}} = f(Y) = g \circ T(X)$, where $\mathcal{R}_{\mathcal{B}} = f$ is defined by the model response (in our case, it is a thermo-hydro-mechanical response evaluated by finite element calculations over a spatial domain

3 Crack distributions and openings can be directly post-treated based on the strain field (Bouhjiti *et al.* 2018a).

of interest $\mathcal{B}$). More specifically, the input random vectors of the models **T**, **H**, **M**, **F**, of respective dimensions n_T, n_H, n_M, n_F are denoted $Y_T^{1\leq i\leq n_T}, Y_H^{1\leq i\leq n_H}, Y_M^{1\leq i\leq n_M}$, $Y_F^{1\leq i\leq n_F}$. The response $\mathcal{R}_\mathcal{B}$ therefore requires the identification of $N = n_T + n_H + n_M + n_F$ variables.

The objective is to characterize, as much as possible, $\mathcal{R}_\mathcal{B}$ and the various THM+L intermediate responses by approximating their means, standard deviations (or covariance matrices), PDFs or CDFs. To this end, initial and boundary conditions are assumed to be well defined and regular enough for the problem to be well posed, as defined by Hadamard[4].

3.3. Random field-based modeling of material properties

This section introduces random fields (r.f.) and how uncertain parameters **y** can be modeled by discretized r.f., autocorrelated or not, on a finite element mesh.

3.3.1. *Random fields*

A *random* field $\mathbb{H}(\mathbf{z}, \omega)$ is defined as a family of random variables indexed by a continuous parameter $\mathbf{z} \in \mathcal{B}$ where $\mathcal{B}$ is an opening of $\mathbb{R}^D$ representing either the geometry of the system considered in space (D varies between 1 and 3) or time (in which case D = 1) or both (D = 4). When there is an index variable of time, it is referred to rather as a stochastic process. Therefore, for a given $\mathbf{z_0} \in \mathcal{B}$, $\mathbb{H}(\mathbf{z_0}, \omega)$ is a random variable and for a given event ω_0, $\mathbb{H}(\mathbf{z},\omega_0)$ is a realization (trajectory) of the random field.

A field is said to be *homogeneous* (stationary for a stochastic process) if it is translation-invariant in the indexing domain (space and/or time): its statistical moments remain constant at all points. In the case where only the first two moments are constant (mean and variance), the field is said to be stationary on average of order 2.

3.3.2. *Generation methods for discretized random fields*

In order to assign discrete values representative of a random field $\mathbb{H}$, several discretization procedures are used and consist of approximating $\mathbb{H}$ by $\widehat{\mathbb{H}}$, defined as a discretized n-vector random variable vector χ of random variables $\{\chi_{1\leq i\leq n}\}$:

---

4 There is a solution. It is unique. It depends continuously on its parameters within the context of a reasonable topology.

$$\mathbb{H}(\mathbf{z}) = \widehat{\mathbb{H}}(\mathbf{z}) + o\left(\left\|\widehat{\mathbb{H}} - \mathbb{H}\right\|\right)$$
$$\widehat{\mathbb{H}}(\mathbf{z}) = \chi = \{\chi_i, i = 1, \dots, n\} \quad [3.7]$$

In the case of probabilistic modeling by a random field, performing a discrete or functional approximation is required (Baroth 2005).

3.3.2.1. *Discrete approximations*

The components of the random field $\chi = \{\chi_{1 \leq i \leq n}\}$ are obtained either based on observations at specific points (integration points, centers of gravity or element nodes) or on spatial averages of the elements resulting, for example, from a FE mesh. Among these methods, one can cite:

– midpoint methods or integration point methods where the random field is represented by the set of its values at the centroids or integration points of the elements $\mathbf{z}_c$ (Der Kiureghian and Ke 1988; Yamazaki *et al.* 1988; Deodatis 1989; Matthies *et al.* 1997). Such a representation introduces discontinuities between elements and tends to overestimate variability within the element (Lemaire 2009):

$$\mathbb{H}(\mathbf{z} = \mathbf{z}_c) = \widehat{\mathbb{H}}(\mathbf{z} = \mathbf{z}_c) = \chi \quad [3.8]$$

– nodal point methods (Hisada *et al.* 1983) such as the polynomial interpolation method (Matthies *et al.* 1997), the shape function method (Vanmarcke 2010) and the optimal linear estimation method (Li *et al.* 1993), known as OLE, which represents a particular application of kriging techniques (Ditlevsen 1991):

$$\widehat{\mathbb{H}}(\mathbf{z}) = \sum_{i=1}^{n} N_i(z)\, \mathbb{H}(\mathbf{z}_i) = \sum_{i=1}^{n} N_i(z)\, \chi_i \quad [3.9]$$

– the turning band method (Matheron 1973) where isotropic random functions on $\mathbb{R}^D$ are built from a process on $\mathbb{R}$ and by making use of an orthogonal random measure on the half sphere unit of $\mathbb{R}^D$. This measure is then estimated using conventional quadrature methods. It is useful to specify that the simulated turning band covariance is not stationary and that the quantification of the bias introduced by the discretization of the reduced domain $\mathbb{R}$ is not obvious. However, it has been shown that this bias tends toward zero when spatial discretization also tends toward zero (Freulon and de Fouquet 1991):

– local averaging methods (LAMs) where each of the components χ_i of the vector random χ is the spatial average of the field $\mathbb{H}$ on an element $\mathcal{B}_i$ (Cornell 1971; Zhu and Ren 1986; Der Kiureghian and Ke 1988; Vanmarcke 2010):

$$\chi_i = \int_{\mathcal{B}_i} \frac{\mathbb{H}(\mathbf{z})}{|\mathcal{B}_i|} d\mathcal{B}_i \qquad \mathbf{z} \in \mathcal{B}_i \quad [3.10]$$

– weighted integral methods (Graham and Deodatis 2011) where the discretization of the Gaussian field depends on the shape functions of the FE discretization.

In all the discrete approximation methods, one must try to generate a random field $\hat{\mathbb{H}}$ formed by a finite number of random variables, and shape or interpolation functions, allowing the complete field $\hat{\mathbb{H}}$ to be generated on Ω. In other words, one is looking for the projection of $\mathbb{H}$ on the subspace of the r.f. of Ω, generated by the family of n products of a random variable χ_i of (Ω, F, P) and a shape function ϕ_j from Ω (in the case of a Gaussian process $\mathbb{H}$, some χ_i of Gaussian probability density will of course be chosen). An approximation $\hat{\mathbb{H}}$ of $\mathbb{H}$ is then written as:

$$\hat{\mathbb{H}}(\mathbf{z}) = \sum_{i=1}^{n} \phi_i(z)\, \mathbb{H}(\mathbf{z}_i) = \sum_{i=1}^{n} \phi_i(z)\, \chi_i\, ;\ z \in \Omega \qquad [3.11]$$

or:

$$\hat{\mathbb{H}}(\mathbf{z}) = \mu_{\mathbb{H}} + \sum_{i=1}^{n} \phi_i(z)\, \tilde{\chi}_i\ ;\ z \in \Omega \qquad [3.12]$$

with the χ_i of the Gaussian random variables, the $\tilde{\chi}_i$ of the centered Gaussian random variables and the ϕ_i of the maps of Ω in $\mathbb{R}$, whose expressions depend on the approximation method being considered; these are indicator functions in the case of the midpoint method and LAM and are interpolation or shape functions for the nodal point method and integration point method (Baroth 2005).

3.3.2.2. *Functional approximations*

The random field is represented by series of random variables and deterministic spatial functions. The approximation of the field is then obtained by truncating these series to a finite order P verifying, for example, an acceptable relative error[5] on the variance (Sudret and Der Kiureghian 2000). Several discretization methods are presented in the literature (Ghanem and Spanos 1991; Li *et al.* 1993; Dietlevsen and Madsen 1996; Matthies *et al.* 1997; Sudret and Der Kiureghian 2000). Among others, there is the Karhunen–Loève method, as well as spectral methods (development of orthogonal series called orthogonal series expansion [OSE], as a development of optimal linear expansion [EOLE]). These approximations are well

5 The definition of a target relative error is often made arbitrarily. However, it should be emphasized that, on the one hand, the target precision should mainly depend on the sensitivity of the model response to the variance error introduced by the random fields, and, on the other hand, that the spatial discretization of the finite elements should be defined in such a way so as not to degrade the quality of the random field once projected.

adapted in the context of the Monte Carlo method simulation. The field $\mathbb{H}$ can be written in the following form:

$$\hat{\mathbb{H}}(\mathbf{z}) = \mu_{\mathbb{H}} + \sum_{i=1}^{\infty} \phi_i(z)\,\xi_i \;;z \in \Omega \qquad [3.13]$$

where the ξ_i are independent centered Gaussian random variables and the ϕ_i are functions of Ω in $\mathbb{R}$ (ϕ_i are orthogonal in the OSE case and are derived from a spectral decomposition of the values at the nodes in the EOLE case). An approximation of $\mathbb{H}$ by a field $\hat{\mathbb{H}}$ by truncation to a certain order P the previous series is obtained as follows:

$$\hat{\mathbb{H}}(\mathbf{z}) \approx \mu_{\mathbb{H}} + \sum_{i=1}^{P} \phi_i(z)\,\xi_i \;;z \in \Omega \qquad [3.14]$$

This type of method consists of projecting the random field onto a certain functional basis independent of the FE mesh size. The two most common functional bases are that of the normalized Hermite polynomials and the eigenfunctions of the covariance operator $C_{\mathbb{H}(z)\mathbb{H}(z')}$ of the field $\mathbb{H}$. The determination of the functions ϕ_i consists of finding the couples (λ_i, ϕ_i) such that $\forall z \in \Omega$:

$$\int_{\Omega} C_{\mathbb{H}(z)\mathbb{H}(z')}\varphi_i(z')dz' = \lambda_i\varphi_i(z) \qquad [3.15]$$

The solution to [3.15] is not always analytically obvious; this requires the use of adapted numerical resolution methods (quadratures or other) (Ghanem and Spanos 1991; Phoon *et al.* 2002; Li *et al.* 1993; Zhang and Ellingwood 1994). Finally, the random field $\mathbb{H}$ takes the following form:

$$\begin{aligned} \mathbb{H}(\mathbf{z}) &= \mu_{\mathbb{H}} + \sum_{i=1}^{\infty}\sqrt{\lambda_i}\phi_i(z)\,\tilde{\xi}_i \\ \mathbb{H}(\mathbf{z}) &\approx \hat{\mathbb{H}}(\mathbf{z}) = \mu_{\mathbb{H}} + \sum_{i=1}^{P}\sqrt{\lambda_i}\phi_i(z)\,\tilde{\xi}_i \end{aligned} \;;z \in \Omega \qquad [3.16]$$

where the $\tilde{\xi}_i$ are independent copies of a random Gaussian variable ξ.

3.3.3. *Random fields and autocorrelations*

The covariance function between two spatial parameters $\mathbf{z}$ and $\mathbf{z}'$ of $\mathcal{B}$ is written as:

$$\begin{aligned} & C_{\mathbb{H}(z)\mathbb{H}(z')} = \sigma_{\mathbb{H}}^2\,\rho_{\mathbb{H}}(u = |z - z'|) \\ & \rho(\mathbb{H}(\mathbf{z},\omega), \mathbb{H}(\mathbf{z}',\omega)) = \rho_{\mathbb{H}}(\mathbf{z},\mathbf{z}') = \rho_{\mathbb{H}}(\mathbf{z} - \mathbf{z}') \end{aligned} \qquad [3.17]$$

with $\rho_{\mathbb{H}}$ the autocorrelation function of the field $\mathbb{H}$, depending only on the relative distance $u = |z - z'|$ and a scale of fluctuation l_{flu} designating the degree of dependency of two field values in two different positions. It is related to the autocorrelation function $\rho_{\mathbb{H}}$ by the relation (Vanmarcke 2010):

$$l_{flu} = \int_{-\infty}^{+\infty} \rho_{\mathbb{H}}(u)\, du = 2\int_{0}^{+\infty} \rho_{\mathbb{H}}(u)\, du \qquad [3.18]$$

For a given physical quantity X, the scale of fluctuation l_{flu} and the spatial correlation function $\rho_{\mathbb{H}}$ are identified based on variograms (Baroth *et al.* 2011). The relation between the covariance matrix and the semivariogram is defined by:

$$\begin{aligned} \gamma(h)_{\substack{= \\ \text{discrete}}} \frac{1}{2\,N_h} \sum\nolimits_{h=|z_i - z_j|} \left(x(z_i) - x(z_j)\right)^2 \\ C_{\mathbb{H}(z)\mathbb{H}(z')} = C_{\mathbb{H}(z)\mathbb{H}(z)} - \gamma(h = |z - z'|) \end{aligned} \qquad [3.19]$$

with N_h the number of couples of the distance h, γ the value of the semi-variogram and $x(z)$ the realization of the measure X at a given point z.

In Baroth (2005) and Sudret (2007), several expressions (limited to the stationary variograms in [3.19]) are cited to describe $\rho_{\mathbb{H}}$. The couple $(l_{flu}, \rho_{\mathbb{H}})$ that allows the best fit of the observed experimental variogram:

$$\begin{aligned} \rho_{\mathbb{H}}^{lin}\left(v = \frac{u}{l_{flu}}\right) &= \max(1 - v; 0) \\ \rho_{\mathbb{H}}^{exp}(u, l_{flu}) &= e^{-2\,v} \\ \rho_{\mathbb{H}}^{gau}(u, l_{flu}) &= e^{-\pi\, v^2} \\ \rho_{\mathbb{H}}^{sin}(u, l_{flu}) &= \frac{\sin(\pi\, v)}{\pi\, v} \end{aligned} \qquad [3.20]$$

3.3.4. *Application: contribution to modeling the cracking of reinforced concrete works by self-correlated r.f.*

3.3.4.1. *Context presentation*

In addition to their structural role, containment facilities of nuclear power reactors have a second functional role which aims to protect the environment from radioactive substances under accidental operational conditions. In double-walled vessels (see Figure 3.2), static containment is provided by the inner wall, made of prestressed reinforced concrete. French regulations (*Autorité de sûreté nucléaire*, ASN) require a maximum dry air leakage rate through this wall equivalent to 1.5% of the mass of air occupying the inner enclosure per day, and under a pressure gradient of 4.2 bar. Such verifications are carried out periodically (generally every ten years) by the operator Electricité de France (EDF), which sets a more penalizing threshold of 1.1% in order to take into account the "degrading" effects of concrete aging between two successive leakage checks. Among the effects that induce a loss of air tightness, one can mention:

– concrete cracking due to, on the one hand, the heterogeneity of the structural rigidity of concrete and tensile stresses induced by thermo-hydraulic gradients or external mechanical loads, and, on the other hand, by the random distribution of the mechanical properties of concrete, resulting in random spacing and length of cracks on site;

– concrete drying (thermo-activated phenomenon) inducing an increase in dry air permeability and an amplification of visco-elastic strains and shrinkages;

– prestressing losses due to creep and shrinkage of the concrete, as well as cable relaxation. One of the direct consequences is an increased risk of cracking and crack openings during pressurization of the vessel;

– local defects existing at the level of additional concreting operations or other causes of additional leaks, although marginal compared to the leaks through cracks or concrete mass, at the structural scale.

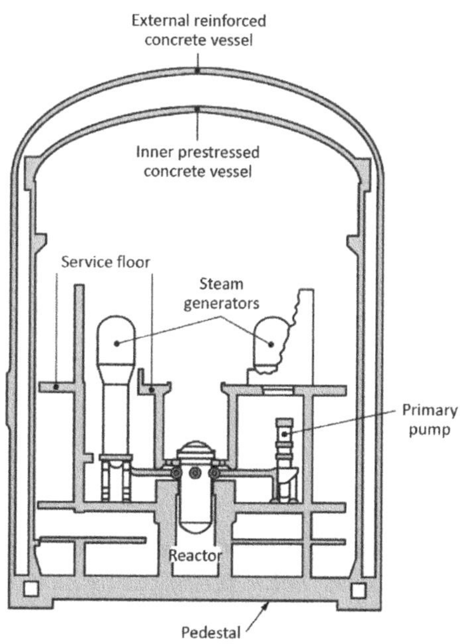

Figure 3.2. *Containment principle for double-walled nuclear vessels*

In this application, an experimental double-walled containment (VeRCoRs[6] project in Figure 3.3), on a scale of 1:3, is modeled. To do so, calculations are performed using finite elements and at the representative structural volume (RSV) scale of each structural part of the containment (see Figure 3.3). This has a twofold advantage:

– to allow a relatively fine spatial discretization to objectively describe local phenomena;

– to limit the calculation time compared to complete 3D simulations. This point is important for a numerically inexpensive probabilistic coupling.

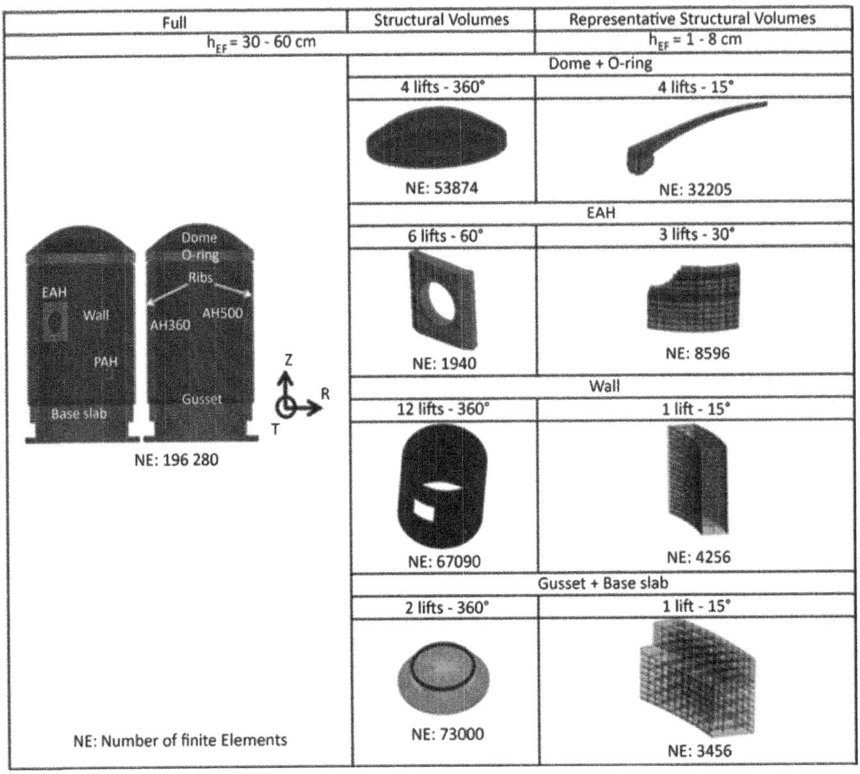

Figure 3.3. *Diagram of the internal wall of the VeRCoRs model, its structural parts: gusset, wall, equipment access hatch (EAH), dome and their associated RSV (NE: number of finite elements). For a color version of this figure, see www.iste.co.uk/gogu/uncertainties.zip*

6 https://fr.xing-events.com/EDF-vercors-project.html.

However, RSV-based modeling incurs additional challenges concerning the definition of:

– their dimensions, which allow an objective description of the spatial variability of the concrete properties and those of the THM loads;

– conditions at kinematic boundaries, often idealized, to take into account the effect of the remaining structural volume not represented and their evolution over time;

– the interactions between RSVs and the transition from RSVs to the whole structure.

It is also necessary to point out that, due to downscaling, the regulatory dry air leakage rates of the VeRCoRs model are higher than those of 1:1 vessels. Therefore, the 1.5% and 1.1% previously mentioned should be replaced by 13.5% and 10.1%, respectively.

Therefore, the objective of this section is not to demonstrate the physical validity of the models used (the reader may refer to the works in Bouhjiti (2018) for this purpose), but to evaluate, from a methodological point of view, the contribution of modeling the spatial variability of concrete properties with respect to the prediction of the real behavior of large structures.

3.3.4.2. *Effect of autocorrelated r.f.*

In this section, THM calculations are considered merely in early stages, that is, during the hydration phase of the concrete (especially during the first 100 days after pouring). The gusset RSV is of particular interest because the risk of cracking is much higher compared to other structural parts. This higher risk of cracking is due to the temperature and hydration gradients in the thickness of the concrete and, above all, to the presence of the invert (massive volume), which limits thermal and chemical shrinkages in concrete. During this early phase, the gusset undergoes strong tensile forces which are sources of cracking that can affect the structural tightness of the containment vessel.

The early THM model has about 50 input parameters and its resolution requires about 15 h. The deterministic response of the model, and in the absence of spatial randomness, allows two boundary cracks per RSV, as shown in Figure 3.4 (*2 boundary cracks*). As previously explained, one difficulty with RSV is to properly define the boundaries and their conditions. In this case, the local absence of transversal rebars creates a deterministic weakness along the boundaries. Nevertheless, this response is far from being representative of all the cracking modes and their frequency observed on site (0–3 cracks per RSV). A deterministic calculation is therefore insufficient to reproduce the observed behavior of concrete; in particular, it overlooks the effect of the spatial variability of concrete properties.

In order to improve our modeling and make it more realistic, one can consider the spatial variability of the Young's modulus of concrete (this is a parameter that defines the elasto-damageable properties of concrete in the model).

Unfortunately, variograms are not currently available for this parameter at the structural scale. One can therefore explore the various stationary distributions in [3.20]. According to the assumptions adopted in other works (Larrard *et al.* 2010), the random field associated with the Young's modulus is considered to have a lognormal nature with a coefficient of variation of 10% and a metric scale of fluctuation. For practical reasons (non-intrusiveness and compatibility with the computational code used, namely Code_Aster®) and ease of implementation, the field (see Figure 3.5) is generated discretely at the centroids of the finite element (*MidPoint Method*).

The spatial correlation is then established with the EOLE method by addressing only the first 102 terms with a relative error on the target variance of 10^{-1} (see Figure 3.6 – according to our sensitivity analysis, concrete cracking is not very sensitive to this order of magnitude of error on the simulated variance).

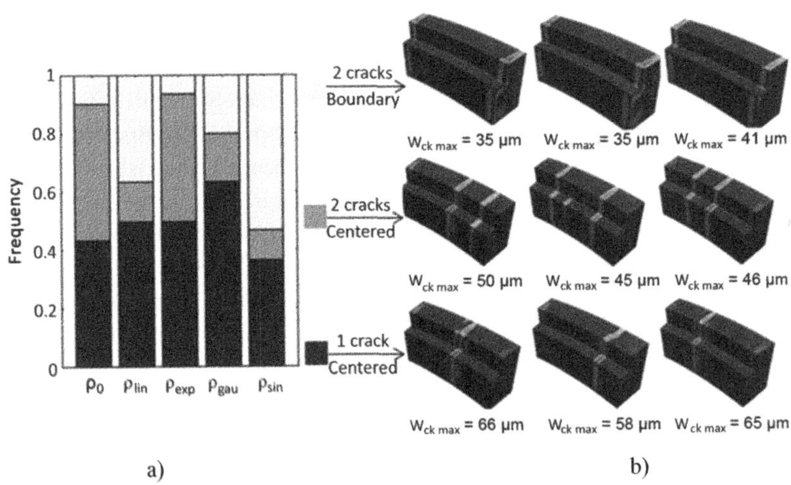

Figure 3.4. *Evolution of cracking profile frequencies according to the spatial correlations under consideration of the Young's modulus: (a) frequency of cracking profiles versus spatial autocorrelation function (constant, linear, exponential, Gaussian, sinusoidal); (b) cracking profiles (number and openings of cracks) for 12 days. For a color version of this figure, see www.iste.co.uk/gogu/uncertainties.zip*

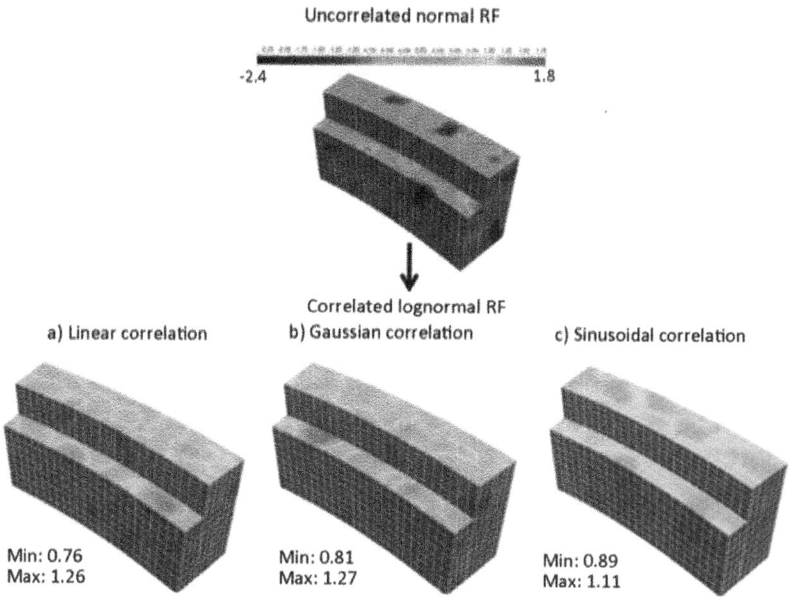

Figure 3.5. *Effect of spatial correlation on the distribution of the r.v. of a lognormal field realization (mean: 1, coefficient of variation: 10%, scale of fluctuation: 1 m). For a color version of this figure, see www.iste.co.uk/gogu/uncertainties.zip*

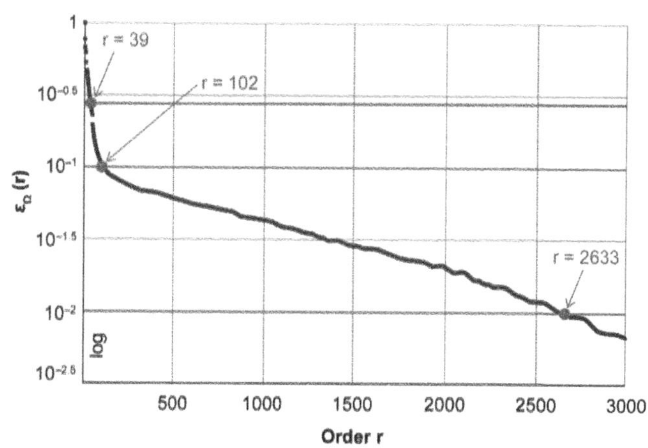

Figure 3.6. *Evolution of the relative error on variance $\varepsilon_{\Omega=Gousset}$ according to order r of the EOLE expansion of the lognormal field (before spatial correlation). For a color version of this figure, see www.iste.co.uk/gogu/uncertainties.zip*

Due to the introduction of this spatial randomness, and considering only 30 random field[7] realizations, two additional cracking profiles could be identified: one and two cracks centered by RSV (*1, 2 cracks centered* in Figure 3.4). The random field also has a non-negligible effect on crack spacing and crack openings. Although the results obtained are consistent with on-site observations, it should be noted that 0- and 3-crack profiles could not be identified. This is due to:

– the insufficient number of realizations under consideration. However, a rigorous application of the Monte Carlo method seems difficult to achieve;

– overlooking the variability of the other THM parameters of concrete; this would require experimental identification of the correlation matrix. However, such measurements are still very costly for industrial applications and deserve, first, treatment at the material scale, in order to discern the highly correlated parameters from those that are less correlated.

Finally, this cracking simulation exercise on the scale of a large structure illustrates:

– the high sensitivity of the cracking profiles to the spatial correlation of the damage parameters;

– the relative inefficiency of r.f. in predicting all probable cracking profiles. Therefore, other methods should be explored for efficiency gains; in particular, methods based on response surfaces.

3.4. Modeling uncertainty propagation using response surface methods

This section introduces SFEM for describing uncertainty propagation in chained THM+L models. Our objective is to directly access PDFs of the quantities of interest and not only statistical quantities or failure probabilities. The latter should merely be a postprocessing phase.

3.4.1. *Probabilistic coupling strategies*

Among the stochastic methods used in the literature, the following are noteworthy.

7 This is equivalent to 19 days of numerical computations.

3.4.1.1. *Monte Carlo or quasi-Monte Carlo-based approaches*

These approaches theoretically allow for an almost complete probabilistic description of the variables of interest (Metropolis and Ulam 1949; Morokoff and Caflish 1995; Kroese *et al.* 2014). They consist of establishing a fairly high number N of random draws of the input parameters and simulating the model response for each of these draws. Access to the CDF (and to the PDF) is then direct; the same applies to the first statistical moments [3.21], failure probabilities [3.22] and global sensitivity indices, as defined by Sobol' (2001).

These can be used to identify the contribution of each parameter to the observed variance of the model over the entire definition domain, and to analyze possible correlations between parameters for non-additive models ($\mathcal{R}_{\mathcal{B}}(a+b) \neq \mathcal{R}_{\mathcal{B}}(a) + \mathcal{R}_{\mathcal{B}}(b)$):

$$\overline{\mathcal{R}_{\mathcal{B}}} = \frac{\sum_{i=1}^{N} \mathcal{R}_{\mathcal{B},i}}{N} \quad \sigma_{\mathcal{R}_{\mathcal{B}}}^2 = \frac{\sum_{i=1}^{N} (\mathcal{R}_{\mathcal{B},i} - \overline{\mathcal{R}_{\mathcal{B}}})}{N-1} \qquad [3.21]$$

$$P_f(\mathcal{R}_{\mathcal{B},f}) = \frac{Card(\mathcal{R}_{\mathcal{B}}/g(\mathcal{R}_{\mathcal{B},1\leq i\leq N})=1)}{N} \text{ with } \begin{matrix} g = 1 \text{ if failing} \\ g = 0 \text{ otherwise} \end{matrix} \qquad [3.22]$$

Monte Carlo-type methods are therefore robust, *a priori* applicable to any physical problem regardless of the degree of complexity. Nevertheless, from a practical point of view, their applicability is limited (Baroth 2005; Sudret 2007) by:

– the numerical cost required for highly nonlinear problems (for THM+L computations);

– the large number of computations N to access failure probabilities of the order of $10^{2-log_{10}N}$;

– the low speed of convergence $o(1/\sqrt{N})$ and $o(1/N)$ for the Monte Carlo and quasi-Monte Carlo methods, respectively.

This has driven the development of alternative methods for partial probabilistic characterization at lower cost, to the detriment of a complete description: reliability methods (FORM, SORM, etc.) for failure probabilities, quadrature methods for statistical moments and spectral methods or response surfaces for metamodels (for the purpose of replacing expensive models and facilitating the use of Monte Carlo). Particular interest is given to these metamodeling methods later on.

3.4.1.2. *Spectral approaches to metamodeling*

Given the physical nature of the problem discussed in this chapter, the variances of the THM+L responses of interest are assumed to be finite and square-integrable.

This is sufficient to accept the existence of a Hilbertian representation of the response $\mathcal{R}_{\mathcal{B}}$ (Soize and Ghanem 2004):

$$\mathcal{R}_{\mathcal{B}}(X) = \sum_{k=0}^{\infty} a_k \Psi_k(X) \quad [3.23]$$

with $\Psi_{k\in\mathbb{N}}$ a series of random variables forming the basis of the Hilbertian space, and $a_{k\in\mathbb{N}}$ the coordinates of the response $\mathcal{R}_{\mathcal{B}}$ in this basis.

In the case of random and independent variables $X = \{X_1, X_2, \dots, X_N\}$, the marginal PDF and the Hilbertian basis are constructed as follows (tensor product):

$$f_X(x) = \prod_{i=1}^{N} f_{X_i}(x), \quad \Psi_k(X) = \prod_{i=1}^{N} \Psi_{k_i}(X_i) \quad [3.24]$$

with $\{\Psi_{k_i\in\mathbb{N}}\}$ a Hilbertian basis associated with the random variable X_i of PDF f_{X_i}.

Specifically, the basis $\{\Psi_{k_i\in\mathbb{N}}\}$ is defined in such a way as to verify the conditions for orthonormality:

$$< \Psi_{k_i}^{l}, \Psi_{k_i}^{m} >_i = \int_{D_{X_i}} \Psi_{k_i}^{l}(x)\Psi_{k_i}^{m}(x)\, f_{X_i}(x)dx = \delta_{lm} \quad [3.25]$$

$<,>$ is an internal product associated with the probabilized Hilbertian subspace D_{X_i} of basis $\{\Psi_{k_i\in\mathbb{N}}\}$. δ_{lm} is the Kronecker symbol equal to 1 when l = m and 0 otherwise.

This condition is, therefore, verified for the basis $\{\Psi_{k\in\mathbb{N}}\}$ also on space $D_X = \prod_{i=1}^{N} D_{X_i}$, which allows the coordinates y_k to be directly deduced:

$$a_k = \int_{D_X} \mathcal{R}_{\mathcal{B}}(x)\, \Psi_k(x) f_X(x)dx \quad [3.26]$$

The decomposition [3.23] contains an infinite number of terms. In practice, only a finite number is considered. The truncated version then corresponds to an approximation of the model response [3.27].

The order of truncation is to be defined according to a given condition (numerical cost, convergence criteria, error on statistic moments, etc.):

$$\mathcal{R}_{\mathcal{B}}(X) \approx \widehat{\mathcal{R}_{\mathcal{B}}} = \sum_{k=0}^{Q<+\infty} a_k \Psi_k(X) \quad [3.27]$$

3.4.2. *Polynomial chaos method*

Several bases $\{\Psi_{k\in\mathbb{N}}\}$ verify expressions [3.23]–[3.27]. In particular, we choose to impose a polynomial form to the bases Ψ_k. This is then referred to as the polynomial chaos expansion (PCE) method (Xiu and Karniadis 2002). The orthonormality condition [3.25], and subject to the independence of random parameters, makes it possible to identify, for example, the Hermitian polynomial basis for normal distributions, and the Legendre polynomial basis for uniform distributions (Baroth 2005). The definition of the PCE then requires:

– the definition of the order of truncation Q in expression [3.27], which is not *a priori* known. In practice, the criterion is arbitrarily defined so as to have a maximal polynomial order q of the polynomial function $\widehat{\mathcal{R}_\mathcal{B}}$, allowing an accurate, physical representativeness of the metamodel. In Baroth *et al.* (2011), orders from 3 to 5 are suggested. The number of terms to be identified is therefore $1 + \mathrm{Q} = \binom{\mathrm{N}+\mathrm{q}}{\mathrm{q}} = \frac{(\mathrm{N}+\mathrm{q})!}{\mathrm{N}!\mathrm{q}!}$. Other methods, based more on convergence criteria for measuring the quality of fitting (Blatman 2010), have been developed. Each *k*th term is thus included or excluded from the sum according to their contribution to the improvement of the overall adjustment;

– the identification of the coefficients $\{\mathrm{a}_{\mathrm{kl'}}\}$ in [3.27] can be achieved in two different ways. The first is known as the regression method (Berveiler *et al.* 2006) where the computational points of the model $\mathcal{R}_\mathcal{B}$ are defined in accordance with the experimental or factorial design methods (Cavazzuti 2012; Montgomery 2012) or with the quadrature methods. For the problem to be well posed, as many calls to the model $\mathcal{R}_\mathcal{B}$ as number of terms $\{\mathrm{a}_{\mathrm{kqu}}\}$ are required ($\mathrm{N}_{\mathrm{simulation}} \geq \frac{(\mathrm{N}+\mathrm{q})!}{\mathrm{N}!\mathrm{q}!}$). Therefore, for the Gaussian quadrature method, for example, at least $\sqrt[\mathrm{N}]{\frac{(\mathrm{N}+\mathrm{q})!}{\mathrm{N}!\mathrm{q}!}}$ integration points per parameter would be needed (isotropic scheme). The fit of the approximate response $\widehat{\mathcal{R}_\mathcal{B}}$ can be done based on the method of least squares. Nonetheless, this procedure does not take into account the orthonormality of the Hilbertian basis. The second method is the projection method (Baroth *et al.* 2006, 2007) and relies on numerical methods for the evaluation of the integral [3.26], namely the Gauss quadrature. It helps to recall that to estimate the integral of a polynomial function of order $2\mathrm{n} - 1$, n integration points are required. Therefore, for an approximation $\widehat{\mathcal{R}_\mathcal{B}}$ [3.27] of order q and for elements of the polynomial basis $\{\Psi_{\mathrm{k}\in\mathbb{N}}\}$ of order less than or equal to q, the number of integration points required is q+1 per random variable. The number of calls to the model $\mathcal{R}_\mathcal{B}$ is $(\mathrm{q}+1)^{\mathrm{N}}$ ($\geq \frac{(\mathrm{N}+\mathrm{q})!}{\mathrm{N}!\mathrm{q}!}$). As such, the quadrature method is, in principle, more expensive than the regression method. Nevertheless, this cost can be greatly reduced by resorting to adaptive approaches

(Gary *et al.* 2001) or sparse integration methods such as the Smolyak's method (1963) or a combination of both.

For both methods, and in cases where the coefficients a_k are time dependent, a Karhunen–Loève-type expansion (Ghanem and Spanos 1991) can be envisaged (Hawchar *et al.* 2017).

Once the approximation of the metamodel $\widehat{\mathcal{R}_{\mathcal{B}}}$ is built and the error $\left\|\widehat{\mathcal{R}_{\mathcal{B}}} - \mathcal{R}_{\mathcal{B}}\right\|$ accepted, probabilistic, reliability or sensitivity analyses can be carried out using conventional less greedy Monte Carlo methods (given the explicit nature of $\widehat{\mathcal{R}_{\mathcal{B}}}$). However, it is worth noting that:

– if the orthonormality condition is respected, the global sensitivity indices are easily inferred through direct postprocessing of the coefficients a_k (Sudret 2008);

– the identification of the point of failure is conditioned by its membership to the definition domain of the approximation $\widehat{\mathcal{R}_{\mathcal{B}}}$. Indeed, the latter is valid only on a limited domain $\overline{D_X}$ defined by the lower and upper bounds of the integration points for each parameter X_i. The representativeness of the approximation $\widehat{\mathcal{R}_{\mathcal{B}}}$ outside this area is not guaranteed.

3.4.2.1. *Application: contribution to the modeling of cracking of reinforced concrete works at an early age using metamodeling*

In this section, we consider the same issues as those described in section 3.3.4. However, we address these differently this time by ensuring probabilistic coupling using spectral approximation methods and not the use of several random field realizations.

Random variable (CoV = 10%, log law)		Number of cracks by RSV	Frequencies	
			Experience (visual observations)	Metamodel
Young's modulus	Monte Carlo → 500 draws per parameter	0	46%	48%
Tensile strength		1	37%	30%
Endogenous shrinkage coefficient		2	12%	17%
Thermal expansion coefficient		3	5%	5%

Table 3.1. *List of independent random parameters and distributions of the resulting cracking profiles*

We therefore propose to construct a cracking response surface at the gusset RSV scale (see Figure 3.5) based on the random parameters presented in Table 3.1 (Young's modulus, tensile strength, endogenous shrinkage coefficient and thermal expansion coefficient). We thus consider more uncertainties in comparison with the calculations presented in section 3.3.4, where only the spatial variability of the Young's modulus is taken into account. The reader should note that, in this application, spatial randomness is maintained to allow for random localization of cracking and its propagation. However, only one realization is considered for all parameter sets in order to define the metamodel of the number of cracks in the gusset at the RSV scale.

Given the discrete (their number can only be an integer) and limited nature of the cracks (for energy considerations, their number is limited to a finite maximum value), a continuous description such as [3.27] is of little relevance. We thus propose to search for a discrete formulation of our response surface (possible interpolation with the Lagrange polynomial basis) by exploiting discretization techniques resulting from experimental designs (Cavazzuti 2012; Montgomery 2012). In particular, we propose an adaptive central composite design (ACCD) scheme whose boundaries are defined in such a way as to allow the identification of all physically acceptable cracking modes. Therefore, we impose two conditions for the enrichment of the four-dimensional experimental design (four parameters in Table 3.1); the addition of a central point is conditioned:

– by the existence of two different cracking profiles at its edges. As a matter of fact, on a given segment, if the edges correspond to the same number of cracks, all the points of the segment should have the same number of cracks;

– by satisfying a discretization criterion of the given experimental design (for example, a maximum number of points per dimension).

As an example, we set a stopping condition as the selection of a maximum of 30 computation points per dimension. It should be noted that in the absence of an adaptive scheme, the experimental design has $30^4 = 810{,}000$ points to be identified. However, with the help of some additional physical simplifications, which we shall not detail here (Bouhjiti 2018), this number is reduced to only 286. The two previous enrichment conditions then result in reducing this number by more than 50%. The reader will note that the identified cracking profiles are in higher numbers compared to the random field method (0 to 12 cracks per RSV in Figure 3.7 – which includes on-site observations of 0–3 cracks). For visualization purposes, the discrete response surface was projected onto the axis of the "tensile strength" parameter in Figure 3.8. Similar results are available for the other axes, which allows the definition of an explicit response surface for the number of cracks as a function of the parameters selected (Young's modulus, endogenous shrinkage coefficient and coefficient of thermal expansion).

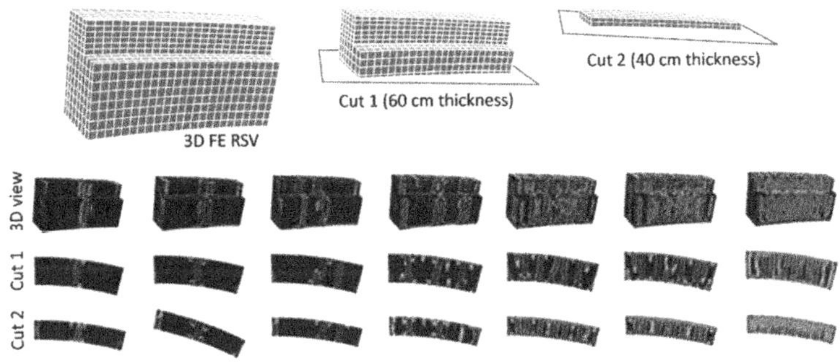

Figure 3.7. *Examples of 12-day cracking profiles involved in the ACCD experimental design. For a color version of this figure, see www.iste.co.uk/gogu/uncertainties.zip*

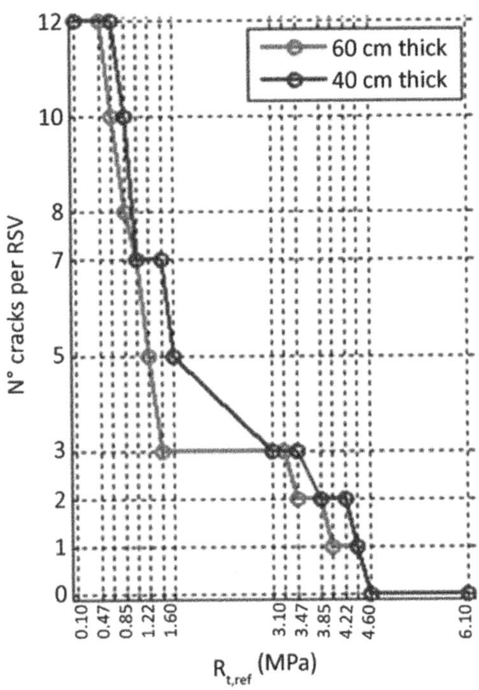

Figure 3.8. *Evolution of the number of cracks per gusset RSV with the tensile strength of concrete. For a color version of this figure, see www.iste.co.uk/gogu/uncertainties.zip*

Now that the response surface is available and accessible at low cost, the Monte Carlo method can be used in order to access the frequency of each cracking mode per gusset RSV. This work has thus allowed us to make a blind prediction of a cracking distribution *in situ* that is comparable to visual observations (Table 3.1).

The effectiveness and efficiency of this approach is therefore more significant compared to the generation of a limited series of random field realizations of a variable parameter. Nevertheless, the importance of modeling the correct spatial variability in the model for better physical representativeness must be stressed.

3.4.2.2. *Application: contribution to the modeling of the THM+L uncertainty propagation in a reinforced concrete structure by polynomial chaos methods*

In this section, we consider the entire THM+L chain from the early phase to the end of the operational phase (planned for 6 years) of the VeRCoRs mock-up. More specifically, we study the overall leakage rate of the structure – the response $\mathcal{R}_{\mathcal{B}}$ in [3.6].

The latter is obtained by superimposing the simulated flow rates for each RSV in Figure 3.3, which allows us to write:

$$Q_{tot} = 24 * Q_{GUS} + 268 * Q_{WALL} + 4 * Q_{EAH} + 24 * Q_{DOME} \qquad [3.28]$$

The THM+L model includes about 60 input parameters and its resolution requires about 20 h. However, for this application, only the physical parameters influencing the tightness of the concrete are considered to be random (totaling 12 in Table 3.2[8]).

In the absence of data on the subject, these random parameters are considered to be independent and lognormally distributed. The objective is to be able to express Q_{tot} in the form of 12-r.v. stochastic processes. One chooses to search for a PCE of maximum order $q = 3$ (Baroth *et al.* 2011). This will therefore require the identification of 220 terms a_k per RSV.

In this application, the Gauss quadrature method is used isotropically with 4 integration points per parameter. This results in 2.6×10^5 calls to the model per RSV. The Hermite polynomial basis is used given the lognormal nature of the parameter distributions in Table 3.2.

8 Only the spatial variability of the Young's modulus is considered. Other parameters are considered as being spatially homogeneous.

One should note that, because of *a priori* local sensitivity analysis, the random parameters listed in Table 3.2 have been identified as the most influential in the model (compared to the rest of inputs). This identification step is not detailed herein for the sake of conciseness.

Parameter	CoV (%)	Distribution
Source term	40	Log
Thermal enthalpy	40	Log
Water diffusivity	20	Log
Initial water content	20	Log
Young's modulus	10	Log
Tensile strength	10	Log
Endogenous shrinkage coefficient	10	Log
Thermal expansion coefficient	10	Log
Coefficient of drying shrinkage	10	Log
Prestressing	50	Log
Pressurization	50	Log
Intrinsic permeability	50	Log

Table 3.2. *List of independent random parameters and their distributions*

Finally, the metamodel $\widehat{\mathcal{R}_{\mathcal{B}}}$ obtained approximates the actual response $\mathcal{R}_{\mathcal{B}}$ of the overall containment leakage rate with a relative error of the order of 18% (average error calculated between the metamodel response and that of the THM-F numerical model, at the integration points of the r.v. in Table 3.2). Compared to the accuracy of on-site measurements, this relative error is largely satisfactory for our numerical model. We therefore decide to accept it. Then, by applying the Monte Carlo method to the metamodel $\mathcal{R}_{\mathcal{B}}$, and with 500 draws per parameter, we can access the CDFs associated with the overall leakage rate, at the risk of exceeding the regulatory leakage criterion and its evolution over time (between 1 and 6 years of operation in Figure 3.9), as well as the contribution of each parameter to the total observed variance of $\widehat{\mathcal{R}_{\mathcal{B}}}$ (see Table 3.3).

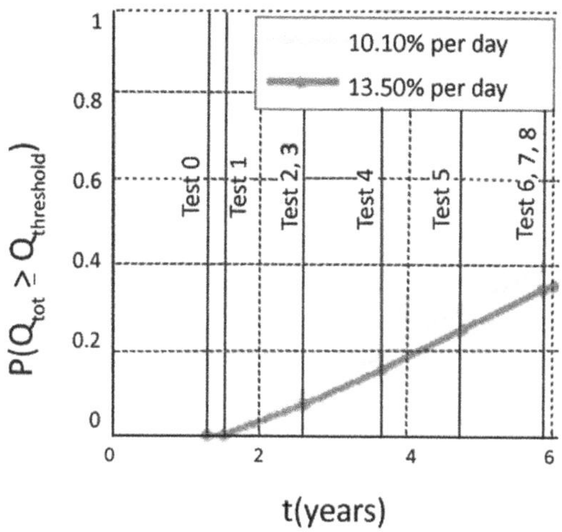

Figure 3.9. *Evaluation of the risk of exceeding the regulatory criterion overall leakage in dry air over time during pressurization tests. For a color version of this figure, see www.iste.co.uk/gogu/uncertainties.zip*

Parameter	Contribution to the total variance of $\widehat{\mathcal{R}_B}$ Change between 1 year → 6 years of operation
Source term	–
Thermal enthalpy	–
Water diffusivity	13% → 1%
Initial water content	13% → 1%
Young's modulus	9% → 22%
Tensile strength	9% → 22%
Endogenous shrinkage coefficient	4% → 10%
Thermal expansion coefficient	4% → 10%
Coefficient of drying shrinkage	–
Prestressing	–
Pressurization	23% → 25%
Intrinsic permeability	24% → 8%

Table 3.3. *Overall sensitivity analysis, as defined by Sobol' (2001), applied to the metamodel* $\widehat{\mathcal{R}_B}$

3.5. Conclusion

This chapter introduced the steps for the probabilistic modeling of a thermo-hydro-mechanical problem with leakage analysis. Emphasis was placed on the use of discretized r.f. within a finite element model (FEM) context, as well as on the study of sensitivity with uncertain parameters. This study is developed with the practical case of the stochastic FEM of a reinforced and prestressed concrete model vessel of nuclear plants. The objective is to evaluate, from a methodological point of view, the added value of modeling the uncertainty of concrete properties on our ability to predict its realistic (by nature, random) behavior at the structural scale.

In a first approach, spatial variability of the Young's modulus was introduced using lognormal random field realizations at the centroids of the finite elements at the scale of an RSV. These realizations allow different cracking profiles to be distinguished and are sensitive to the spatial correlation of the random parameters. However, it was not possible to predict all of the physically possible cracking patterns. Indeed, a higher number of realizations (and numerical simulations) might be required. To keep the numerical cost reasonable, a complementary approach, based on discrete response surface methods, was explored. Eventually, by using a chaotic expansion metamodel of concrete cracking phenomena, involving 12 influential random variables, it was possible to account more satisfactorily for the diversity of cracking patterns and, ultimately, to accurately characterize the evolution of the structural tightness of the vessel over time and its variation.

This work was co-financed by EDF-SEPTEN/DTG/CIH within the framework of the PERENITI Industrial Chair, hosted by the Grenoble INP Partnership Foundation and the ENDE and MACENA projects (ProjetIA-11-RSNR-0009 and ProjetIA-11-RSNR-0012).

3.6. References

Baroth, J. (2005). Analyse par éléments finis stochastiques de la propagation d'incertitudes dans un modèle mécanique non linéaire. PhD Thesis, Blaise Pascal University, France.

Baroth, J., Bodé, L., Bressolette, P., Fogli, M. (2006). SFEM using Hermite polynomials: An approach for solving nonlinear problems with uncertain parameters. *Computer Methods in Applied Mechanics and Engineering*, 195, 6479–6501.

Baroth, J., Bressolette, P., Chauviere, C., Fogli, M. (2007). An efficient SFEM using Lagrange polynomials: Application to nonlinear mechanical problems with uncertain parameters. *Computer Methods in Applied Mechanics and Engineering*, 196(46–58), 4419–4429.

Baroth, J., Schoefs, F., Breysse, D. (2011). *Construction Reliability: Safety, Variability and Sustainability*. ISTE Ltd, London and John Wiley & Sons, New York.

Berveiller, M., Sudret, B., Lemaire, M. (2006). Stochastic finite element: A nonintrusive approach by regression. *European Journal of Computational Mechanics*, 15(1–3), 81–92.

Blatman, G. (2007). Chaos polynomial creux et adaptatif pour la propagation d'incertitudes et l'analyse de sensibilité. PhD Thesis, Blaise Pascal University, France.

Bouhjiti, D. (2018). Analyse probabiliste de la fissuration et du confinement des grands ouvrages en béton armé et précontraint. PhD Thesis, Grenoble Alpes University, France.

Bouhjiti, D., Baroth, J., Briffaut, M., Dufour, F., Masson, B. (2018a). Statistical modeling of cracking in large concrete structures under thermo-hydro-mechanical loads: Application to nuclear containment buildings. Part 1: Random field effects (reference analysis). *Nuclear Engineering Design*, 333, 196–223.

Bouhjiti, D., Boucher, M., Briffaut, M., Dufour, F., Baroth, J., Masson, B. (2018b). Accounting for realistic thermo-hydro-mechanical boundary conditions whilst modeling the ageing of concrete in nuclear containment buildings: Model validation and sensitivity analysis. *Engineering Structures*, 166, 314–338.

Cavazzuti, M. (2012). *Optimization Methods: From Theory to Design of Scientific and Technological Aspects in Mechanics*. Springer, Berlin.

Cornell, A.C. (1971). First Order uncertainty analysis of soils deformation and stability. *Proceedings of the First Conference on Applications of Statistics and Probability to Soil and Structural Engineering*. University of Hong Kong Press, Hong Kong, 130–144.

Deodatis, G. (1989). Stochastic FEM sensitivity analysis of nonlinear dynamic problems. *Probabilistic Engineering Mechanics*, 4(3), 135–141.

Der Kiureghian, A. and Ke, J.B. (1988). The stochastic finite element method in structural reliability. *Probabilistic Engineering Mechanics*, 3(2), 83–91.

Der Kiureghian, A. and Liu, P. (1986). Structural reliability under incomplete probability information. *Engineering Mechanics*, 112, 85–104.

Ditlevsen, O. (1991). Random field interpolation between point by point measured properties. In *Computational Stochastic Mechanics*, Spanos, P.D., Brebbia, C.A. (eds). Springer, London.

Ditlevsen, O. and Madsen, H. (1996). *Structural Reliability Methods*. John Wiley & Sons, Chichester.

Ditlevsen, O. and Madsen, H. (2005). Probabilistic information. In *Structural Reliability Methods*, Ditlevson, O., Madsen, H, (eds). Internet edition 2.2.5, July [Online]. Available at: https://citeseerx.ist.psu.edu/viewdoc/download?doi=10.1.1.121.3682&rep=rep1&type=pdf.

Freulon, X. and de Fouquet, C. (1991). Remarques sur la pratique des bandes tournantes en 3D. *Cahier de géostatique*, Fascicule 1, 101–117.

Gary, W.G., Dong, Z., Aitchison, P. (2001). Adaptive response surface method. A global optimization scheme for computation-intensive design problems. *Engineering Optimization*, 3(6), 707–734.

Ghanem, R. and Spanos, P. (1991). *Stochastic Finite Elements: A Spectral Approach*. Dover Publications, New York.

Graham, L. and Deodatis, G. (2011). Response and eigenvalue analysis of stochastic finite element systems with multiple correlated material and geometric properties. *Probabilistic Engineering Mechanics*, 16(1), 11–29.

Hawchar, L., El Soueidy, C.P., Schoefs, F. (2017). Principal component analysis and polynomial chaos expansion for time-variant reliability problems. *Reliability Engineering and System Safety*, 167, 406–416.

Hisada, T., Nagakiri, S., Nagakiri, T. (1983). Stochastic finite element analysis of uncertain intrinsic stress caused by structural misfits. *SMIRT Conference*, Chicago.

Kroese, D.P., Brereton, T., Taimre, T., Botev, Z.I. (2014). Why the Monte Carlo method is so important today. *WIREs Computational Statistics*, 6(6), 386–392.

de Larrard, T., Colliat, J.B., Benboudjema, F., Torrenti, J.M., Nahas, G. (2010). Effect of the Young's modulus variability on the mechanical behaviour of a nuclear containment vessel. *Nuclear Engineering and Design*, 240(12), 4051–4060.

Lemaire, M. (2009). *Structural Reliability*. ISTE Ltd, London and John Wiley & Sons, New York.

Li, C., Der Kiureghian, A., Ke, J.B. (1983). Optimal discretization of random fields. *Engineering Mechanics*, 119(6), 1136–1154.

Matheron, G. (1973). The intrinsic random functions and their applications. *Advances in Applied Probability*, 5, 435–468.

Matthies, H.G., Chistoph, E.B., Bucher, C., Soares, C.G. (1997). Uncertainties in probabilistic numerical analysis of structures and solids-stochastic finite elements. *Structural Safety*, 19(3), 283–336.

Metropolis, N. and Ulam, S. (1949). The Monte Carlo method. *American Statistical Association*, 44, 335–341.

Montgomery, D. (2012). *Design and Analysis of Experiments*. John Wiley & Sons, New York.

Morokoff, W.J. and Caflish, R.E. (1995). Quasi-Monte Carlo integration. *Journal of Computational Physics*, 122–2, 218–230.

Nataf, A. (1962). Détermination des distributions dont les marges sont données. *Comptes rendus hebdomadaires des séances de l'Académie des sciences*, 255, 42–43.

Nelson, R. (1999). *An Introduction to Copulas*. Springer, Berlin.

Phoon, K., Huang, S., Quek, S. (2002). Implementation of Karhunen-Loève expansion for simulation using a wavelet-Galerkin scheme. *Probabilistic Engineering Mechanics*, 17(3), 293–303.

Smolyak, S. (1963). Quadrature and interpolation formulas for tensor products of certain classes of functions. *Soviet Mathematics, Doklady*, 4, 420–243.

Sobol', M.I. (2001). Global sensitivity indices for nonlinear mathematical models and their Monte Carlo estimates. *Mathematica and Computers in Simulation*, 55(1–5), 271–280.

Soize, C. and Ghanem, R. (2004). Physical systems with random uncertainties: Chaos representations with arbitrary probability measure. *Scientific Computing*, 26(2), 395–410.

Stefanou, G. (2009). The stochastic finite element method: Past, present and future. *Computer Methods in Applied Mechanics and Engineering*, 198, 1031–1051.

Sudret, B. (2007). Uncertainty propagation and sensitivity analysis in mechanical models. Contributions to structural reliability and stochastic spectral methods. HDR, Blaise Pascal University, France.

Sudret, B. (2008). Global sensitivity analysis using polynomial chaos expansion. *Reliability Engineering & System Safety*, 93(7), 964–979.

Sudret, B. and Der Kiureghian, A. (2000). Stochastic finite element methods and reliability. A state of the art report. Report, UCB/SEMM/2000-08, University of California, Berkeley.

Vanmarcke, E. (2010). *Random Fields: Analysis and Synthesis.* World Scientific, Singapore.

Xiu, D. and Karniadis, G. (2002). The Wiener-Askey polynomial chaos for stochastic differential equations. *Journal of Scientific Computing*, 24(2), 619–644.

Yamazaki, F., Shinozuka, M., Dasgupta, G. (1988). Neumann expansion for stochastic finite element analysis. *Engineering Mechanics*, 114(8), 1335–1354.

Zhang, J. and Ellingwood, B. (1994). Orthogonal series expansions of random fields in reliability analysis. *Engineering Mechanics*, 120(12), 2660–2677.

Zhu, W.Q. and Ren, Y.J. (1986). Local average of random field and stochastic finite element method. *Acta aeronautica et astronautica sinica*, 7(6), 604–611.

4

Reduction of Uncertainties in Multidisciplinary Analysis Based on a Polynomial Chaos Sensitivity Study

Sylvain DUBREUIL[1], Nathalie BARTOLI[1], Christian GOGU[2] and Thierry LEFEBVRE[1]

[1] *ONERA/DTIS, University of Toulouse, France*

[2] *Clément Ader Institute, Paul Sabatier University, Toulouse, France*

4.1. Introduction

Presently, the design and study of a new system calls upon the intensive use of numerical simulation. The latest advances in numerical resolution methods, combined with access to large numerical computational resources, make the use of a *digital twin* very advantageous and economical. In order to simulate the behavior of complex systems, so-called multiphysics or multidisciplinary simulations are often used. For example, the study of a new concept in aeronautics can involve many coupled disciplines, such as aerodynamics, structural mechanics, propulsion, acoustics, automatic control, mission analysis, etc. These different modules of disciplinary analysis are then coupled within a multidisciplinary, often nonlinear, analysis (MDA), which needs to be solved in order to obtain a system in equilibrium. Within the context of this chapter, it is assumed that this nonlinear system is solved by non-intrusively coupling the different disciplinary modules, the approach adopted then makes use of a fixed-point iterative Gauss–Seidel or Jacobi-type algorithm. Once the equilibrium of the system is reached, different

Mechanical Engineering under Uncertainties,
coordinated by Christian GOGU.

quantities of interest can be calculated, such as fuel consumption, range or operating costs of the aircraft. When MDA is used inside an optimization loop, these different quantities also depend on design variables. Two types of variables will therefore have to be considered: coupling variables, the solution of the MDA coupled system, and design variables that can be fixed or managed by an optimization algorithm.

As far as the MDA is concerned, its resolution relies on an iterative algorithm that necessitates calling disciplinary modules at every iteration; the computational time can therefore become prohibitive. It should also be noted that within a multi-disciplinary design optimization (MDO) framework, this resolution is carried out for several sets of parameters in the design space, which considerably increases computation time. In order to solve this problem, the use of approximated disciplinary models with negligible computational costs can prove helpful. These models can be given by means of conventional approximation tools (for example, regression and interpolation by Gaussian processes) or be derived from a simplification of the modeled physical phenomenon (for example, linearization and model reduction). In any case, we shall assume in the following that the error of these approximated disciplinary models is modeled through probabilistic uncertainty. The construction of this probabilistic model may be inherent to the selected model, built from a database (see Jones *et al.* 1998) or obtained through expert opinion.

The next step concerns propagating the uncertainties associated with each of these models within the nonlinear system that is the MDA. The uncertainty of each model is modeled by random fields on the space of the design and coupling variables. This choice involves reformulating the problem as a random nonlinear system. The solution of this system will thus be defined as the joint probability distribution associated with the random vector of the converged coupling variables. The solution of a random MDA involving model uncertainties is relatively complex. In Liang *et al.* (2015), a semi-decoupled approach is proposed, and in Dubreuil *et al.* (2016) a direct resolution of this random MDA is performed by introducing a polynomial chaos approximation of the coupling variables. Nevertheless, this last approach can prove to be numerically costly. Within the context of this chapter, an approximate solution using a simplification of the disciplinary Gaussian processes, combined with a Monte Carlo method, is implemented.

Since the quantities of interest are functions of the random coupling variables, they will also be modeled by random variables. Sensitivity analysis then aims to identify and classify the variables responsible for the variation of the quantity of interest under study. To this end, the influence of the uncertainty associated with disciplinary Gaussian processes on the uncertainty of this quantity of interest must be evaluated. The approach proposed here then consists of approximating this quantity of interest through polynomial chaos expansion (see Ghanem and Spanos 1991) in order to deduce an analytical estimation of the Sobol' indices (see Sobol' 2001) from the chaos coefficients based on the work of Sudret (2008) and Blatman and Sudret (2010).

Following the sensitivity analysis, the last step consists of enriching the disciplinary metamodel(s) that have been identified to reduce the variance on the quantity of interest. In the end, the proposed methodology provides a means to quantify and reduce the uncertainty on the quantity of interest, resulting from MDA, due to the use of disciplinary surrogate models. The gains in terms of computational time are illustrated on an aero-structural coupling of an aircraft wing.

The chapter is organized as follows: section 4.2 presents the formalism and resolution of a MDA with model uncertainty. Section 4.3 describes sensitivity analysis using Sobol' indices calculated from a polynomial chaos expansion of the quantity of interest. Finally, section 4.4 presents the application test case related to the aeroelastic equilibrium of an aircraft wing. The results obtained in terms of uncertainty reduction are described. Section 4.5 concludes this chapter.

4.2. MDA with model uncertainty

4.2.1. *Formalism*

We consider here a conventional MDA problem, involving n_d disciplines coupled together. Such an analysis can be written in the form of a system of n_d equations:

$$y_i(z, y_{c^{(i)}}) = f_i(z, y_{c^{(i)}}),\ i = 1, \cdots, n_d \qquad [4.1]$$

where $y_i(z, y_{c^{(i)}}) \in \mathbb{R}$, $z \in \mathcal{Z}$ represents the vector of design variables (with $\mathcal{Z} \subset \mathbb{R}^n$), $y_{c^{(i)}}$ represents the vector of coupling variables for the discipline i; $c^{(i)}$ represents the set of indices identifying the interdisciplinary coupling variables, in other words $\#(c^{(i)}) \leq (n_d - 1)$ and $i \notin c^{(i)}$. Finally, f_i is the solver associated with the discipline i (for example, a numerical model solver).

Let us first underline the following points:

– with the notations previously introduced, the disciplines i and j are said to be coupled through a loop (sometimes also called strong coupling) if $i \in c^{(j)}$ and $j \in c^{(i)}$. In the following, we consider that equation [4.1] contains at least one loop;

– the disciplinary solvers f_i, $i = 1,\cdots,n$ are considered as being “black boxes”. The accurate model behind f_i is assumed to be unknown to the user of the MDA (it should be noted that f_i could also be a linear or nonlinear model with respect to both design and coupling variables). This is representative of a MDA context, where the designer is only a user of disciplinary models, f_i, which in turn are provided by experts of each discipline;

– the solution of equation [4.1] is the set y of size n_d. Since disciplinary solvers f_i, $i = 1,\cdots,n_d$ are regarded as coupled black boxes, this solution is computed by an iterative algorithm. This approach is sometimes called a partitioned solver in the literature. In the following, the existence of a single solution for equation [4.1] is assumed on the basis of physical considerations;

– the system of equations [4.1] was written for $y_i(z, y_{c^{(i)}}) \in \mathbb{R}$ in order to simplify the notations. In the following, we shall address a problem involving a vector output with $y_i(z,\ y_{c^{(i)}}) \in \mathbb{R}^{n_i}$ representative of the solution of a finite element solver (discretized displacement field).

While this partitioned approach is very convenient as it enables a multidisciplinary system to be solved by coupling disciplinary solvers in a non-intrusive way; the associated computational cost can, however, become problematic if the solvers $f_i, i = 1,\cdots,n_d$ are computationally expensive. Moreover, this problem is amplified if the computation of the solution has to be performed at several points z in the design domain $\mathcal{Z}$. Within this framework, it can be interesting to use disciplinary metamodels for the solvers $f_i, i = 1,\cdots,n_d$. In the following, we denote by $\tilde{Y}_i(z, y_{c^{(i)}})$ the metamodel of the discipline i and we assume that the uncertainty of this metamodel is under a probabilistic form, that is to say that $\forall\,(z, y_{c^{(i)}})$ $\in \mathcal{Z} \times C^{(i)}, \tilde{Y}_i(z, y_{c^{(i)}})$ is a random variable, with $C^{(i)}$ the space of the coupling variables for the discipline i.

The modeling uncertainty is therefore dependent on the design and coupling variables. An exhaustive presentation of methods for modeling and quantifying uncertainties falls outside the scope of this chapter, but we refer the reader to Chapter 1. Note that this modeling uncertainty is purely epistemic, as it is related to the lack of knowledge in the disciplinary metamodel. It can therefore be reduced by improving the knowledge of the metamodel and, in this chapter, we present an effective approach to reduce this uncertainty in a relevant manner according to a given quantity of interest.

For determining the modeling uncertainty in metamodels, we shall adopt a fairly common modeling approach consisting of the construction of Gaussian process-based (GP) surrogate models (or metamodels). This type of surrogate model has two advantages, in particular:

– once constructed, the cost of evaluating the surrogate model is negligible;

– an estimate of the modeling uncertainty is intrinsically provided by the Gaussian process model.

In practice, Gaussian processes are constructed based on a design of experiments (denoted by DoE_{f_i}) on space $\mathcal{Z}\times C^{(i)}$. The model is then constructed by conditioning a GP on DoE_{f_i}. The parameters of this GP are estimated, typically using maximum likelihood. Readers wishing to access more details on the construction of Gaussian processes can refer to Rasmussen and Williams (2006). In the framework of the MDA, if $y_i \in \mathbb{R}$, the disciplinary solver f_i can thus be replaced by the following GP:

$$\tilde{Y}_i(z, y_{c^{(i)}}) = \mu_{f_i}(z, y_{c^{(i)}}) + \varepsilon_i(z, y_{c^{(i)}}) \qquad [4.2]$$

where $\mu_{f_i}(z, y_{c^{(i)}})$ is the mean value and $\varepsilon_i(z, y_{c^{(i)}})$ is a GP of null mean and covariance function conditioned on DoE_{f_i}. Therefore, for a couple $(z^{(0)}, y^{(0)}_{c^{(i)}})$ $\in \mathbb{R}^n \times C^{(i)}$, which does not belong to DoE_{f_i} the surrogate model can be written as:

$$\tilde{Y}_i(z^{(0)}, y^{(0)}_{c^{(i)}}) = \mu_{f_i}(z^{(0)}, y^{(0)}_{c^{(i)}}) + \sigma_{f_i}(z^{(0)}, y^{(0)}_{c^{(i)}})\xi_i \qquad [4.3]$$

where $\mu_{f_i}(z^{(0)}, y^{(0)}_{c^{(i)}})$ is the mean value, $\sigma_{f_i}(z^{(0)}, y^{(0)}_{c^{(i)}})$ is the standard deviation and ξ_i is a standard Gaussian random variable. It should be noted that the expressions of ε_i, μ_{f_i} and σ_{f_i} are entirely determined by the type of Gaussian process being used. In the following, we shall use Gaussian processes with a constant mean and a square exponential covariance function, but the approach described remains valid for any other type of Gaussian process.

Note that in many cases, the output of disciplinary simulations is not a scalar but a large vector. This is the case, for example, in our application to aeroelastic coupling, which will be presented in section 4.4, where the two disciplinary outputs are a displacement field and a circulation vector. Therefore, we shall now formulate the problem in the more general case where the output of solver f_i is a vector of dimension n_i. In this case, with $y_i \in \mathbb{R}^{n_i}$ we assume that the disciplinary metamodel will be of the form:

$$\begin{aligned}\tilde{Y}_i(z,y_{c(i)}) &= \sum_{j=1}^{N_i}\left(\mu_{f_{i_j}}(z,y_{c(i)})+\varepsilon_{i_j}(z,y_{c(i)})\right)Y_{i_j} \\ &= \sum_{j=1}^{N_i}\mu_{f_{i_j}}(z,y_{c(i)})Y_{i_j}+\sum_{j=1}^{N_i}\varepsilon_{i_j}(z,y_{c(i)})Y_{i_j}\end{aligned} \qquad [4.4]$$

with $Y_{i_j}, j=1,\cdots,N_i \in \mathbb{R}^{n_i}$ a basis of N_i vectors to be defined (see section 4.4) and $\mu_{f_{i_j}}(z,y_{c(i)})+\varepsilon_{i_j}(z,y_{c(i)}), j=1,\cdots,N_i$ independent Gaussian processes with values in $\mathbb{R}$. Thereby, $\tilde{Y}_i(z,y_{c(i)})$ is a Gaussian process with values in $\mathbb{R}^{n_i}$ by linear combination of Gaussian processes with:

$$\mu_{f_i}(z,y_{c(i)}) = \sum_{j=1}^{N_i}\mu_{f_{i_j}}(z,y_{c(i)})Y_{i_j} \qquad [4.5]$$

and:

$$\varepsilon_i(z,y_{c(i)}) = \sum_{j=1}^{N_i}\varepsilon_{i_j}(z,y_{c(i)}))Y_{i_j} \qquad [4.6]$$

In particular, for a couple $(z^{(0)},y_{c(i)}^{(0)}) \in \mathcal{Z}\times C^{(i)}$ that does not belong to DoE_{f_i}, the surrogate model will be written as

$$\tilde{Y}_i(z^{(0)},y_{c(i)}^{(0)}) = \sum_{j=1}^{N_i}\left(\mu_{f_{i_j}}(z^{(0)},y_{c(i)}^{(0)})+\sigma_{f_{i_j}}(z^{(0)},y_{c(i)}^{(0)})\xi_{i_j}\right)Y_{i_j} \qquad [4.7]$$

where $\mu_{f_{i_j}}(z^{(0)}, y^{(0)}_{c(i)})$ are mean values, $\sigma_{f_{i_j}}(z^{(0)}, y^{(0)}_{c(i)})$ the standard deviations and ξ_{i_j} independent standard Gaussian random variables.

In the following, we shall expand computations using the more general form given by equation [4.4]. Nonetheless, it is possible to reduce the problem to a discipline with scalar output $y_i \in \mathbb{R}^{n_i}$ with $n_i = N_i = 1$ and $Y_{i,1} = 1$.

4.2.2. *Solving the random MDA*

The disciplinary surrogate model given by equation [4.4] is now introduced in the MDA (see equation [4.1]), which leads us to the nonlinear random system composed of $\sum_{i=1}^{n_d} n_i$ equations, $\forall z \in \mathcal{Z}$:

$$\tilde{Y}_i(z, \tilde{Y}_{c(i)}) = \sum_{j=1}^{N_i} \left(\mu_{f_{i_j}}(z, \tilde{Y}_{c(i)}) + \varepsilon_{i_j}(z, \tilde{Y}_{c(i)}) \right) Y_{i_j}, i = 1, \cdots, n_d \quad [4.8]$$

where $\tilde{Y}_{c(i)}$ represents the random vector associated with the coupling variables that affect discipline i and $\tilde{Y}_i \in \mathbb{R}^{n_i}$. It should be noted that since the basis vectors $Y_{i_j} \in \mathbb{R}^{n_i}$ are constant, the unknowns of this system are only the $n_{unk} = \sum_{i=1}^{n_d} N_i$ coefficients, grouped in the vector, $\forall z \in \mathcal{Z}$:

$$V_{unk}(z) = \left\{ \mu_{f_{i_j}}(z, \tilde{Y}_{c(i)}) + \varepsilon_{i_j}(z, \tilde{Y}_{c(i)}), i = 1, \cdots, n_d, j = 1, \cdots, N_i \right\} \quad [4.9]$$

The solution of this nonlinear system is defined by the joint probability distribution of $V^\star_{unk}(z)$ defined by:

$$V^\star_{unk}(z) = \left\{ \left(\mu_{f_{i_j}}(z, \tilde{Y}^\star_{c(i)}) + \varepsilon_{i_j}(z, \tilde{Y}^\star_{c(i)}) \right)^\star, i = 1, \cdots, n_d, j = 1, \cdots, N_i \right\} \quad [4.10]$$

yielding, through linear combination, the random vectors of the converging coupling variables, grouped in the list $\tilde{Y}^\star(z) = \left[\tilde{Y}^\star_i(z), i = 1, \cdots, n_d \right]$ with $\tilde{Y}^\star_i(z) \in \mathbb{R}^{n_i}$ and such that, $\forall z \in \mathcal{Z}$:

$$\tilde{Y}_i^\star(z,\tilde{Y}_{c^{(i)}}^\star(z)) = \sum_{j=1}^{N_i}\left(\mu_{f_{i_j}}(z,\tilde{Y}_{c^{(i)}}^\star(z)) + \varepsilon_{i_j}(z,\tilde{Y}_{c^{(i)}}^\star(z))\right)^\star Y_{i_j}, i=1,\cdots,n_d \qquad [4.11]$$

It is important to note here that the joint probability distribution of $V_{unk}^\star(z)$ (see equation [4.10]) is not Gaussian since the MDA is a nonlinear system. However, samples of the vector $V_{unk}^\star(z)$ and thus of $\tilde{Y}^\star(z)$ (by linear combination, see equation [4.11]) can be obtained by performing several realizations of the GP surrogate models and solving the MDA for each draw. This sampling approach does not raise any numerical cost issues because draws from the GP surrogate models have a negligible computational cost.

Moreover, in this work, we are only interested in the variation of the solutions of equation [4.8], hence our choice to model the solution by the random vector of the coefficients $V_{unk}^\star(z)$. The number of solutions for a given realization is also a random number, but this point falls beyond the scope of this chapter and is not necessary to our study. However, readers interested in such developments should consult Azaïs and Wschebor (2005).

Solving the system of nonlinear random equations given by equation [4.8] is not a simple task. A direct strategy would be to approach the random fields ε_{i_j} by spatial or spectral discretization, and then solve the nonlinear random equation system parameterized by a finite number of random variables. This number depends on the correlation length of ε_{i_j} and, depending on its value, it can induce an expensive computational time. This approximation has been numerically verified in Dubreuil *et al.* (2016) using the Karhunen–Loève decomposition and can result in high numerical costs if the number of random variables needed to obtain an accurate approximation of the random fields is large.

A less costly approach, developed in Dubreuil *et al.* (2016), and called the perfectly dependent GP is used here. In order to estimate the probability distribution of $V_{unk}^\star(z)$, the idea developed in this section is to solve a simpler system of equations whose solution is close to that of equation [4.8]. Indeed, the objective of this uncertainty propagation step is to model the variation of the solutions of the MDA due to the use of disciplinary metamodels. The variance of the solution of the MDA is directly related to the variances of the disciplinary GPs. Therefore, we propose the use of simplified GP models for the propagation of uncertainties by considering a perfectly dependent disciplinary GP.

Therefore, the linear combination of GP given by equation [4.4] is simplified by:

$$\tilde{Y}_i'(z, y_{c(i)}) = \sum_{j=1}^{N_i} \left(\mu_{f_{i_j}}(z, y_{c(i)}) + \sigma_{f_{i_j}}(z, y_{c(i)}) \xi_{i_j} \right) Y_{i_j}$$

where the variances are given by:

$$\sigma_{f_{i_j}}(z, y_{c(i)}) \xi_{i_j} = \varepsilon_{i_j}(z, y_{c(i)}) \quad \forall\, z, y_{c(i)} \in \mathcal{Z} \times \mathbb{R}^{\#c(i)}$$

with ξ_{i_j} a reduced centered Gaussian random variable. It should be noted that this allows one to exactly approximate the mean and variance of the field ε_{i_j} (but not its covariance at all) using a single random variable ξ_{i_j}. This ultimately amounts to constructing a perfectly dependent GP for each component of the vector V_{unk}.

In the end, the following system has to be solved:

$$\tilde{Y}_i'^{\star}(z, \tilde{Y}_{c(i)}'^{\star}) = \sum_{j=1}^{N_i} \left(\mu_{f_{i_j}}(z, \tilde{Y}_{c(i)}'^{\star}) + \sigma_{f_{i_j}}(z, \tilde{Y}_{c(i)}'^{\star}) \xi_{i_j} \right)^{\star} Y_{i_j}, i = 1, \cdots, n_d \quad [4.12]$$

whose solution is given in the form of the random vector:

$$V_{unk}'^{\star}(z) = \left\{ \left(\mu_{f_{i_j}}(z, \tilde{Y}_{c(i)}'^{\star}) + \sigma_{f_{i_j}}(z, \tilde{Y}_{c(i)}'^{\star}) \xi_{i_j} \right)^{\star}, i = 1, \cdots, n_d, j = 1, \cdots, N_i \right\}$$

Given that the variance of the simplified disciplinary GPs involved in equation [4.12] and those of the disciplinary GPs involved in equation [4.8] are equal, the probability distribution of $V_{unk}'^{\star}(z)$ is assumed to be a correct estimate of the estimation of $V_{unk}^{\star}(z)$. This approximation was numerically verified in Dubreuil *et al.* (2016). Moreover, from a numerical point of view, equation [4.12] is much simpler to solve than equation [4.8] because each GP ε_{i_j} is discretized with only one random variable ξ_{i_j}. In the following, the random vector containing all the random variables $\xi_{i_j}, i = 1,\cdots,n_d, j = 1,\cdots,N_i$ is denoted by $\Xi = \{\Xi_i, i = 1,\cdots,n_d\}$ with $\Xi_i = \{\xi_{i_j}, j = 1,\cdots,N_i\}$.

Random MDA has already been the focus of studies and several methods have been proposed to approach the solution of such a system (see, for example, Arnst *et al.* 2012; Sankararaman and Mahadevan 2012; Jiang *et al.* 2015; Dubreuil *et al.* 2016). In the present study, the simplified GP models involved in the MDA are inexpensive, so the Monte Carlo method can be used directly. The point is to solve equation [4.12] for a given realization $\Xi^{(\ell)}$ of Ξ, for $\ell = 1,\cdots,n_{MC}$, with n_{MC} the number of Monte Carlo simulations performed. Let us point out that for a given realization $\Xi^{(\ell)}$ any nonlinear solver can be used to solve these deterministic systems of equations and in what follows, the nonlinear Jacobi method is retained. In addition, it should be noted that these systems of equations can only be solved in a point-to-point manner inside the design space, that is, for a fixed value $z^{(0)}$ of z.

The main objective of the following section is to assess the influence of uncertainty related to the modeling of coupling variables by GPs, on a given quantity of interest.

4.2.3. *Approximation of the quantity of interest using sparse polynomial chaos*

The previous section has shown that the use of disciplinary GPs in MDA leads to determining random coefficients $V^{\star}_{unk}(z)$ and thus to obtaining the random coupling variables denoted by $\tilde{Y}^{\star}(z) = \left[\tilde{Y}_i^{\star}(z), i = 1,\cdots,n_d, \tilde{Y}_i^{\star}(z) \in \mathbb{R}^{n_i}\right]$.

Consequently, the quantity of interest, which is a function of the coupling variables, is also modeled as a random variable or, more precisely, as a random field on the design space $\mathcal{Z}$, which is expressed as:

$$Y_{QoI}(z, \tilde{Y}^{\star}_{c^{(QoI)}}(z)), \forall z \in \mathcal{Z} \qquad [4.13]$$

where the notation $c^{(QoI)}$ designates the coupling variables involved in the expression of the quantity of interest. The method proposed to characterize this random variable is based on a polynomial chaos approximation (Ghanem and Spanos 1991) of $Y_{QoI}(z, \tilde{Y}^{\star}_{c(QoI)}(z))$. Assuming that the quantity Y_{QoI} is a second-order random variable, it can be written in a general context (see Cameron and Martin 1947) that:

$$Y_{QoI}(z,\Xi) = \sum_{i=0}^{\infty} a_i(z)\phi_i(\Xi) \quad \forall z \in \mathcal{Z} \qquad [4.14]$$

where the convergence is in L_2 norm, $\{\phi_i\}_{i\in\mathbb{N}}$ are the polynomials of an orthogonal basis, with respect to the probability density of Ξ, and $a_i(z)$ are coefficients to be determined. In order to simplify the notations, the dependency in z will be omitted in the following.

For the calculation of the coefficients, several approaches are possible and the regression method, developed in Berveiller *et al.* (2006) and improved in Blatman and Sudret (2011), will be used in the following. In order to reduce the number of coefficients to be determined, the sparse polynomial chaos approach, described in Blatman and Sudret (2011) and implemented in the OpenTURNS library (Baudin *et al.* 2017), is used. In our study, the approximation of the quantity of interest is written as:

$$\hat{Y}_{QoI}(\Xi) = \sum_{\boldsymbol{\alpha}\in A} a_{\boldsymbol{\alpha}} H_{\boldsymbol{\alpha}}(\Xi) \qquad [4.15]$$

with $\Xi = \left\{\Xi_1, \cdots, \Xi_{n_d}\right\}$ where each vector $\Xi_i, i = 1, \cdots, n_d$ is a vector whose components are the independent standard normal random variables used to discretize the field ε_i (see equation [4.6]), that is $\Xi_i = \left\{\xi_{i_1}, \cdots, \xi_{i_{N_i}}\right\}$. The approximation basis is composed of the multivariate Hermite polynomials $H_{\boldsymbol{\alpha}}(\Xi)$ constructed by tensor product of the dimension 1 Hermite polynomials and normalized such that:

$$\mathbb{E}[H_{\boldsymbol{\alpha}}(\Xi) H_{\boldsymbol{\beta}}(\Xi)] = \delta_{\boldsymbol{\alpha\beta}}$$

with $\mathbb{E}$ the expectation, $\delta_{\alpha\beta}$ the Kronecker symbol where $\boldsymbol{\alpha} = \left\{\alpha_1, \cdots, \alpha_{n_d}\right\}$ is the list containing the multi-indices $\boldsymbol{\alpha_i}, i = 1, \cdots, n_d$ such that $\boldsymbol{\alpha_i} = \left\{\alpha_{i_1}, \cdots, \alpha_{i_{N_i}}\right\}$.

Finally, the indices α_{i_j} indicate the degree of the variable ξ_{i_j} in the polynomial H_α of degree d (that is $\sum_{i=1}^{n_d} |\boldsymbol{\alpha}_i| = d$). Let the set $\mathcal{A}$ be a subset of $\mathcal{B}$ that contains all the indices $\boldsymbol{\alpha}$ up to degree d, that is $card(\mathcal{B}) = \frac{(d+n_{unk})!}{d!n_{unk}!}$ with $n_{unk} = \sum_{i=1}^{n_d} N_i$. The objective of the sparse approach is to reduce, as much as possible, the set of subindices $\mathcal{A}$ in order to obtain $card(\mathcal{A}) << card(\mathcal{B})$. For this purpose, the LARS (least-angle regression) algorithm (see Efron *et al.* 2004) is employed. In this non-intrusive approach, the unknown coefficients $a_{\boldsymbol{\alpha}}$ are calculated iteratively by solving a least squares problem, and by selecting the most correlated polynomial with the residual at each iteration. The details of the calculation of the coefficients $a_{\boldsymbol{\alpha}}$ can be found in Blatman and Sudret (2011).

In the end, using the orthogonality of the basis vectors, once the coefficients have been calculated, the quantity $\hat{Y}_{QoI}(\Xi)$ is completely determined from equation [4.15]. This polynomial chaos expansion allows for an analytical estimation of the mean and variance of $Y_{QoI}(\Xi)$ given by:

$$\mathbb{E}[Y_{QoI}(\Xi)] \approx \mathbb{E}[\hat{Y}_{QoI}(\Xi)] = a_{\overline{\mathbf{0}}} \qquad [4.16]$$

and:

$$\mathrm{Var}[Y_{QoI}(\Xi)] \approx \mathrm{Var}[\hat{Y}_{QoI}(\Xi)] = \sum_{\boldsymbol{\alpha} \in \mathcal{A} \setminus \{\overline{\mathbf{0}}\}} a_{\boldsymbol{\alpha}}^2 \qquad [4.17]$$

where Var denotes the variance and $\overline{\mathbf{0}}$ is the list containing the multi-indices $\alpha_i = \mathbf{0}$, $i = 1, \cdots, n_d$ of zero components.

4.3. Sensitivity analysis and uncertainty reduction

4.3.1. *Introduction*

The previous sections have shown how the uncertainties associated with each disciplinary metamodel propagated to the quantity of interest. This propagation leads us to model the quantity of interest by a random variable Y_{QoI} approximated by its development on polynomial chaos $\hat{Y}_{QoI}$ The variability of Y_{QoI} is thus due to the variability of the disciplinary metamodels. It is then recalled that $Y_{QoI} = f_{QoI}(Y^{\star}_{c(QoI)}(\boldsymbol{\varepsilon}))$ with $\boldsymbol{\varepsilon} = \{\varepsilon_i, i = 1, \cdots, n_d\}$ the random fields modeling the uncertainties of the disciplinary metamodels. The objective of the sensitivity analysis is then to identify the input random fields ε_i responsible for the variation of the output Y_{QoI} and to quantify their influence on this variation.

Among the different approaches for performing a sensitivity analysis, methods based on analysis of variance are frequently retained since they propose to quantify the relationship between the variance due to an input, and the total variance of the response. Formally, the Sobol' sensitivity indices (Sobol' 2001) are defined as follows. Let $Y_{QoI} = f_{QoI}(Y^{\star}_{c(QoI)}(\boldsymbol{\varepsilon}))$ the first-order Sobol' index of the input ε_i be defined by:

$$S_{\varepsilon_i} = \frac{\mathrm{Var}(\mathbb{E}(Y_{QoI} \mid \varepsilon_i))}{\mathrm{Var}(Y_{QoI})}$$

In the present case where the inputs are the random fields ε_i (namely functional inputs and not scalar random variables), several works have focused on the estimation of indices S_{ε_i} (Iooss and Ribatet 2009). As previously explained, in our framework, it is assumed that each field ε_i can be discretized in the form:

$$\varepsilon_i(z, y_{c(i)}) = \sum_{j=1}^{N_i} \sigma_{f_{i_j}}(z, y_{c(i)}) \xi_{i_j} Y_{i_j}$$

where ξ_{i_j} are independent random variables. Following the idea proposed in Jacques *et al.* (2006), we shall *group* the scalar variables used in the discretization of the field ε_i. In Jacques *et al.* (2006), this approach is used to deal with groups of dependent variables. The first-order Sobol' index of the input ε_i is then defined by:

$$S_{\varepsilon_i} = \frac{\mathrm{Var}(\mathbb{E}(Y_{QoI} \mid \xi_{i_j}, j = 1, \cdots, N_i))}{\mathrm{Var}(Y_{QoI})} \qquad [4.18]$$

Higher order indices can also be defined to quantify the influence of interactions between different input variables. Similarly, the total index of the variable ε_i can be used to quantify the influence of the variable ε_i alone, as well as in interaction with all other input variables. It is defined by:

$$S_{\varepsilon_i}^T = 1 - \frac{\mathrm{Var}(\mathbb{E}(Y_{QoI} \mid \xi_{\sim i_j}, j = 1,\cdots,N_i))}{\mathrm{Var}(Y_{QoI})} \qquad [4.19]$$

where the notation $\sim i$ denotes all input variables except i.

These indices can be estimated by Monte Carlo (see, for example, Iooss and Ribatet 2009; Jiang *et al.* 2016) but the computational cost can then be significant. In the following, we shall make use of the sparse polynomial chaos expansion $\hat{Y}_{QoI}$ (presented in section 4.2.3) in order to estimate these indices analytically.

NOTE.– Other sensitivity indices are available for identifying the variables that influence the variation of the model output. In particular, there are indices using the entire probability distribution of the response (and not just the variance), developed in Borgonovo (2007) and Krzykacz-Hausmann (2001). Readers interested in the use of these indices in multidisciplinary analysis can find an example of their application in Jiang *et al.* (2016). In what follows, variance-based sensitivity indices have been chosen because they provide relevant information for reducing uncertainty and can be easily estimated as will be shown in the following section.

4.3.2. *Sobol' indices approximated by polynomial chaos*

As recalled in section 4.2.3, the approximation of the quantity of interest by sparse polynomial chaos is of the form:

$$\hat{Y}_{QoI}(\Xi) = \sum_{\boldsymbol{\alpha} \in A} a_{\boldsymbol{\alpha}} H_{\boldsymbol{\alpha}}(\Xi)$$

with $\Xi = \left\{\Xi_1, \cdots, \Xi_{n_d}\right\}$ where each vector $\Xi_i, i = 1, \cdots, n_d$ is a vector whose components are the independent standard normal random variables retained for discretizing the field ε_i, that is $\Xi_i = \left\{\xi_{i_1}, \cdots, \xi_{i_{N_i}}\right\}$. From the coefficients, we have an analytical expression of the mean and variance via equations [4.16] and [4.17]. In addition, the works of Sudret (2008) and Blatman and Sudret (2010) show that, in the case of scalar inputs, an estimate of the Sobol' indices can be obtained analytically from the coefficients $a_{\boldsymbol{\alpha}}$. We can easily adapt this work to the framework of our study where each functional input ε_i is described by a group of variables ξ_{i_j}.

Applying equations [4.18] and [4.19] to the polynomial chaos decomposition given by equation [4.15], we obtain the first-order Sobol' index of the input ε_i:

$$\hat{S}_{\varepsilon_i} = \frac{1}{\mathrm{Var}(\hat{Y}_{QoI})} \sum_{\boldsymbol{\alpha} \in \mathcal{A}_i} a_{\boldsymbol{\alpha}} \qquad [4.20]$$

where:

$$\mathcal{A}_i = \left\{ \boldsymbol{\alpha} \in \mathcal{A} \setminus \{\bar{0}\}, \boldsymbol{\alpha}_{\sim i} = \mathbf{0} \right\}$$

and the associated total Sobol' index:

$$\hat{S}^T_{\varepsilon_i} = 1 - \frac{1}{\mathrm{Var}(\hat{Y}_{QoI})} \sum_{\boldsymbol{\alpha} \in \mathcal{A}_i^T} a_{\boldsymbol{\alpha}} \qquad [4.21]$$

where:

$$\mathcal{A}_i^T = \left\{ \boldsymbol{\alpha} \in \mathcal{A} \setminus \{\bar{0}\}, \boldsymbol{\alpha}_i = \mathbf{0} \right\}$$

The estimates of the Sobol' indices given by equations [4.20] and [4.21] can then be used to determine which disciplinary metamodel should be enriched. As explained in section 4.2.1, disciplinary metamodels are constructed from designs of experiments DoE_{f_i} where disciplinary models $f_i(z, y_{c^{(i)}})$ are evaluated. These disciplinary models depend on design variables z and also on coupling variables $y_{c^{(i)}}$. In order to enrich a disciplinary metamodel, the disciplinary solver f_i must therefore be evaluated at a point $(z^{(new)}, y^{(new)}_{c^{(i)}})$. The previous sensitivity analysis is performed for a fixed value of the design variables $z^{(new)}$ that will therefore be chosen to sample the disciplinary model. On the other hand, the determination of

$y_{c^{(i)}}^{(new)}$ is more complex since without solving the MDA we do not know the values of $y_{c^{(i)}}^{(new)}$ the solution of the MDA at $z^{(new)}$. In order to solve this problem, we propose, in the following, to solve the MDA using the mean values of the disciplinary metamodels and to use the value of the coupling variables obtained for $y_{c^{(i)}}^{(new)}$.

4.4. Application to an aeroelastic test case

4.4.1. *Presentation*

The method presented in the previous sections will now be applied to a test case involving the static aeroelastic equilibrium of an aircraft wing. To this end, we assume a linear model for the elastic deformation of the structure and a potential fluid model for obtaining the aerodynamic loads. The calculation of the displacement field is achieved using the finite element method (FEM), while the calculation of the aerodynamic circulation (and thus of the aerodynamic loads) is achieved with the vortex lattice method (VLM). Therefore, the coupling of these two solvers using a Gauss–Seidel-type algorithm allows the displacement field and the circulation field that balance the wing to be found. In practice, we use mesh interpolation by radial basis functions (see, for example, Boer *et al.* 2007) to transfer information between the two meshes. With the notations adopted so far, the multidisciplinary system to solve is thus: find the displacement field $U(z,\Gamma)$ and the circulation vector $\Gamma(z,U)$ that solve:

$$\begin{cases} U(z,\Gamma) & = f_1(z,\Gamma) \\ \Gamma(z,U) & = f_2(z,U) \end{cases} \quad [4.22]$$

where f_1 and f_2, respectively, symbolize the FEM and VLM solvers. In the following, we shall use, as the only design parameter, the angle of attack of the wing $z \in [1°,10°]$. Figures 4.1(a) and (b) respectively present the FEM and VLM models of the wing under study.

The following results illustrate the advantages of taking into account the coupling between these two disciplines by comparing certain quantities of interest, with and without taking into account the multidisciplinary coupling. These results are given for z = 8° and the quantities of interest are normalized by the maximal value. More specifically, Figure 4.2 shows the displacement field, and Figure 4.3 shows the value of the circulation.

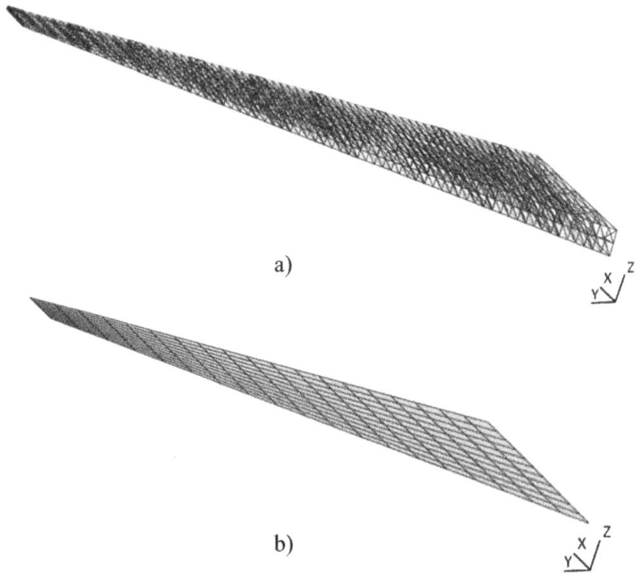

Figure 4.1. *(a) Finite element mesh of the structural part of the wing (skin, spars and ribs); (b) VLM model mesh of the wing (mean plane of the wing). For a color version of this figure, see www.iste.co.uk/gogu/uncertainties.zip*

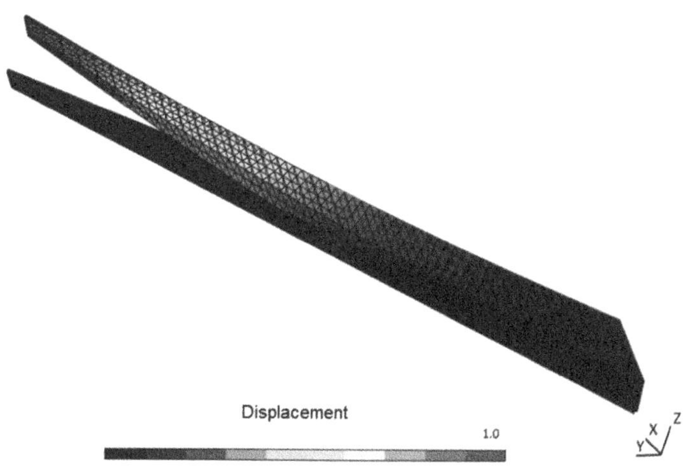

Figure 4.2. *Displacement field after convergence of the multidisciplinary analysis. For a color version of this figure, see www.iste.co.uk/gogu/uncertainties.zip*

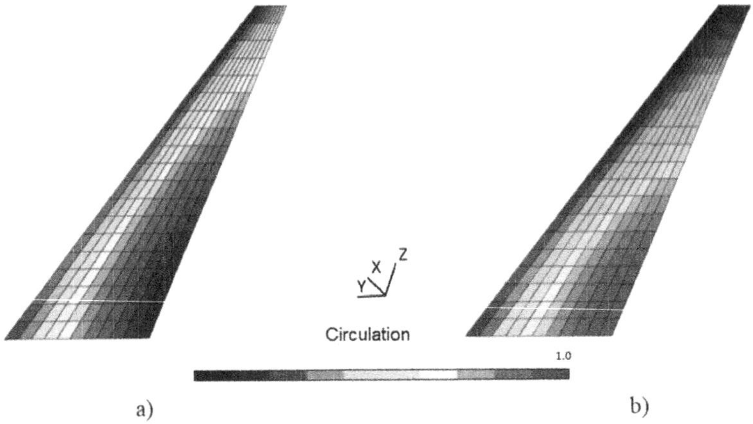

Figure 4.3. *Circulation (a) before and (b) after convergence of the multidisciplinary analysis. For a color version of this figure, see www.iste.co.uk/gogu/uncertainties.zip*

Finally, Figure 4.4 shows the quantity of interest $Y_{QoI}(z)$ that we shall study in the following. This is the lift-to-drag ratio, the ratio between the wing's lift coefficient $Cz(z)$ and the drag coefficient $Cx(z)$ defined by:

$$Y_{QoI}(z) = \frac{Cz}{Cx}$$

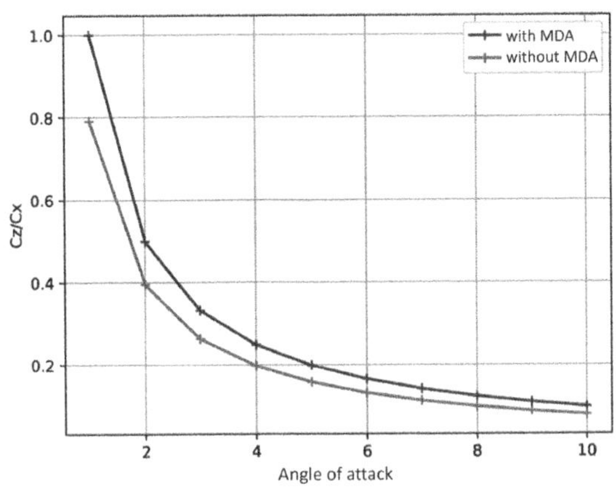

Figure 4.4. *Lift-to-drag ratio $Y_{QoI}(z)$ as a function of the angle of attack z, with or without taking multidisciplinary coupling into account. For a color version of this figure, see www.iste.co.uk/gogu/uncertainties.zip*

Figure 4.4 shows the benefit of MDA for evaluating lift-to-drag ratio. Nevertheless, the coupled calculation must be performed for each value of z, which is numerically very expensive. In order to reduce this computational time, we propose to build metamodels of both disciplines.

4.4.2. Construction of disciplinary metamodels

Disciplinary metamodels are of the form:

$$\hat{U}(z) = \sum_{i=1}^{N_U} \hat{u}_i(z) U_i \qquad [4.23]$$

and:

$$\hat{\Gamma}(z) = \sum_{i=1}^{N_\Gamma} \hat{\gamma}_i(z) \Gamma_i \qquad [4.24]$$

where the bases $U_i, i = 1, \cdots, N_U$ and $\Gamma_i, i = 1, \cdots, N_\Gamma$ are constructed by singular value decomposition from a sample of solutions (also called design of experiments) where the MDA of the disciplinary models has been solved. The coefficients $u_i, i = 1, \cdots, N_U$ and $\gamma_i, i = 1, \cdots, N_\Gamma$ are obtained by projecting these solutions onto the bases $U_i, i = 1, \cdots, N_U$ and $\Gamma_i, i = 1, \cdots, N_\Gamma$, respectively. We then construct interpolations by Gaussian processes of each coefficient denoted by $\hat{u}_i, i = 1, \cdots, N_U$ and $\hat{\gamma}_i, i = 1, \cdots, N_\Gamma$. This approach is known as POD+I (proper orthogonal decomposition + interpolation) in the literature (see, for example, Guo and Hesthaven 2018; Aversano *et al.* 2019).

Therefore, the metamodels $\hat{U}$ and $\hat{\Gamma}$ are Gaussian by linear combination of Gaussian variables and the MDA system to be solved becomes $\forall z \in \mathcal{Z}$:

$$\begin{aligned} \hat{U}(z, \hat{\gamma}_j, j = 1, \cdots, N_\Gamma) &= \sum_{i=1}^{N_U} \hat{u}_i(z, \hat{\gamma}_j, j = 1, \cdots, N_\Gamma) U_i \\ \hat{\Gamma}(z, \hat{u}_i, i = 1, \cdots, N_U) &= \sum_{j=1}^{N_\Gamma} \hat{\gamma}_j(z, \hat{u}_i, i = 1, \cdots, N_U) \Gamma_j \end{aligned} \qquad [4.25]$$

In practice, for the construction of the bases $U_i, i = 1, \cdots, N_U$ and $\Gamma_i, i = 1, \cdots, N_\Gamma$; the MDA is solved for five values of z (1.0, 2.5, 4.5, 7.25, 10.0). Since the MDA is solved using an iterative algorithm, we thus have several points for each value of z for the construction of the designs of experiments used to determine the Gaussian processes $\hat{u}_i, i=1,\cdots,N_U$ and $\hat{\gamma}_i, i=1,\cdots,N_\Gamma$. Using the notations of equation [4.6], we shall denote in the following by $\varepsilon_1 = \sum_{i=1}^{N_U} \varepsilon_{\hat{u},i} U_i$ the model uncertainty associated with $\hat{U}$ and $\varepsilon_2 = \sum_{i=1}^{N_\Gamma} \varepsilon_{\hat{\gamma},i} \Gamma_i$ the model uncertainty associated with $\hat{\Gamma}$.

In practice, after solving the five MDAs, we obtain a design of experiments of size 50. The bases $U_i, i=1,\cdots,N_U$ and $\Gamma_i, i=1,\cdots,N_\Gamma$ are truncated so that the sum of the retained singular values represents more than 99.999% of the total sum of the singular values, which leads to bases of size $N_U = 4$ and $N_\Gamma = 3$. The uncertainty quantification problem is therefore parameterized by $N_U + N_\Gamma = 7$ independent standard Gaussian random variables. Using the notations of section 4.2.2, we obtain $\Xi = \{\Xi_1, \Xi_2\}$ with $\Xi_1 = \{\xi_1, \xi_2, \xi_3, \xi_4\}$ and $\Xi_2 = \{\xi_5, \xi_6, \xi_7\}$.

We can now solve the random MDA for different realizations of the random vector Ξ. As an example, Figure 4.5 shows 20 random realizations of the lift-to-drag ratio, obtained by solving the random MDA given by equation [4.25]. In this figure, the reference curve is obtained by solving the MDA using the disciplinary solvers (FEM and VLM).

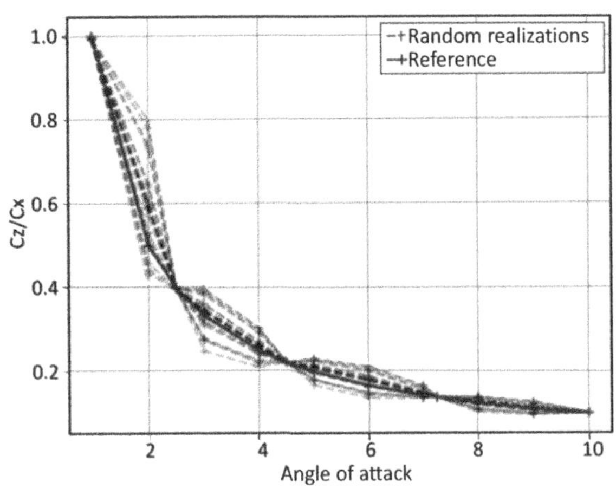

Figure 4.5. *Twenty random realizations of the lift-to-drag ratio Y_{QoI} (z) obtained by solving the random MDA of equation [4.25] as a function of the angle of attack z. The reference curve is obtained by solving the MDA using the disciplinary solvers. For a color version of this figure, see www.iste.co.uk/gogu/uncertainties.zip*

It is interesting to note that the approximation illustrated in Figure 4.5 is not obtained by constructing a Gaussian process from values of Y_{QoI} associated with a design of experiments but by solving equation [4.25] for different realizations of the random vector Ξ. This implies, in particular, that $Y_{QoI}(z)$ is not a Gaussian random variable. However, as in Gaussian interpolation, we do obtain a null variance at points of the design of experiments, namely 1.0, 2.5, 4.5, 7.25 and 10.0 because the disciplinary metamodels were constructed by solving the MDA at these five points.

We shall now focus on sensitivity analysis and on the reduction of uncertainty of $Y_{QoI}(z)$ for $z = 2°$.

4.4.3. *Sensitivity analysis and uncertainty reduction*

As described in section 4.2.3, we begin by computing a polynomial chaos approximation of the quantity of interest $Y_{QoI}(z)$, denoted $\hat{Y}_{QoI}(z)$, for the angle of attack $z = 2°$. To this end, we use a basis of multivariate Hermite polynomials of dimension $n_{unk} = 7$. The maximum degree of the polynomials being used is set to $d = 3$, which leads to $card(\mathcal{B}) = 120$. In order to determine the coefficients $a_{\boldsymbol{\alpha}}$ of the approximation by polynomial chaos (see equation [4.15]), we use a design of experiments of size 300, that is, the random MDA of equation [4.25] is solved for 300 random realizations of the vector Ξ.

Of these 300 points, 250 are used for determining the coefficients $a_{\boldsymbol{\alpha}}$ by least angle regression and 50 are used for validation (the 50 points used for validation are randomly drawn from the 300-sized sample). The validation of the polynomial chaos approximation is evaluated by calculating the coefficient of determination R^2 with the training sample (of size 250 and denoted by $Y_{QoI}^{(tra)}$ in the following) and the mean relative error E_{rel} with the validation sample (of size 50 and denoted $Y_{QoI}^{(val)}$ hereafter). The quantities R^2 and E_{rel} are thus obtained by:

$$R^2 = \frac{\sum_{i=1}^{250} (\hat{Y}_{QoI}^{(tra)}(\Xi_i) - \overline{Y}_{QoI}^{(tra)})^2}{\sum_{i=1}^{250} (Y_{QoI}^{(tra)}(\Xi_i) - \overline{Y}_{QoI}^{(tra)})^2}$$

$$E_{rel} = \frac{1}{50}\sum_{i=1}^{50}\frac{|\hat{Y}_{QoI}^{(val)}(\Xi_i) - Y_{QoI}^{(val)}(\Xi_i)|}{|Y_{QoI}^{(val)}(\Xi_i)|}$$

with $\overline{Y}_{QoI}^{(tra)}$ the average of $Y_{QoI}^{(tra)}$ with 250 points.

The evaluation of the coefficients leads to a sparse basis with card(A) = 16 and we obtain $R^2 = 0.996$ and $E_{rel} = 0.009$, which allows us to be confident about the quality of the approximation obtained by polynomial chaos. In order to illustrate the quality of the approximation by polynomial chaos, Figure 4.6 presents the histogram obtained from the 300 values of $Y_{QoI}(\Xi_i), i = 1, \cdots, 300,$ as well as the one obtained from the approximation $\hat{Y}_{QoI}(\Xi_i), i = 1, \cdots, 300.$ This figure also illustrates that the random variable Y_{QoI} is not Gaussian, as explained earlier.

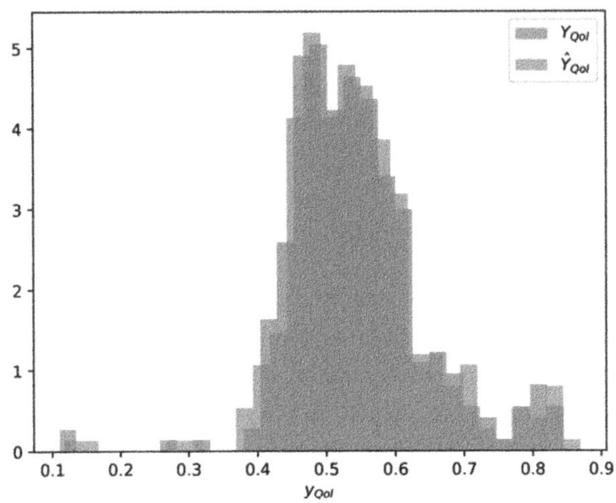

Figure 4.6. *Histogram of the quantity of interest Y_{QoI} based on the sample of size 300 used for the construction and validation of the polynomial chaos approximation and histogram obtained with the polynomial chaos approximation $\hat{Y}_{QoI}$ from the same sample. For a color version of this figure, see www.iste.co.uk/gogu/uncertainties.zip*

We now consider the estimation of the Sobol' sensitivity indices. Although the previous results allow us to be confident about the quality of the polynomial chaos approximation, they do not make it possible to quantify the estimation error of the Sobol' indices. As a matter of fact, it is difficult to link the quality of the estimator of the Sobol' indices by polynomial chaos to the quality of the approximation itself. For this purpose, we make use of a *bootstrap* resampling approach (see Efron 1979), proposed in Dubreuil *et al.* (2014), for estimating the coefficient of variation of the indices. Table 4.1 gives the mean values of the Sobol' indices as well as the estimate of the coefficient of variation obtained after 300 bootstrap resamplings.

Variable	Mean $\hat{S}_{\varepsilon_i}$	Coeff. var. $\hat{S}_{\varepsilon_i}$	Mean $\hat{S}^T_{\varepsilon_i}$	Coeff. var. $\hat{S}^T_{\varepsilon_i}$
ε_1	5.2×10^{-3}	11.0%	1.5×10^{-2}	8.3%
ε_2	9.8×10^{-1}	0.12%	9.9×10^{-1}	0.06%

Table 4.1. *Estimators of the Sobol' indices calculated by polynomial chaos, the estimators for the means and coefficients of variation are calculated by 300 bootstrap repetitions*

The results in Table 4.1 show that the variance of the quantity of interest Y_{QoI} is almost solely due to the variance of the field ε_2 ($\mathbb{E}[\hat{S}_{\varepsilon_2}] \approx 9.8 \times 10^{-1}$) discretized by the random vector $\Xi_2 = \{\xi_5, \xi_6, \xi_7\}$ and modeling the error of the aerodynamic metamodel by calculating the circulation $\hat{\Gamma}$ (see equation [4.24]).

We shall therefore enrich the aerodynamic disciplinary metamodel $\hat{\Gamma}(z, \hat{u}_i, i = 1, \cdots, N_U)$ by calculating $\Gamma(z, \hat{u}_i, i = 1, \cdots, N_U)$ with the VLM model and then, once the solution is projected onto the basis $\Gamma_j, j = 1, \cdots, N_\Gamma$, reconstruct the Gaussian process approximations of the coefficients $\hat{\gamma}_j, j = 1, \cdots, N_\Gamma$. The point $(z, \hat{u}_i,\ i = 1, \cdots, N_U)$ at which the VLM model is evaluated is chosen such that $z = 2°$ and $\hat{u}_i, i = 1, \cdots, N_U$ is the solution of the random MDA (see equation [4.25]) using the mean value of the fields ε_1 and ε_2 obtained by imposing $\Xi = \bar{0}$.

Figure 4.7 presents the histograms of the quantity of interest before and after enrichment for the angle of attack $z = 2°$. This figure clearly illustrates the reduction in uncertainty in the quantity of interest Y_{QoI} obtained by reducing the uncertainty of the aerodynamic disciplinary metamodel $\hat{\Gamma}$. More precisely, before enrichment, the coefficient of variation of Y_{QoI} is estimated at 18%; after enrichment, it is estimated at 3% (estimated on samples of size 300).

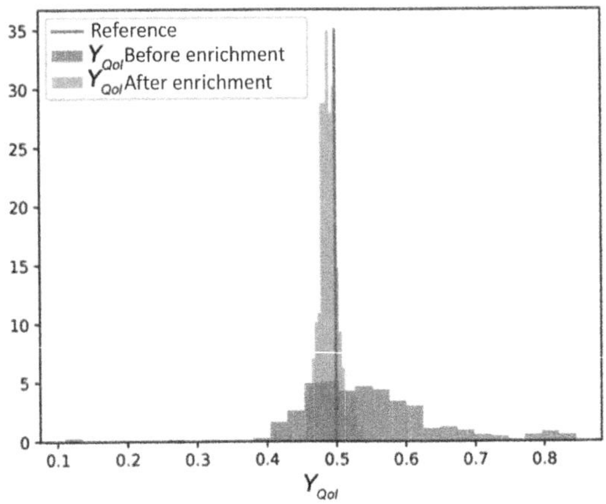

Figure 4.7. *Histogram of the quantity of interest Y_{QoI} before and after enrichment of the discipline $\hat{\Gamma}$. The angle of attack is set to $z = 2°$. The reference is obtained by solving the MDA with the discipline solvers (FEM and VLM). For a color version of this figure, see www.iste.co.uk/gogu/uncertainties.zip*

In order to confirm the interest of sensitivity analysis to detect the disciplinary metamodel to be enriched, Figure 4.8 presents the histogram of the quantity of interest obtained by enrichment of the metamodel structures $\hat{U}$. In this case, the estimation of the coefficient of variation after enrichment is still 17% compared to 18% before enrichment (and 3% after enriching the aerodynamic metamodel $\hat{\Gamma}$). These results are not surprising and indeed confirm those of the sensitivity analysis given in Table 4.1.

It is important to note that the quantity of interest is dependent on the coupling of the two disciplines (as shown in Figure 4.4) but that (almost) only the uncertainty of the metamodel calculating the circulation (that is, the aerodynamic model) is responsible for the variation in the quantity of interest. Indeed, since the quantity of interest studied is the lift-to-drag ratio, obtained by postprocessing the circulation Γ, it would be tempting to conclude that the result of the sensitivity analysis was predictable, which is not the case. In order to illustrate this last point, we now propose to study a quantity of interest obtained by postprocessing the displacement U. As an example, we choose Y_{QoI} the wingtip twist angle, the other parameters of the study remain unchanged. For this example, we use fourth-degree polynomials for the polynomial chaos approximation in order to have a relative error on the validation sample smaller than 5%. Figure 4.9 shows the histogram of the wingtip twist angle before enriching the disciplinary metamodels and its approximation by polynomial chaos. The coefficient of variation of this new quantity of interest is estimated to be 34%. The polynomial chaos approximation yields $R^2 \approx 0.98$ and $E_{rel} \approx 0.044$. In the same way as the previous quantity of interest, the quality of the estimation of the Sobol' indices by polynomial chaos is evaluated by bootstrap resampling of size 300. Table 4.2 presents the results obtained.

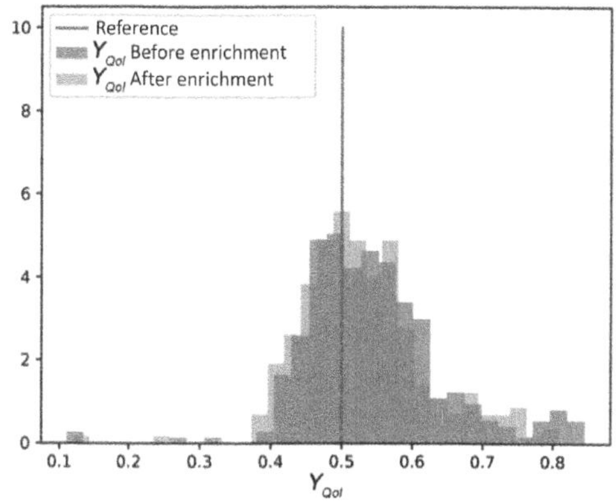

Figure 4.8. Histogram of the quantity of interest Y_{QoI} before and after enrichment of discipline $\hat{U}$. The angle of attack is set to z = 2°. The reference is obtained by solving the MDA with the discipline solvers (FEM and VLM). For a color version of this figure, see www.iste.co.uk/gogu/uncertainties.zip

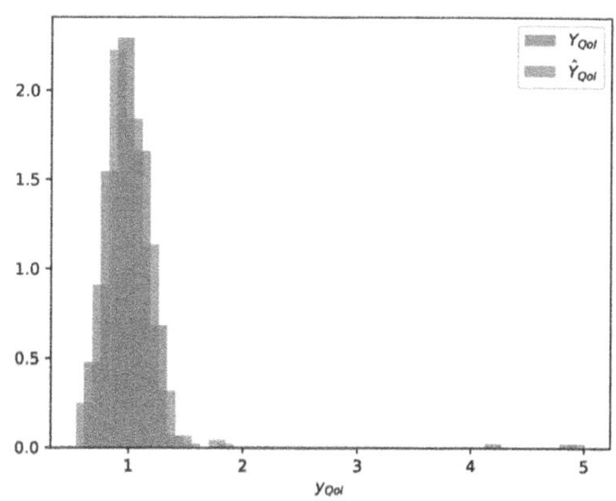

Figure 4.9. Histogram of the quantity of interest, namely the spin angle Y_{QoI}, based on the sample of size 300, used for the construction and validation of the polynomial chaos approximation and histogram obtained with the polynomial chaos approximation $\hat{Y}_{QoI}$ with the same sample. For a color version of this figure, see www.iste.co.uk/gogu/uncertainties.zip

Variable	Mean $\hat{S}_{\varepsilon_i}$	Coeff. var. $\hat{S}_{\varepsilon_i}$	Mean $\hat{S}^T_{\varepsilon_i}$	Coeff. var. $\hat{S}^T_{\varepsilon_i}$
ε_1	1.6×10^{-1}	13.3%	3.5×10^{-1}	18.5%
ε_2	6.5×10^{-1}	9.7%	8.4×10^{-1}	2.5%

Table 4.2. *Wing tip twist. Estimators of the Sobol' indices calculated by polynomial chaos, the estimators of the means and coefficients of variation are calculated by 300 bootstrap repetitions*

The results in Table 4.2 show that the variation in the twist angle is also largely influenced by the uncertainty in the aerodynamic metamodel calculating the circulation. However, it can be noted that the influence of the uncertainty of the metamodel of the structure discipline approximating the displacement field is less negligible than in the previous case and, in addition, the interaction effects between the two sources of uncertainty are also greater (significant difference between the first-order indices and the total indices).

The influence of the uncertainty of the metamodel of Γ being dominant, we enrich the metamodel $\hat{\Gamma}$, as before. After this enrichment, the coefficient of variation of the quantity of interest is only 7% compared to 34% before enrichment. Figure 4.10 presents the histograms of the quantity of interest before and after enrichment of the metamodel $\hat{\Gamma}$ to be compared with the reference value (obtained by solving the MDA with the FEM and VLM solvers for $z = 2°$). The reduction of uncertainty appears clearly in this figure and it should also be noted that, as in the case of the lift-to-drag ratio previously studied, the quantity of interest, here the variation of the twist angle, seems to converge toward the reference value, which confirms the relevance of the disciplinary metamodel-based approach.

Finally, as before, we propose to illustrate the result obtained by also enriching the other disciplinary metamodel, in this case the structural metamodel $\hat{U}$. The histogram of the quantity of interest before and after enriching $\hat{U}$ is given in Figure 4.11. In this case, the coefficient of variation Y_{QoI} only decreases from 34% before enrichment to 28% after enrichment of $\hat{U}$ (versus 7% after enrichment of $\hat{\Gamma}$). This result is not surprising and confirms the results of the sensitivity analysis (see Table 4.2). Therefore, even for a structure-related quantity of interest (wing tip twist angle), the metamodel to be enriched, leading to the greatest reduction in uncertainty, is the aerodynamic metamodel estimating the aerodynamic circulation Γ.

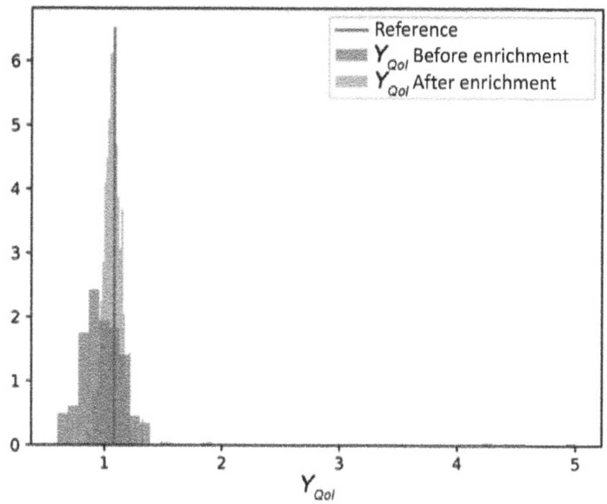

Figure 4.10. *Histogram of the new quantity of interest Y_{QoI} (wingtip spin angle) before and after enrichment of discipline $\hat{\Gamma}$. The angle of attack is set to z = 2°. The reference is obtained by solving the MDA with the discipline solvers (FEM and VLM). For a color version of this figure, see www.iste.co.uk/gogu/uncertainties.zip*

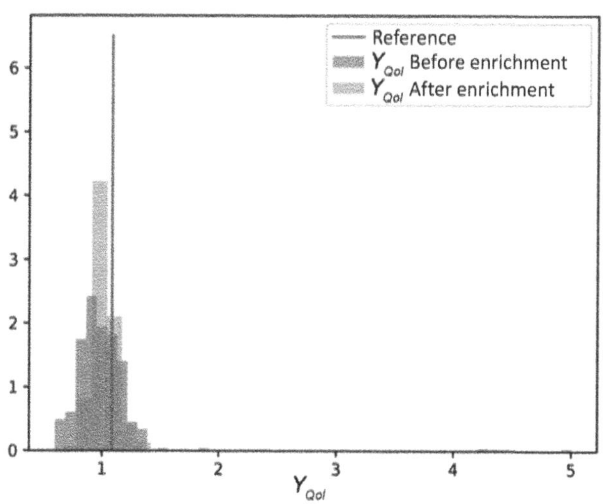

Figure 4.11. *Histogram of the new quantity of interest Y_{QoI} (wingtip spin angle) before and after enrichment of discipline $\hat{U}$. The angle of attack is set to z = 2°. The reference is obtained by solving the MDA with the discipline solvers (FEM and VLM). For a color version of this figure, see www.iste.co.uk/gogu/uncertainties.zip*

This example illustrates the advantage of the approach proposed within the framework of the construction of approximations for MDA. In fact, in addition to the adaptive construction classically used in Gaussian process approximation, sensitivity analysis to the uncertainties of the approximation makes it possible to determine which disciplinary metamodel(s) is/are to be improved, in order to reduce the uncertainty of the quantity of interest under study. This approach allows a reduction in the numerical costs of constructing the approximation when it is not essential to improve all the disciplines. Moreover, we have seen, by way of two different quantities of interest – the lift-to-drag ratio resulting from postprocessing the aerodynamic circulation Γ and the wingtip twist angle resulting from the postprocessing of the displacement U – that only sensitivity analysis could determine which metamodel to enrich, sometimes producing counterintuitive results; this difficulty is however perfectly explained by the nonlinear nature of the MDA.

4.5. Conclusion

This chapter presents an original application of sensitivity analysis of a MDA to the uncertainties of approximation of each discipline. The sensitivity analysis of the quantity of interest of the MDA is performed by estimating the Sobol' indices using polynomial chaos expansion. The calculation of these indices then allows to enrichment only of the disciplinary metamodel(s) responsible for most of the variation of the quantity of interest. The enrichment approach of a disciplinary model in the coupling variables space and design variables space is also presented. The application of this approach to an aeroelastic test case illustrates the reduction of uncertainty in the quantity of interest after the enrichment of the metamodel calculating the aerodynamic forces, identified as being the preponderant source of uncertainty using the estimation of the Sobol' indices.

The perspectives of these works concern their application to multidisciplinary optimization, as well as their generalization to the framework of MDA with disciplinary models of different fidelity. For example, it could be interesting to weight the results of sensitivity analysis based on the computational cost associated with the enrichment of disciplines, that is the use of high-fidelity computational codes.

4.6. References

Arnst, M., Ghanem, R., Phipps, E., Red-Horse, J. (2012). Dimension reduction in stochastic modeling of coupled problems [Online]. *International Journal for Numerical Methods in Engineering*, 92(11), 940–968. Available at: http://dx.doi.org/10.1002/nme.4364.

Aversano, G., Bellemans, A., Li, Z., Coussement, A., Gicquel, O., Parente, A. (2019). Application of reduced-order models based on PCA and Kriging for the development of digital twins of reacting circulation applications [Online]. *Computers and Chemical Engineering*, 121, 422–441. Available at: http://www.sciencedirect.com/science/article/pii/S0098135418305891.

Azaïs, J.-M. and Wschebor, M. (2005). On the roots of a random system of equations. The theorem of shub and smale and some extensions [Online]. *Foundations of Computational Mathematics*, 5(2), 125–144. Available at: https://doi.org/10.1007/s1020 8-004-0119-0.

Baudin, M., Dutfoy, A., Iooss, B., Popelin, A.-L. (2017). OpenTURNS: An industrial software for uncertainty quantification in simulation. In *Handbook of Uncertainty Quantification*, Ghanem, R., Higdon, D., Owhadi, H. (eds). Springer, Cham.

Berveiller, M., Sudret, B., Lemaire, M. (2006). Stochastic finite elements: A non-intrusive approach by regression. *European Journal of Computational Mechanics*, 15(1–3), 81–92.

Blatman, G. and Sudret, B. (2010). Efficient computation of global sensitivity indices using sparse polynomial chaos expansions [Online]. *Reliability Engineering & System Safety*, 95(11), 1216–1229. Available at: http://www.sciencedirect.com/science/article/pii/S0951 832010001493.

Blatman, G. and Sudret, B. (2011). Adaptive sparse polynomial chaos expansion based on least angle regression [Online]. *Journal of Computational Physics*, 230(6), 2345–2367. Available at: http://www.sciencedirect.com/science/article/pii/S002199911000 6856.

de Boer, A., van Zuijlen, A., Bijl, H. (2007). Review of coupling methods for non-matching meshes [Online]. *Computer Methods in Applied Mechanics and Engineering*, 196(8), 1515–1525. Available at: http://www.sciencedirect.com/science/article/pii/S004578250 6002817.

Borgonovo, E. (2007). A new uncertainty importance measure [Online]. *Reliability Engineering & System Safety*, 92(6), 771–784. Available at: http://www.sciencedirect.com/science/article/pii/S0951832006000883.

Cameron, R.H. and Martin, W.T. (1947). The orthogonal development of non-linear functionals in series of Fourier-Hermite functionals [Online]. *Annals of Mathematics*, 48(2), 385–392. Available at: http://www.jstor.org/stable/1969178.

Dubreuil, S., Berveiller, M., Petitjean, F., Salaün, M. (2014). Construction of bootstrap confidence intervals on sensitivity indices computed by polynomial chaos expansion [Online]. *Reliability Engineering & System Safety*, 121, 263–275. Available at: http://www.sciencedirect.com/science/article/pii/S0951832013002688.

Dubreuil, S., Bartoli, N., Gogu, C., Lefebvre, T. (2016). Propagation of modeling uncertainty by polynomial chaos expansion in multisciplinary analysis. *Journal of Mechanical Design*, 138(11), 111411.

Efron, B. (1979). Bootstrap methods: Another look at the jackknife [Online]. *Annals of Statistics*, 7(1), 1–26. Available at: https://doi.org/10.1214/aos/1176344552.

Efron, B., Hastie, T., Johnstone, I., Tibshirani, R. (2004). Least angle regression [Online]. *Annals of Statistics*, 32(2), 407–499. Available at: https://doi.org/10.1214/009053604000000067.

Ghanem, R.G. and Spanos, P.D. (1991). *Stochastic Finite Elements: A Spectral Approach.* Springer, New York.

Guo, M. and Hesthaven, J.S. (2018). Reduced order modeling for nonlinear structural analysis using Gaussian process regression [Online]. *Computer Methods in Applied Mechanics and Engineering*, 341, 807–826. Available at: http://www.sciencedirect.com/science/article/pii/S0045782518303487.

Iooss, B. and Ribatet, M. (2009). Global sensitivity analysis of computer models with functional inputs [Online]. *Reliability Engineering & System Safety*, 94(7), 1194–1204. Available at: http://www.sciencedirect.com/science/article/pii/S0951832008002299.

Jacques, J., Lavergne, C., Devictor, N. (2006). Sensitivity analysis in presence of model uncertainty and correlated inputs [Online]. *Reliability Engineering & System Safety*, 91(10), 1126–1134. Available at: http://www.sciencedirect.com/science/article/pii/S0951832005002231.

Jiang, Z., Li, W., Apley, D.W., Chen, W. (2015). A spatial-random-process based multidisciplinary system uncertainty propagation approach with model uncertainty. *Journal of Mechanical Design*, 137(10), 101402.

Jiang, Z., Chen, W., German, B.J. (2016). Multidisciplinary statistical sensitivity analysis considering both aleatory and epistemic uncertainties [Online]. *AIAA Journal*, 54(4), 1326–1338. Available at: https://doi.org/10.2514/1.J054464.

Jones, D.R., Schonlau, M., Welch, W.J. (1998). Efficient global optimization of expensive black-box functions [Online]. *Journal of Global Optimization*, 13(4), 455–492. Available at: http://dx.doi.org/10.1023/A:1008306431147.

Krzykacz-Hausmann, B. (2001). Epistemic sensitivity analysis based on the concept of entropy. *Third International Symposium on Sensitivity Analysis of Model Output*, Madrid, June 18–20, 31–35.

Liang, C., Mahadevan, S., Sankararaman, S. (2015). Stochastic multidisciplinary analysis under epistemic uncertainty. *Journal of Mechanical Design*, 137, 021404.

Rasmussen, C.E. and Williams, C.K.I. (2006). *Gaussian Processes for Machine Learning. Adaptive Computation and Machine Learning.* MIT Press, Cambridge.

Sankararaman, S. and Mahadevan, S. (2012). Likelihood-based approach to multidisciplinary analysis under uncertainty. *Journal of Mechanical Design*, 134(3), 031008.

Sobol', I.M. (2001). Global sensitivity indices for nonlinear mathematical models and their Monte Carlo estimates [Online]. *Mathematics and Computers in Simulation*, 55(1), 271–280. Available at: http://www.sciencedirect.com/science/article/pii/S0378475400002706.

Sudret, B. (2008). Global sensitivity analysis using polynomial chaos expansions [Online]. *Reliability Engineering & System Safety*, 93(7), 964–979. Available at: http://www.sciencedirect.com/science/article/pii/S0951832007001329.

PART 2

Taking Uncertainties into Account: Reliability Analysis and Optimization under Uncertainties

5

Rare-event Probability Estimation

Jean-Marc BOURINET
SIGMA Clermont, Clermont-Ferrand, France

5.1. Introduction

The scope of this chapter is restricted to *time-invariant* reliability problems as defined in the structural reliability literature (see, e.g. (Ditlevsen and Madsen 2007; Lemaire *et al.* 2010)), also known as *static* simulation problems or models by some authors (see, e.g. (Homem-de-Mello and Rubinstein 2002; Cancela *et al.* 2009)), in which time is not an explicit variable.

We consider a finite set of uncertain scalar parameters modeled by a random vector $\mathbf{X} = (X_1, \dots, X_n)^{\mathrm{T}}$ defined by its joint continuous probability density function (PDF) $f_{\mathbf{X}} : \mathbb{R}^n \to \mathbb{R}_{\geq 0}, \mathbf{x} \mapsto f_{\mathbf{X}}(\mathbf{x})$, whose support[1] $\mathcal{D}_{f_{\mathbf{X}}}$ is denoted by $\mathcal{X}$ and where $\mathbf{x} = (x_1, \dots, x_n)^{\mathrm{T}}$.

We assume that the performance of a given physical system is defined by a deterministic scalar function $g : \mathcal{D}_g \subseteq \mathbb{R}^n \to \mathbb{R}, \mathbf{x} \mapsto g(\mathbf{x})$, known as the *limit-state function* (LSF) or *performance function*, where $\mathcal{D}_g$ is the domain of definition of g, assumed to verify $\mathcal{D}_g \supseteq \mathcal{D}_{f_{\mathbf{X}}} = \mathcal{X}$. As an illustration, if we are interested in the exceedance of a given threshold level y_{th} by some response $s(\mathbf{x})$ of the system, the LSF takes the form $g(\mathbf{x}) = y_{\mathrm{th}} - s(\mathbf{x})$ for any given $\mathbf{x} \in \mathcal{D}_g$.

1 The support $\mathcal{D}_{f_{\mathbf{X}}}$ of a PDF $f_{\mathbf{X}}$ is the set of points over which this PDF is not equal to zero: $\mathcal{D}_{f_{\mathbf{X}}} = \{\mathbf{x} : f_{\mathbf{X}}(\mathbf{x}) > 0\}$.

Mechanical Engineering under Uncertainties,
coordinated by Christian GOGU.

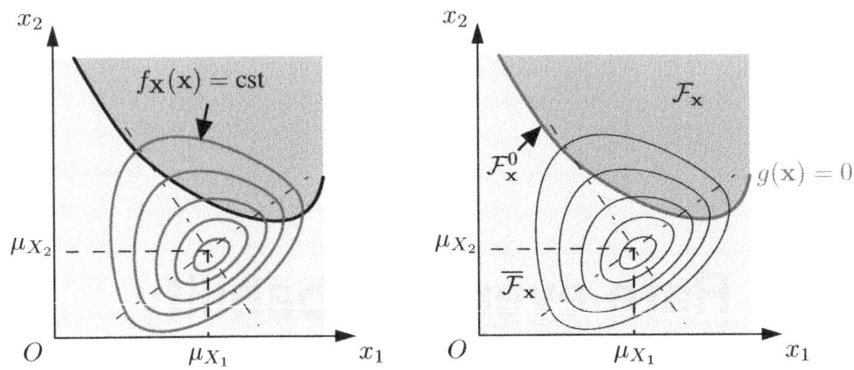

(a) Joint PDF $f_{\mathbf{X}}$ (b) LSS and safe/failure domains

Figure 5.1. *Joint PDF $f_{\mathbf{X}}$, limit-state surface $\mathcal{F}_{\mathbf{x}}^0$, safe domain $\overline{\mathcal{F}}_{\mathbf{x}}$ and failure domain $\mathcal{F}_{\mathbf{x}}$. For a color version of this figure, see www.iste.co.uk/gogu/uncertainties.zip*

The hypersurface $\mathcal{F}_{\mathbf{x}}^0 = \{\mathbf{x} \in \mathcal{D}_g : g(\mathbf{x}) = 0\}$ is called the *limit-state surface* (LSS). It divides $\mathcal{D}_g$ into a *failure domain*, conventionally defined as $\mathcal{F}_{\mathbf{x}} = \{\mathbf{x} \in \mathcal{D}_g : g(\mathbf{x}) \leq 0\}$ and a *safe domain* $\overline{\mathcal{F}}_{\mathbf{x}}$ defined as the complement of $\mathcal{F}_{\mathbf{x}}$ in $\mathcal{D}_g$ (see Figure 5.1(b)).

It is important to point out that the LSF g can only be evaluated pointwise and, as a consequence, the LSS $\mathcal{F}_{\mathbf{x}}^0$ cannot be explicitly defined. The LSF g will be assumed to be expensive to evaluate for any realization $\mathbf{x} \in \mathcal{X}$ of $\mathbf{X}$. Each evaluation of g may require the solution of a large system of equations, e.g. finite element models in structural mechanics. The formulation is restricted here to a single function g but this function may represent a combination of several failure criteria in more general settings, such as those studied in system reliability (union or intersection of failure events, or any more general systems based on the union of minimal cutsets).

The probability p_{f} that the system fails w.r.t. the above-defined LSF g is given by the following n-fold integral:

$$p_{\mathrm{f}} = \mathbb{P}\left(E\right) = \int_{g(\mathbf{x}) \leq 0} f_{\mathbf{X}}(\mathbf{x})\, \mathrm{d}\mathbf{x} = \int_{\mathcal{X}} \mathbb{1}_{\mathcal{F}_{\mathbf{x}}}\left(\mathbf{x}\right)\, f_{\mathbf{X}}(\mathbf{x})\, \mathrm{d}\mathbf{x}, \qquad [5.1]$$

where $E = \{g(\mathbf{X}) \leq 0\}$ is the failure event, $\mathbb{1}_{\mathcal{D}}\left(\cdot\right)$ denotes the *indicator function* of domain $\mathcal{D}$ such that $\mathbb{1}_{\mathcal{D}}\left(\mathbf{x}\right) = 1$ if $\mathbf{x} \in \mathcal{D}$, $\mathbb{1}_{\mathcal{D}}\left(\mathbf{x}\right) = 0$ otherwise, and $\mathrm{d}\mathbf{x} = \mathrm{d}x_1 \ldots \mathrm{d}x_n$.

The *reliability* R of the system is defined as the probability of the complementary event of E: $R = \mathbb{P}\left(\overline{E}\right) = 1 - p_\mathrm{f}$. We will assume that the *failure probability* p_f is small ($p_\mathrm{f} \ll 1$), i.e. the failure event E is rare. Such a situation is frequent in many domains, where the safety level is required to be high.

5.1.1. *Mapping to the multivariate standard normal space*

This section addresses two important interrelated points in reliability assessment: could equation [5.1] be solved in another space than $\mathcal{X}$ by means of a suitable mapping, to be defined in order to make calculations more tractable (this will enable us to introduce the so-called *standard normal space*) and, if so, how can we practically define this mapping such that the new reliability assessment problem to solve is set up in accordance with the statistical information known about the joint PDF $f_\mathbf{X}$. We will assume that both the marginal and conditional CDFs of $\mathbf{X}$ are strictly continuous, i.e. there should not be probability masses associated with any outcome of $\mathbf{X}$.

A central idea of structural reliability is to rewrite the reliability problem defined in equation [5.1] in the standard Gaussian context. To do so, a diffeomorphism[2] $T : \mathcal{X} \to \mathbb{R}^n, \mathbf{x} = (x_1, \dots, x_n)^\mathrm{T} \mapsto \mathbf{u} = (u_1, \dots, u_n)^\mathrm{T} = T(\mathbf{x})$ (referred to as *isoprobabilistic transformation* in the following) is constructed such that the following equalities (in distribution) hold:

$$\mathbf{U} = T(\mathbf{X}) \Leftrightarrow \mathbf{X} = T^{-1}(\mathbf{U}), \qquad [5.2]$$

where $\mathbf{U} = (U_1, \dots, U_n)^\mathrm{T}$ is a n-dimensional standard Gaussian vector whose joint PDF $\varphi_n(\cdot)$ is given, for any $\mathbf{u} = (u_1, \dots, u_n)^\mathrm{T} \in \mathbb{R}^n$, by:

$$\varphi_n(\mathbf{u}) = \frac{1}{(2\pi)^{n/2}} \exp\left(-\frac{\|\mathbf{u}\|^2}{2}\right), \qquad [5.3]$$

where $\|\mathbf{u}\|^2 = \mathbf{u}^\mathrm{T}\mathbf{u} = u_1^2 + \dots + u_n^2$ and $\|\cdot\|$ represents the canonical Euclidean norm on $\mathbb{R}^n$.

In the image space $\mathbb{R}^n$ referred to as *standard normal space*, the LSF is a mapping $G : \mathbb{R}^n \to \mathbb{R}, \mathbf{u} \mapsto G(\mathbf{u})$ such that, for any $\mathbf{u} \in \mathbb{R}^n$:

$$G(\mathbf{u}) = (g \circ T^{-1})(\mathbf{u}). \qquad [5.4]$$

---

2 A diffeomorphism is a globally bijective (i.e. surjective and injective, or onto and one-to-one) mapping, which is continuously differentiable and has a continuously differentiable inverse.

The reformulated problem is obtained by performing the change of variable $\mathbf{x} = T^{-1}(\mathbf{u})$ in equation [5.1], which is justified by the fact that T is a diffeomorphism. Its expression writes:

$$p_f = \mathbb{P}(E) = \int_{G(\mathbf{u})\leq 0} \varphi_n(\mathbf{u})\,\mathrm{d}\mathbf{u} = \int_{\mathbb{R}^n} \mathbb{1}_{\mathcal{F}_\mathbf{u}}(\mathbf{u})\,\varphi_n(\mathbf{u})\,\mathrm{d}\mathbf{u}, \qquad [5.5]$$

where $\mathcal{F}_\mathbf{u} = \{\mathbf{u} \in \mathbb{R}^n : G(\mathbf{u}) \leq 0\}$ is the failure domain in the standard normal space and $\mathrm{d}\mathbf{u} = \mathrm{d}u_1 \ldots \mathrm{d}u_n$.

This formulation in the standard normal space $\mathbb{R}^n$, also referred to as $\mathbf{u}$-space, presents the following properties, which are exploited by the FORM and SORM approximation methods presented later in section 5.2:

– the $\mathbf{u}$-space is normalized (the same standard normal distribution for each component U_i) and rotationally symmetric;

– if we consider a given point P^* in the $\mathbf{u}$-space specified by its coordinates $\mathbf{u}^* = (u_1^*, u_2^*)^\mathrm{T}$, the joint PDF φ_n is characterized by an exponential decay with the square of the distance from the origin (i.e. $\|\mathbf{u}\|^2$) along the radial direction (see the blue cross-section in Figure 5.2). An exponential decay also appears in the orthoradial direction when we move away from P^* in the hyperplane $\{\mathbf{u} \in \mathbb{R}^n : \beta - \boldsymbol{\alpha}^\mathrm{T}\mathbf{u} = 0\}$ where $\beta = \boldsymbol{\alpha}^\mathrm{T}\mathbf{u}^*$ (see the red cross-section in Figure 5.2);

– the probability in the half-space $\{\mathbf{u} \in \mathbb{R}^n : \beta - \boldsymbol{\alpha}^\mathrm{T}\mathbf{u} \leq 0\}$ is equal to $\Phi(-\beta)$ where $\Phi(\cdot)$ is the univariate standard normal CDF.

The transformation T from the physical input space $\mathcal{X}$ to the standard normal space $\mathbb{R}^n$ is constructed in accordance with the level of information that is given about the random vector $\mathbf{X}$. Complete knowledge of the joint PDF $f_\mathbf{X}$ is rare, and some simplifying assumptions are often made due to the available amount of statistical data. The marginal distributions of $\mathbf{X}$ are assumed to be known through their PDFs f_{X_i} or CDFs F_{X_i} for $i = 1, \ldots, n$. Statistical independence between some or all the X_i-components of $\mathbf{X}$ is often encountered, and this specific case is first presented. The case of known linear correlations between X_i-components is then considered. This latter case is often a practical choice in the case of limited statistical information. More general situations, based on other measures of dependence, e.g. defined in terms of a given copula, may be considered. Such models are not addressed within the scope of this chapter. The transformations corresponding to the following cases are considered in section 5.1.3:

– statistical independence between the X_i-components and knowledge of their marginal distributions (marginal PDFs f_{X_i} or CDFs F_{X_i} for $i = 1, \ldots, n$);

– knowledge of linear correlations between X_i-components in addition to their marginal distributions (marginal PDFs f_{X_i} or CDFs F_{X_i} for $i = 1, \ldots, n$);

– complete knowledge of the joint PDF $f_{\mathbf{X}}$.

These transformations are presented in the context of *copulas*, which constitute a useful and general tool to express the dependence between random inputs. To this end, some basic definitions and important results are given in section 5.1.2.

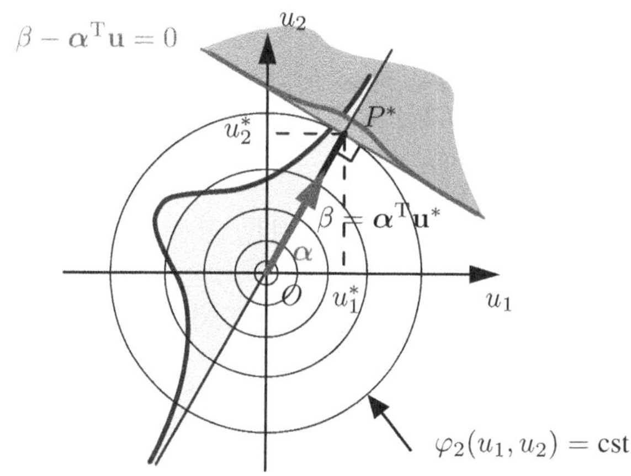

Figure 5.2. *Standard normal space. For a color version of this figure, see www.iste.co.uk/gogu/uncertainties.zip*

5.1.2. *Copulas and correlation*

This section introduces copulas for their practical use in the modeling of dependencies between random inputs. For a detailed presentation of copulas and their mathematical basis, the reader is invited to refer to general textbooks (see, e.g. (Nelsen 1999)). From a practical point of view, a copula C is a joint CDF on $[0,1]^n$ with uniformly distributed marginals (on $[0,1]$). From Sklar's theorem (Sklar 1959) recalled below, it is clear that a copula represents the complementary information needed to define the joint CDF of a random vector $\mathbf{X}$ in addition to the information given by its marginals (here marginal CDFs F_{X_i}).

THEOREM 5.1 (Sklar's Theorem).– Let $F_{\mathbf{X}}$ be a n-dimensional joint CDF with given marginal CDF F_{X_i} for $i = 1, \ldots, n$. Then there exists a n-dimensional copula C such that, for all $\mathbf{x} = (x_1, \ldots, x_n)^{\mathrm{T}} \in [-\infty, +\infty]^n$:

$$\begin{aligned} F_{\mathbf{X}}(\mathbf{x}) &= F_{\mathbf{X}}(x_1, \ldots, x_n) \\ &= C\left(F_{X_1}(x_1), \ldots, F_{X_n}(x_n)\right). \end{aligned} \qquad [5.6]$$

In addition, if the marginal CDFs $F_{X_1}, \ldots, F_{X_n}$ are continuous, then C is unique.

As a corollary of Sklar's theorem, if we consider a given n-dimensional joint CDF $F_{\mathbf{X}}$ with continuous marginal CDFs, we can express the copula C as follows, for any $\mathbf{u} = (u_1, \ldots, u_n)^{\mathrm{T}} \in [0,1]^n$:

$$C(u_1, \ldots, u_n) = F_{\mathbf{X}}\left(F_{X_1}^{-1}(u_1), \ldots, F_{X_n}^{-1}(u_n)\right), \qquad [5.7]$$

where $F_{X_i}^{-1}$ denotes the inverse of the marginal CDF F_{X_i} for $i = 1, \ldots, n$.

From Sklar's theorem, it is also possible to express the joint PDF $f_{\mathbf{X}}$ for a given copula C, knowing the marginal PDFs f_{X_i} of $\mathbf{X}$ for $i = 1, \ldots, n$:

$$\begin{aligned} f_{\mathbf{X}}(\mathbf{x}) &= f_{\mathbf{X}}(x_1, \ldots, x_n) \\ &= c\left(F_{X_1}(x_1), \ldots, F_{X_n}(x_n)\right) \prod_{i=1}^{n} f_{X_i}(x_i), \end{aligned} \qquad [5.8]$$

where $c(u_1, \ldots, u_n) = \dfrac{\partial^n C(u_1, \ldots, u_n)}{\partial u_1 \ldots \partial u_n}$ is the density of copula C.

Several copulas are available for modeling dependencies, e.g. normal, Student, Frank, Clayton and Gumbel, among others. Only the definitions of the independent and normal copulas are given in this chapter, since they are used in the construction of the isoprobabilistic transformation T in section 5.1.3. We also recall a few important results concerning copulas (Lebrun and Dutfoy 2009a, p. 578). These results are useful in the following for the definition of T.

DEFINITION 5.1.– *The n-dimensional* independent copula *is given by:*

$$C(u_1, \ldots, u_n) = C_{ind}(u_1, \ldots, u_n) = \prod_{i=1}^{n} u_i. \qquad [5.9]$$

DEFINITION 5.2.– *Let $\Phi_n(\cdot; \mathbf{R}_0)$ denote the n-dimensional standard normal CDF with linear correlation matrix $\mathbf{R}_0$. Then*

$$\begin{aligned} C(u_1, \ldots, u_n) &= C_{\mathcal{N}_n}(u_1, \ldots, u_n; \mathbf{R}_0) \\ &= \Phi_n\left(\Phi^{-1}(u_1), \ldots, \Phi^{-1}(u_n); \mathbf{R}_0\right) \end{aligned} \qquad [5.10]$$

is the n-dimensional normal copula *(or Gaussian copula) parameterized by the linear correlation matrix $\mathbf{R}_0$.*

THEOREM 5.2.– Let $\mathbf{X} = (X_1, \ldots, X_n)^{\mathrm{T}}$ be a vector of continuous random variables with copula C. If $\alpha_1, \ldots, \alpha_n$ are strictly increasing functions on the respective supports of X_i components, then C is also the copula of $(\alpha_1(X_1), \ldots, \alpha_n(X_n))^{\mathrm{T}}$.

CDF of random vector $(X_1, \ldots, X_k)^{\mathrm{T}}$

Let $\mathbf{X} = (X_1, \ldots, X_n)^{\mathrm{T}}$ be a vector of continuous random variables defined by its copula C and its marginal CDFs F_{X_i} for $i = 1, \ldots, n$. The CDF of the k-dimensional random vector $(X_1, \ldots, X_k)^{\mathrm{T}}$ is given for $k \leq n$ by:

$$F_{X_1,\ldots,X_k}(x_1, \ldots, x_k) = C_{1,\ldots,k}\left(F_{X_1}(x_1), \ldots, F_{X_k}(x_k)\right), \tag{5.11}$$

where $C_{1,\ldots,k}(u_1, \ldots, u_k) = C(u_1, \ldots, u_k, 1, \ldots, 1)$.

Conditional CDF of $X_k|X_1, \ldots, X_{k-1}$

Let $\mathbf{X} = (X_1, \ldots, X_n)^{\mathrm{T}}$ be a vector of continuous random variables defined by its copula C and its marginal CDFs F_{X_i} for $i = 1, \ldots, n$. The CDF of the conditional random variable $X_k|X_1, \ldots, X_{k-1}$ is given for $k \leq n$ by:

$$\begin{aligned} F_{X_k|X_1,\ldots,X_{k-1}}(x_k|x_1, \ldots, x_{k-1}) = \\ C_{k|1,\ldots,k-1}\left(F_{X_k}(x_k)|F_{X_1}(x_1), \ldots, F_{X_{k-1}}(x_{k-1})\right), \end{aligned} \tag{5.12}$$

where $C_{k|1,\ldots,k-1}(u_k|u_1, \ldots, u_{k-1}) = \dfrac{\dfrac{\partial^{k-1} C_{1,\ldots,k}(u_1, \ldots, u_k)}{\partial u_1 \ldots \partial u_{k-1}}}{\dfrac{\partial^{k-1} C_{1,\ldots,k-1}(u_1, \ldots, u_{k-1})}{\partial u_1 \ldots \partial u_{k-1}}}$.

In addition to the basics of copula modeling, we present a brief review of some scalar and bivariate measures of dependence (see, e.g. (Nelsen 1999; Lebrun 2013) for more details, including the definitions of measures of concordance, dependence and association). It is important to stress that dependences defined in terms of copulas are more informative than such scalar measures, and as a consequence we cannot, in general, define a dependence structure only in terms of these scalar measures.

A first and usual measure used to describe the dependence between two random variables X_i and X_j is *linear correlation* (or Pearson's correlation), defined in terms of the following coefficient:

$$\begin{aligned} \rho_{ij} &= \rho(X_i, X_j) \\ &= \frac{\mathrm{Cov}\left[X_i, X_j\right]}{\sqrt{\mathrm{Var}\left[X_i\right]}\sqrt{\mathrm{Var}\left[X_j\right]}} \\ &= \frac{1}{\sqrt{\mathrm{Var}\left[X_i\right]}\sqrt{\mathrm{Var}\left[X_j\right]}} \int_0^1 \int_0^1 \left[C\left(u_i, u_j\right) - u_i u_j\right] \mathrm{d}F_{X_i}^{-1}(u_i)\, \mathrm{d}F_{X_j}^{-1}(u_j), \end{aligned} \tag{5.13}$$

where $\mathrm{Cov}\left[X_i, X_j\right] = \mathbb{E}\left[\left(X_i - \mathbb{E}\left[X_i\right]\right)\left(X_j - \mathbb{E}\left[X_j\right]\right)\right]$ is the covariance between X_i and X_j, and $\mathrm{Var}\left[X_i\right]$ and $\mathrm{Var}\left[X_j\right]$ are the respective variances of X_i, X_j (assumed to be finite and non-zero).

From equation [5.13], it is found that linear correlation is not a function of the copula only, but also of the marginal CDFs of X. The direct consequence is that linear correlation is not independent of the marginal distributions. This scalar measure is nevertheless well suited to joint normal or elliptical distributions, provided the second moments exist. The dependence structure of more general distributions cannot, however, be represented by such a unique scalar for each pair of inputs, and using linear correlation in such situations could be misleading. Recourse to better-suited scalar measures is recommended for non-elliptical distributions, such as Spearman's rank correlation ρ_S given in equation [5.14] and Kendall's τ given in equation [5.15], where C denotes the bivariate copula of $(X_i, X_j)^T$:

$$\begin{aligned} \rho_S &= \rho\left(F_{X_i}(X_i), F_{X_j}(X_j)\right) \\ &= 12 \int_0^1 \int_0^1 \left[C(u_i, u_j) - u_i u_j\right] \mathrm{d}u_i \mathrm{d}u_j, \end{aligned} \quad [5.14]$$

$$\tau = 4 \int_0^1 \int_0^1 C(u_i, u_j)\, \mathrm{d}C(u_i, u_j). \quad [5.15]$$

5.1.3. *Isoprobabilistic transformations*

5.1.3.1. *Independent random inputs*

In this first case, we make the following assumptions about the random vector $\mathbf{X} = (X_1, \ldots, X_n)^T$:

– the n marginal distributions of $\mathbf{X}$ are known through their PDFs f_{X_i} or CDFs F_{X_i} for $i = 1, \ldots, n$;

– the X_i-components of $\mathbf{X}$ are mutually independent.

In order to construct the isoprobabilistic transformation T, each component X_i of $\mathbf{X}$ is first mapped into a uniform random variable V_i on $[0, 1]$ using its CDF F_{X_i}. This uniform random variable V_i is then mapped into a standard normal variable U_i using the inverse CDF of the standard normal distribution.

The isoprobabilistic transformation T is therefore defined as the following composed application:

$$\begin{aligned} T = T_{vu} \circ T_{xv} : \mathcal{X} \to \quad & [0,1]^n \quad \to \quad \mathbb{R}^n \\ \mathbf{x} \mapsto \mathbf{v} = & T_{xv}(\mathbf{x}) \mapsto \mathbf{u} = T_{vu}(\mathbf{v}) \end{aligned} \quad [5.16]$$

where $\mathbf{x} = (x_1, \ldots, x_n)^T$, $\mathbf{v} = (v_1, \ldots, v_n)^T$, $\mathbf{u} = (u_1, \ldots, u_n)^T$, and where, for $i = 1, \ldots, n$:

$$v_i = F_{X_i}(x_i) \quad , \quad u_i = \Phi^{-1}(v_i). \quad [5.17]$$

The joint CDF of $\mathbf{X}$ is given, from theorem 5.1, by:

$$\begin{aligned} F_{\mathbf{X}}(\mathbf{x}) = F_{\mathbf{X}}(x_1, \ldots, x_n) &= C_{\mathbf{X}}\left(F_{X_1}(x_1), \ldots, F_{X_n}(x_n)\right) \\ &= \prod_{i=1}^{n} F_{X_i}(x_i) \ , \end{aligned} \qquad [5.18]$$

where $C_{\mathbf{X}}$ is the independent copula C_{ind}, i.e. such that $C_{\text{ind}}(u_1, \ldots, u_n) = \prod_{i=1}^{n} u_i$ (see equation [5.9]).

From equation [5.8], the joint PDF of $\mathbf{X}$ simply reads:

$$\begin{aligned} f_{\mathbf{X}}(\mathbf{x}) = f_{\mathbf{X}}(x_1, \ldots, x_n) &= c_{\mathbf{X}}\left(F_{X_1}(x_1), \ldots, F_{X_n}(x_n)\right) \prod_{i=1}^{n} f_{X_i}(x_i) \\ &= \prod_{i=1}^{n} f_{X_i}(x_i) , \end{aligned} \qquad [5.19]$$

where $c_{\mathbf{X}}$ is the density of the independent copula given by $c_{\text{ind}}(u_1, \ldots, u_n) = 1$.

The joint PDF in the $\mathbf{u}$-space $\mathbb{R}^n$ is obtained as a direct application of the invariance of the copula $C_{\mathbf{X}}$ by the n strictly increasing transformations $\left(\Phi^{-1} \circ F_{X_i}\right)(\cdot)$ for $i = 1, \ldots, n$ (see theorem 5.2). We therefore have $C_{\mathbf{U}} = C_{\mathbf{X}}$ and $c_{\mathbf{U}} = c_{\mathbf{X}}$, and we can write:

$$\begin{aligned} f_{\mathbf{U}}(\mathbf{u}) = f_{\mathbf{U}}(u_1, \ldots, u_n) &= c_{\mathbf{U}}\left(\Phi(u_1), \ldots, \Phi(u_n)\right) \prod_{i=1}^{n} \varphi(u_i) \\ &= \prod_{i=1}^{n} \varphi(u_i) , \end{aligned} \qquad [5.20]$$

where $\varphi(\cdot)$ denotes the univariate standard normal PDF, which confirms that $f_{\mathbf{U}}(\mathbf{u})$ is the n-dimensional standard normal PDF $\varphi_n(\mathbf{u}) = \prod_{i=1}^{n} \varphi(u_i)$.

5.1.3.2. *Nataf transformation*

We now make the following assumptions about the random vector $\mathbf{X} = (X_1, \ldots, X_n)^{\mathrm{T}}$:

– the n marginal distributions of $\mathbf{X}$ are known through their PDFs f_{X_i} or CDFs F_{X_i} for $i = 1, \ldots, n$;

– the linear correlation coefficients $\rho_{ij} = \rho(X_i, X_j)$ of (X_i, X_j)-pairs are known for $i, j = 1, \ldots, n$. $\mathbf{R} = [\rho_{ij}]_{1 \leq i,j \leq n}$ will denote the matrix of linear correlation coefficients.

The isoprobabilistic transformation T under such assumptions is known as the *Nataf transformation* (Nataf 1962). This transformation was introduced into structural reliability by Liu and Der Kiureghian (1986). Its presentation in the framework of copulas is due to Lebrun and Dutfoy (2009b), from which the text of this section is inspired. In the Nataf transformation, each component X_i of $\mathbf{X}$ is first mapped into a uniform random variable V_i on $[0,1]$ using its CDF F_{X_i}. These uniform random variables V_i are then mapped into correlated standard normal variables Z_i using the inverse CDF of the standard normal distribution, i.e. such that $\mathbf{Z} = (Z_1, \dots, Z_n)^{\mathrm{T}} \sim \mathcal{N}(\mathbf{0}, \mathbf{R}_0)$ where $\mathbf{R}_0$ denotes the correlation matrix of the random vector $\mathbf{Z}$. The correlated standard normal variables Z_i are finally mapped into independent standard normal variables U_i using a linear transformation denoted T_{zu}.

The isoprobabilistic transformation T is therefore defined as the following composed application:

$$\begin{array}{rl} T = T_{zu} \circ T_{vz} \circ T_{xv} : \mathcal{X} \to & [0,1]^n \quad \to \quad \mathbb{R}^n \quad \to \quad \mathbb{R}^n \\ & \mathbf{x} \mapsto \mathbf{v} = T_{xv}(\mathbf{x}) \mapsto \mathbf{z} = T_{vz}(\mathbf{v}) \mapsto \mathbf{u} = T_{zu}(\mathbf{z}) \end{array} \qquad [5.21]$$

where $\mathbf{x} = (x_1, \dots, x_n)^{\mathrm{T}}$, $\mathbf{v} = (v_1, \dots, v_n)^{\mathrm{T}}$, $\mathbf{z} = (z_1, \dots, z_n)^{\mathrm{T}}$, $\mathbf{u} = (u_1, \dots, u_n)^{\mathrm{T}}$, and where, for $i = 1, \dots, n$:

$$v_i = F_{X_i}(x_i) \quad , \quad z_i = \Phi^{-1}(v_i). \qquad [5.22]$$

For the construction of T_{zu}, we assume a linear relation between $\mathbf{z}$ and $\mathbf{u}$ in the form $\mathbf{z} = \mathbf{A}\mathbf{u}$. The correlation matrix $\mathbf{R}_0$ can be written as follows:

$$\begin{aligned} \mathbf{R}_0 = [\mathrm{Cov}[Z_i, Z_j]]_{1 \le i,j \le n} &= \mathbb{E}\left[\mathbf{Z}\mathbf{Z}^{\mathrm{T}}\right] - \mathbb{E}[\mathbf{Z}]\,\mathbb{E}\left[\mathbf{Z}^{\mathrm{T}}\right] \\ &= \mathbf{A}\,\mathbb{E}\left[\mathbf{U}\mathbf{U}^{\mathrm{T}}\right]\mathbf{A}^{\mathrm{T}} \\ &= \mathbf{A}\,\mathbf{A}^{\mathrm{T}}. \end{aligned} \qquad [5.23]$$

From equation [5.23], it appears that $\mathbf{A}$ can be chosen as the lower triangular matrix $\mathbf{L}_0$ of the Cholesky decomposition of matrix $\mathbf{R}_0$ (assumed positive definite) such that:

$$\mathbf{R}_0 = \mathbf{L}_0 \mathbf{L}_0^{\mathrm{T}}, \qquad [5.24]$$

which defines the transformation T_{zu}:

$$\begin{aligned} T_{zu} : \mathbb{R}^n &\to \mathbb{R}^n \\ \mathbf{z} &\mapsto \mathbf{u} = T_{zu}(\mathbf{z}) = \mathbf{L}_0^{-1}\mathbf{z} \end{aligned} \qquad [5.25]$$

As described in Lebrun and Dutfoy (2009b), the underlying assumption of the Nataf transformation is that the copula $C_{\mathbf{Z}}$ of the random vector $\mathbf{Z}$ (or equivalently the copula $C_{\mathbf{X}}$ of $\mathbf{X}$) is the n-dimensional normal copula parameterized by the correlation matrix $\mathbf{R}_0$.

The joint CDF of $\mathbf{Z}$ is therefore assumed to be given by:

$$\begin{aligned} F_{\mathbf{Z}}(\mathbf{z}) = F_{\mathbf{Z}}(z_1, \ldots, z_n) &= C_{\mathbf{Z}}(\Phi(z_1), \ldots, \Phi(z_n)) \\ &= C_{\mathcal{N}_n}(\Phi(z_1), \ldots, \Phi(z_n); \mathbf{R}_0) \\ &= \Phi_n(z_1, \ldots, z_n; \mathbf{R}_0), \end{aligned} \qquad [5.26]$$

where $C_{\mathbf{Z}}$ is the n-dimensional normal copula $C_{\mathcal{N}_n}$ parameterized by the correlation matrix $\mathbf{R}_0$ (see equation [5.10]).

From equation [5.8], the joint PDF of $\mathbf{Z}$ reads:

$$\begin{aligned} f_{\mathbf{Z}}(\mathbf{z}) = f_{\mathbf{Z}}(z_1, \ldots, z_n) &= c_{\mathbf{Z}}(\Phi(z_1), \ldots, \Phi(z_n)) \prod_{i=1}^{n} \varphi(z_i) \\ &= c_{\mathcal{N}_n}(\Phi(z_1), \ldots, \Phi(z_n); \mathbf{R}_0) \prod_{i=1}^{n} \varphi(z_i) \\ &= \varphi_n(z_1, \ldots, z_n; \mathbf{R}_0), \end{aligned} \qquad [5.27]$$

where $c_{\mathbf{Z}}$ is the density of a n-dimensional normal copula parameterized by the correlation matrix $\mathbf{R}_0$, whose expression is:

$$c_{\mathcal{N}_n}(u_1, \ldots, u_n; \mathbf{R}_0) = \frac{\varphi_n\left(\Phi^{-1}(u_1), \ldots, \Phi^{-1}(u_n); \mathbf{R}_0\right)}{\prod_{i=1}^{n} \varphi\left(\Phi^{-1}(u_i)\right)}, \qquad [5.28]$$

where $\varphi_n(\cdot; \mathbf{R}_0)$ is the n-dimensional standard normal PDF with linear correlation matrix $\mathbf{R}_0$.

The joint PDF in the $\mathbf{x}$-space $\mathcal{X}$ is obtained as a direct application of the invariance of the copula $C_{\mathbf{Z}}$ by the n strictly increasing transformations

$\left(F_{X_i}^{-1} \circ \Phi\right)(\cdot)$ for $i = 1, \ldots, n$ (see theorem 5.2). We therefore have $C_{\mathbf{X}} = C_{\mathbf{Z}}$ and $c_{\mathbf{X}} = c_{\mathbf{Z}}$. The joint PDF of $\mathbf{X}$ is therefore given by the following expression:

$$\begin{aligned} f_{\mathbf{X}}(\mathbf{x}) = f_{\mathbf{X}}(x_1, \ldots, x_n) &= c_{\mathbf{X}}\left(F_{X_1}(x_1), \ldots, F_{X_n}(x_n)\right) \prod_{i=1}^{n} f_{X_i}(x_i) \\ &= \frac{\varphi_n(z_1, \ldots, z_n; \mathbf{R}_0)}{\prod_{i=1}^{n} \varphi(z_i)} \prod_{i=1}^{n} f_{X_i}(x_i), \end{aligned} \quad [5.29]$$

which is the expression given in equation (11) of Liu and Der Kiureghian (1986).

The $\mathbf{R}_0$ matrix has not been explicited yet. The elements $\rho_{0,ij}$ of this matrix are defined in terms of the given linear correlation coefficients $\rho_{ij} = \rho(X_i, X_j)$ of X_i and X_j components of $\mathbf{X}$. They are obtained from the following equation, for $i, j = 1, \ldots, n$, in which we use the bivariate form of the joint PDF, expressed in equation [5.29]:

$$\begin{aligned} \rho_{ij} &= \mathbb{E}\left[\left(\frac{X_i - \mu_i}{\sigma_i}\right)\left(\frac{X_j - \mu_j}{\sigma_j}\right)\right] \\ &= \int_{\mathcal{X}_j}\int_{\mathcal{X}_i} \left(\frac{x_i - \mu_i}{\sigma_i}\right)\left(\frac{x_j - \mu_j}{\sigma_j}\right) \frac{\varphi_2(z_i, z_j; \rho_{0,ij})}{\varphi(z_i)\varphi(z_j)} f_{X_i}(x_i) f_{X_j}(x_j)\, \mathrm{d}x_i \mathrm{d}x_j \\ &= \int_{\mathbb{R}}\int_{\mathbb{R}} h(z_i, z_j, \mu_i, \mu_j, \sigma_i, \sigma_j)\, \varphi_2(z_i, z_j; \rho_{0,ij})\, \mathrm{d}z_i \mathrm{d}z_j, \end{aligned} \quad [5.30]$$

where $\mathcal{X}_i$ and $\mathcal{X}_j$ denote the respective supports of the ith and jth components of the random vector $\mathbf{X}$, μ_i and μ_j, their respective means, and σ_i and σ_j their respective standard deviations,

where $h(z_i, z_j, \mu_i, \mu_j, \sigma_i, \sigma_j) = \left(\dfrac{F_{X_i}^{-1}(\Phi(z_i)) - \mu_i}{\sigma_i}\right)\left(\dfrac{F_{X_j}^{-1}(\Phi(z_j)) - \mu_j}{\sigma_j}\right)$,

and where

$$\begin{aligned} \varphi_2(z_i, z_j; \rho_{0,ij}) &= \varphi_2\left(z_i, z_j; \begin{bmatrix} 1 & \rho_{0,ij} \\ \rho_{0,ij} & 1 \end{bmatrix}\right) \\ &= \frac{1}{2\pi\sqrt{1 - \rho_{0,ij}^2}} \exp\left(-\frac{z_i^2 - 2\rho_{0,ij} z_i z_j + z_j^2}{2\left(1 - \rho_{0,ij}^2\right)}\right). \end{aligned}$$

The Nataf transformation is valid if equation [5.30] admits solutions for all $i, j = 1, \ldots, n$ and if the linear correlation matrix $\mathbf{R}_0$ is, positive definite. Such a situation is often met in applications of practical interest, as reported by Liu and Der Kiureghian (1986). Finding $\rho_{0,ij}$ solutions of the integral relation defined in equation [5.30] is, in general, tedious, and for this reason approximate formulas for $\rho_{0,ij}$ were derived by Liu and Der Kiureghian (1986) for most common statistical distributions. These formulas are, in general, obtained by least squares fitting and are therefore approximate, except for a few pairs of distributions. An alternative solution consists of calculating $\rho_{0,ij}$ for $i, j = 1, \ldots, n$ and $i > j$ by numerical integration, as implemented in FERUM 4.x (Bourinet *et al.* 2009). Such calculations are performed only once to define the transformation T and with currently available computers they are fast. Coefficients $\rho_{0,ij}$ of the $\mathbf{R}_0$ matrix are obtained by 2D numerical integration as solutions[3] of the following equation for $i, j = 1, \ldots, n$ and $i > j$:

$$\rho_{ij} = \sum_{k=1}^{n_i} \sum_{l=1}^{n_j} \omega_k \omega_l \, h\left(z_k, \mu_i, \sigma_i, z_l, \mu_j, \sigma_j\right) \varphi_2\left(z_k, z_l, \rho_{0,ij}\right), \qquad [5.31]$$

where (z_k, z_l) are the $n_i \times n_j$ integration points and $\omega_k \omega_l$ their respective weights. A Gaussian quadrature rule is applied over the truncated domain $[-6, 6] \times [-6, 6]$. Special attention must be paid to strongly correlated random variables for accurate $\rho_{0,ij}$ values. A practical rule adopted in FERUM 4.x consists of increasing the number of integration points along each dimension with correlation, ranging from $n_i = n_j = 32$ points along each dimension for absolute correlation values lower than 0.9 to $1,024$ points for absolute values larger than 0.9995.

In many engineering problems, the joint PDF $f_{\mathbf{X}}$ is often out of reach and the available statistical information is often limited to marginal distributions of $\mathbf{X}$ and, in some cases, linear correlations between its components. This explains why the Nataf transformation is so popular in the structural reliability literature, even if the assumption of normal copula may be questionable, as mentioned by Lebrun and Dutfoy (2009b).

5.1.3.3. *Rosenblatt transformation*

We now assume that the random vector $\mathbf{X} = (X_1, \ldots, X_n)^{\mathrm{T}}$ is known through its joint distribution, e.g. its joint CDF $F_{\mathbf{X}}$.

In such a case, the isoprobabilistic transformation applied is known as the *Rosenblatt transformation* (Rosenblatt 1952), see the paper from Hohenbichler and Rackwitz (1981) for its introduction in structural reliability. Based on the joint CDF $F_{\mathbf{X}}$, the random vector $\mathbf{X}$ is first mapped into a uniformly distributed random vector

3 It is important to point out that the unknown $\rho_{0,ij}$ in equation [5.31] is in the right-hand side of the equation, which requires an iterative solution for each pair (i, j).

$\mathbf{V} = (V_1, \ldots, V_n)^{\mathrm{T}}$ over $[0,1]^n$ with independent copula (see demonstration by Ditlevsen and Madsen 2007, p. 123). The uniform random variables V_i are then mapped into independent standard normal variables U_i using the inverse CDF of the standard normal distribution.

The isoprobabilistic transformation T is defined as the following composed application:

$$\begin{aligned} T = T_{vu} \circ T_{xv} : \mathcal{X} \rightarrow \quad & [0,1]^n \quad \rightarrow \quad \mathbb{R}^n \\ \mathbf{x} \mapsto \mathbf{v} = & \, T_{xv}(\mathbf{x}) \mapsto \mathbf{u} = T_{vu}(\mathbf{v}) \end{aligned} \qquad [5.32]$$

where $\mathbf{x} = (x_1, \ldots, x_n)^{\mathrm{T}}$, $\mathbf{v} = (v_1, \ldots, v_n)^{\mathrm{T}}$, $\mathbf{u} = (u_1, \ldots, u_n)^{\mathrm{T}}$, and where $u_i = \Phi^{-1}(v_i)$ for $i = 1, \ldots, n$.

One possible choice for the transformations T_{xv} is given in equation [5.33]. Note that there exist $n!$ different choices due to the variable ordering in the conditional expressions.

$$\begin{aligned} v_1 &= F_{X_1}(x_1) \\ v_2 &= F_{X_2|X_1}(x_2|x_1) \\ v_3 &= F_{X_3|X_1,X_2}(x_3|x_1, x_2) \\ &\ldots \\ v_n &= F_{X_n|X_1,\ldots,X_{n-1}}(x_n|x_1, \ldots, x_{n-1}) \end{aligned} \qquad [5.33]$$

It is important to note that the definition of T_{xv} necessitates a knowledge of the conditional CDFs of $X_k|X_1, \ldots, X_{k-1}$ for $k = 2, \ldots, n$. When the joint distribution of $\mathbf{X}$ is defined by its copula C and its marginal CDFs F_{X_i} for $i = 1, \ldots, n$, the Rosenblatt transformation T is computationally tractable by using the conditional expression obtained in equation [5.12]. The definition of the inverse T^{-1} of the Rosenblatt transformation (needed in approximation methods such as FORM and SORM, addressed in section 5.2) may, however, require substantial efforts.

As a final remark, the dependence of the FORM results on the variable ordering in the conditional expressions of equation [5.33] has been studied in several works, e.g. see the example based on the bivariate exponential CDF $F_{\mathbf{X}}(\mathbf{x}) = F_{\mathbf{X}}(x_1, x_2) = 1 - \exp(-x_1) - \exp(-x_2) - \exp(-x_1 - x_2 - x_1 x_2)$ in Ditlevsen and Madsen (2007) and Lemaire *et al.* (2010). As pointed out by Lebrun (2013), it is important to mention that, despite differences in terms of FORM results, the exact failure probability remains unchanged, whatever the order of variables we choose in the Rosenblatt transformation.

5.2. MPFP-based methods

5.2.1. *First-order reliability method*

5.2.1.1. *Problem statement*

The *first-order reliability method* (FORM) aims at finding the point P^* of the LSS $\mathcal{F}_{\mathbf{u}}^0$ that is the closest to the origin O in the standard normal space, i.e. the point in the failure domain $\mathcal{F}_{\mathbf{u}}$ with the largest PDF value (see, e.g. (Ditlevsen and Madsen 2007; Lemaire *et al.* 2010)) (Figure 5.3). This point, known as the *most probable failure point* (MPFP) (also called the *design point*) in structural reliability, is the solution of the following quadratic optimization problem under nonlinear constraint:

$$\mathbf{u}^* = \arg\min_{\mathbf{u}\in\mathbb{R}^n} \frac{1}{2}\mathbf{u}^{\mathrm{T}}\mathbf{u} \text{ s.t. } G(\mathbf{u}) = 0, \qquad [5.34]$$

where $\mathbf{u}^*$ is the coordinate vector of P^* in the standard normal space $\mathbb{R}^n$.

The so-called Hasofer-Lind *reliability index* β_{HL} is given by:

$$\beta_{\mathrm{HL}} = \beta = \boldsymbol{\alpha}^{\mathrm{T}}\mathbf{u}^*, \qquad [5.35]$$

where $\boldsymbol{\alpha}$ is a unit vector such that $\boldsymbol{\alpha} = -\nabla G(\mathbf{u}^*) / \|\nabla G(\mathbf{u}^*)\|$ and ∇ is the gradient operator. In general, the origin O of the standard normal space belongs to the safe domain. As a result, β is positive and represents the Euclidean distance from the MPFP P^* to the origin O of the standard normal space, and $\boldsymbol{\alpha}$ is the unit vector of the (O, P^*) axis (see Figure 5.3).

Under the assumption that the LSF is continuous, smooth and differentiable in the neighborhood of the MPFP, we define the linear approximation of G at the MPFP P^* (first-order Taylor polynomial of G at $\mathbf{u}^*$), which is given by the following expression:

$$G_1(\mathbf{u}) = \nabla G(\mathbf{u}^*)^{\mathrm{T}} (\mathbf{u} - \mathbf{u}^*), \qquad [5.36]$$

in which the term $G(\mathbf{u}^*)$ has been dropped since it is equal to zero (the MPFP P^* belongs to the LSS $\mathcal{F}_{\mathbf{u}}^0$).

The FORM consists of approximating the unknown probability p_{f} defined in equation [5.5] by the following n-dimensional integral:

$$p_{\mathrm{f}}^{\mathrm{FORM}} = \int_{\mathbb{R}^n} \mathbb{1}_{\mathcal{F}_{\mathbf{u}1}}(\mathbf{u})\, \varphi_n(\mathbf{u})\, \mathrm{d}\mathbf{u}, \qquad [5.37]$$

where $\mathcal{F}_{\mathbf{u}1} = \{\mathbf{u} \in \mathbb{R}^n : G_1(\mathbf{u}) \leq 0\} = \{\mathbf{u} \in \mathbb{R}^n : \beta - \boldsymbol{\alpha}^{\mathrm{T}}\mathbf{u} \leq 0\}$.

This integral can be evaluated in a closed form, and a simple calculation gives:

$$p_{\mathrm{f}}^{\mathrm{FORM}} = \Phi(-\beta), \quad [5.38]$$

where Φ denotes the one-dimensional standard normal CDF.

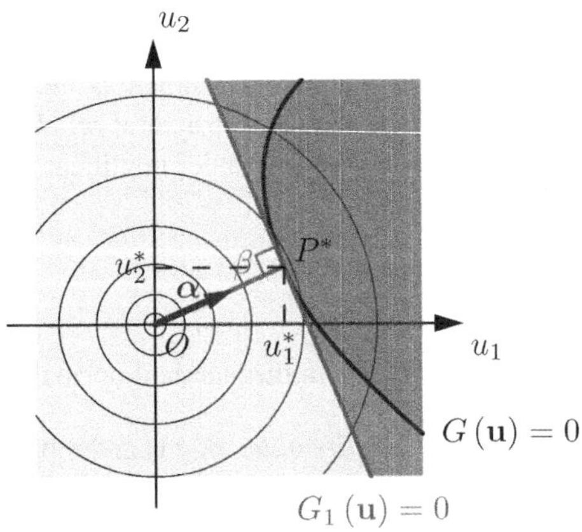

Figure 5.3. *FORM approximation. For a color version of this figure, see www.iste.co.uk/gogu/uncertainties.zip*

5.2.1.2. *Solving the constrained optimization problem*

The main computational task of the FORM is to search for the supposedly unique MPFP using a suitable optimization algorithm. Several methods are available to solve the optimization problem in equation [5.34], including general algorithms, such as the usual sequential quadratic programming (SQP) algorithm, and others which have been specifically tailored to solve equation [5.34], e.g. the HLRF algorithm (Hasofer and Lind 1974; Rackwitz and Fiessler 1978) and the i-HLRF algorithm (Zhang and Der Kiureghian 1994). The reader may refer to Lemaire *et al.* (2010, Chapter 5) for an overview of these algorithms and to Liu and Der Kiureghian (1991) for a comparison of the performances of the SQP, the HLRF and a few other algorithms.

Most of the existing algorithms share the same basis: they are iterative until some convergence criteria are met and, at each iteration of the algorithm, a search direction is determined, along with a step length in that direction. Their performances are usually evaluated based on the following criteria: types of problems that can be solved, convergence to the solution with a prescribed accuracy, convergence rate of the algorithm, ability of the algorithm to scale with the dimension n of the problem.

From a practical viewpoint, all the above-mentioned algorithms require an evaluation of the gradient of the LSF at each iteration. The LSF gradient involves those of one or more outputs of a numerical model (e.g. a FE model) w.r.t. its inputs, considered as random. The gradients (or sensitivities to input parameters) are, in general, not directly accessible from the numerical codes, except in a very few cases and under some often restrictive assumptions, e.g. model linearity. If these sensitivities are available, e.g. by direct differentiation of the equations governing the response(s), they can make FORM very efficient, see Zhang and Der Kiureghian's early paper (1993) for the application of the so-called *direct differentiation method* to the dynamic response of structures with inelastic materials, and the report by Haukaas and Der Kiureghian (2004) for more details on such approaches in the context of FE analyses. In the most general case, these sensitivities are inaccessible (the case of non-intrusive approaches in reliability assessment) and need to be assessed by finite differences, with the dilemma of choosing very small or sufficiently large perturbation steps in the finite difference scheme:

– if the applied perturbations are too small, the numerical model is often unable to evaluate outputs with sufficient accuracy (this could be interpreted as noisy outputs from the code) and the calculated sensitivities are therefore not representative of the level of perturbation applied to the model inputs;

– if large perturbations are selected, the output values are well separated from each other but the calculated sensitivities are no longer representative of first-order derivatives.

The selection of suitable perturbation steps, usually expressed as ratios of random input standard deviations, has considerable consequences on the convergence and convergence rate of the optimization algorithms used to solve equation [5.34]. The tuning of such perturbation steps in a finite difference scheme may require considerable efforts in engineering applications of practical interest.

5.2.1.3. *Remarks about the accuracy of the FORM solution*

It is important to emphasize that the approximation of p_f given in equation [5.38] is under the assumption of a unique MPFP P^* and a linear (or weakly nonlinear) LSS $\mathcal{F}_{\mathbf{u}}^0$. If these assumptions are not fulfilled, the FORM approximation could be a fairly crude one with a large bias w.r.t. the true failure probability.

The uniqueness of the MPFP is often encountered in practice. Note that such an assumption cannot be checked beforehand, and it is also hard to verify it when a FORM solution has been obtained. Some engineering reliability assessment problems may, however, exhibit several zones of similar importance (and therefore multiple MPFPs). In the FORM context, the method proposed by Der Kiureghian and Dakessian (1998) aims at solving such problems. In essence, the method consists of repeating the FORM analysis with a modified LSF, which triggers the search outside the zone where a MPFP has been found. Assuming that K MPFPs have already been identified (the first one being obtained e.g. with the i-HLRF algorithm), the FORM analysis performed next uses the following modified LSF $G_{\text{bulge},K+1}$ in the standard normal space, which consists of adding bulges centered at each of the K obtained MPFPs (see examples of corresponding LSS in Figure 5.4):

$$G_{\text{bulge},K+1}(\mathbf{u}) = G(\mathbf{u}) + \sum_{k=1}^{K} \mathbb{1}_{\mathcal{B}_k}(\mathbf{u})\, s_k \left(r_k^2 - \|\mathbf{u} - \mathbf{u}_k^*\|^2\right)^2, \qquad [5.39]$$

where $\mathbf{u}_k^*$ and β_k are the MPFP and reliability index of the kth MPFP, $\mathcal{B}_k$ is the ball of radius r_k centered on the kth MPFP, i.e. such that $\mathcal{B}_k = \{\mathbf{u} \in \mathbb{R}^n : \|\mathbf{u} - \mathbf{u}_k^*\| \leq r_k\}$, and where r_k is the radius of the bulge and s_k a parameter that controls its height, defined as follows:

$$r_k = \gamma\beta_k \quad , \quad s_k = \frac{\delta\beta_k \|\nabla G(\mathbf{u}_k^*)\|}{\left[(\gamma\beta_k)^2 - (\delta\beta_k)^2\right]^2}, \qquad [5.40]$$

where δ and γ are user-defined parameters such that $0 < \delta < \gamma$.

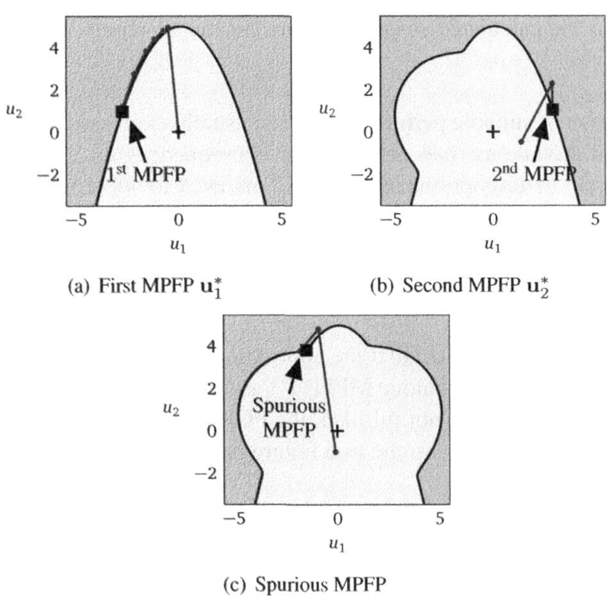

(a) First MPFP $\mathbf{u}_1^*$ (b) Second MPFP $\mathbf{u}_2^*$

(c) Spurious MPFP

Figure 5.4. *Successive use of bulges to find multiple MPFPs (Der Kiureghian and Dakessian 1998, Example 1). Selected parameters: $\delta = 0.75$, $\gamma = 1.1$, $\epsilon = 0.5$. For a color version of this figure, see www.iste.co.uk/gogu/uncertainties.zip*

This procedure is repeated until a spurious MPFP is obtained. Such points appear in the foot of one of the created bulges and correspond to artificially created minimal distance points (see Figure 5.4(c)). To make the method more efficient, Der Kiureghian and Dakessian (1998) suggest starting the FORM algorithm from an initial point $\mathbf{u}_{0,K+1}$, defined as follows to search for the $(K+1)$th MPFP:

$$\mathbf{u}_{0,K+1} = -\epsilon \sum_{k=1}^{K} \mathbf{u}_k^*, \qquad [5.41]$$

where ϵ is a user-defined parameter in the range of 0.2–0.5.

Regarding a potential nonlinearity of the LSS, the FORM approximation may be corrected by a subsequent recourse to one of the second-order reliability methods (SORMs) presented in the next section (either curvature-fitting or point-fitting SORM).

5.2.2. *Second-order reliability method*

5.2.2.1. *Curvature-fitting SORM*

We assume that the LSF is continuous, smooth and at least twice differentiable in the neighborhood of the MPFP obtained by a FORM analysis. In the SORM based on *curvature-fitting* (SORM-cf), we consider the quadratic approximation of G at the MPFP P^* (or second-order Taylor polynomial of G at $\mathbf{u}^*$) given by:

$$G_2(\mathbf{u}) = \nabla G(\mathbf{u}^*)^{\mathrm{T}} (\mathbf{u} - \mathbf{u}^*) + \frac{1}{2} (\mathbf{u} - \mathbf{u}^*)^{\mathrm{T}} \nabla^2 G(\mathbf{u}^*) (\mathbf{u} - \mathbf{u}^*), \qquad [5.42]$$

where ∇^2 denotes the Hessian operator. The SORM-cf approximation of the unknown probability p_{f} defined in equation [5.5] is expressed as the following n-dimensional integral:

$$p_{\mathrm{f}}^{\text{SORM-cf}} = \int_{\mathbb{R}^n} \mathbb{1}_{\mathcal{F}_{\mathrm{u}2}}(\mathbf{u})\, \varphi_n(\mathbf{u})\, \mathrm{d}\mathbf{u}, \qquad [5.43]$$

where $\mathcal{F}_{\mathrm{u}2} = \{\mathbf{u} \in \mathbb{R}^n : G_2(\mathbf{u}) \leq 0\}$.

In order to make the calculation of this integral tractable, the quadratic approximation defined in equation [5.42] is usually expressed in another set of axes by means of the following composed orthonormal transformations (Der Kiureghian 1999).

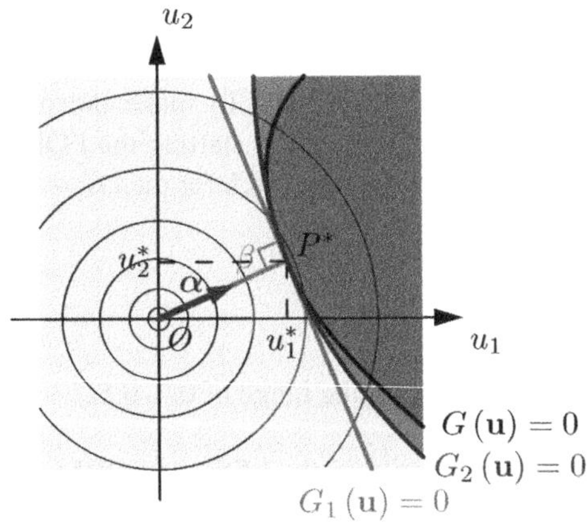

Figure 5.5. *SORM-cf approximation. For a color version of this figure, see www.iste.co.uk/gogu/uncertainties.zip*

The first, $\mathbf{u} \mapsto \mathbf{u}' = (u'_1, \ldots, u'_n)^{\mathrm{T}} = \mathbf{R}_1\mathbf{u}$, is constructed such that the last coordinate axis u'_n coincides with the direction (O, P^*), e.g. by means of a Gram–Schmidt algorithm. In the new space, equation [5.42] is rewritten as follows, in which the factor terms of $(u'_n - \beta)$ are dropped in relation to the remaining terms:

$$\frac{G_2\left(\mathbf{R}_1^{-1}(\mathbf{u}')\right)}{\|\nabla G(\mathbf{u}^*)\|} \approx \beta - u'_n + \frac{1}{2}\widetilde{\mathbf{u}}'^{\mathrm{T}}\widetilde{\mathbf{A}}\widetilde{\mathbf{u}}', \qquad [5.44]$$

where $\widetilde{\mathbf{u}}' = \left(u'_1, \ldots, u'_{n-1}\right)^{\mathrm{T}}$ and $\widetilde{\mathbf{A}}$ is the sub-matrix composed of the first $(n-1)$ rows and columns of $\mathbf{A} = \mathbf{R}_1\nabla^2 G(\mathbf{u}^*)\,\mathbf{R}_1^{\mathrm{T}}/\|\nabla G(\mathbf{u}^*)\|$.

A rotation around axis u'_n is then applied to be in the principal coordinates of the parabolic approximation of the LSS. Let us denote this rotation $\widetilde{\mathbf{u}}' \mapsto \widetilde{\mathbf{u}}'' = \widetilde{\mathbf{R}}_2\widetilde{\mathbf{u}}'$ where $\widetilde{\mathbf{u}}'' = \left(u''_1, \ldots, u''_{n-1}\right)^{\mathrm{T}}$, defined such that $\widetilde{\mathbf{R}}_2\widetilde{\mathbf{A}}\widetilde{\mathbf{R}}_2^{\mathrm{T}} = \mathrm{diag}(\kappa_i)$. The rotation matrix $\widetilde{\mathbf{R}}_2$ represents the eigenmatrix of $\widetilde{\mathbf{A}}$, and the $(n-1)$ principal curvatures $\kappa_1, \ldots, \kappa_{n-1}$ are the corresponding eigenvalues. In this rotated space, equation [5.44] takes the following canonical form:

$$\frac{G_2\left(\mathbf{R}_1^{-1}\left(\mathbf{R}_2^{-1}(\mathbf{u}'')\right)\right)}{\|\nabla G(\mathbf{u}^*)\|} \approx \beta - u''_n + \sum_{i=1}^{n-1}\kappa_i u_i''^2, \qquad [5.45]$$

where $\mathbf{u}'' = (u''_1, \ldots, u''_n)^{\mathrm{T}}$ and $\mathbf{R}_2 = \begin{bmatrix}\widetilde{\mathbf{R}}_2 & \mathbf{0} \\ \mathbf{0}^{\mathrm{T}} & 1\end{bmatrix}$.

Equation [5.43] can then be rewritten in the following form:

$$p_{\mathrm{f}}^{\text{SORM-cf}} \approx \int_{\mathbb{R}^n} \mathbb{1}_{\mathcal{F}_{\mathbf{u}''2}}\left(\mathbf{u}''\right) \varphi_n\left(\mathbf{u}''\right) \mathrm{d}\mathbf{u}'', \qquad [5.46]$$

where $\mathcal{F}_{\mathbf{u}''2} = \left\{\mathbf{u}'' \in \mathbb{R}^n : \beta - u''_n + \sum_{i=1}^{n-1} \kappa_i u_i''^2 \leq 0\right\}$.

Several formulas can be found in the literature to approximate this integral. The best known are those established by Breitung (1984); Tvedt (1988); Hohenbichler and Rackwitz (1988). Only the SORM approximation given by the asymptotic formula of Hohenbichler and Rackwitz is recalled here:

$$p_{\mathrm{f}}^{\text{SORM-cf}} \approx \Phi\left(-\beta\right) \prod_{i=1}^{n-1} \frac{1}{\sqrt{1+\psi\left(\beta\right)\kappa_i}} \quad \left(\beta \to \infty\right), \qquad [5.47]$$

where $\psi\left(\beta\right) = \varphi\left(\beta\right)/\Phi\left(-\beta\right)$, assuming that $\kappa_i > -1/\psi\left(\beta\right)$ for all $i \in \{1, \ldots, (n-1)\}$.

It is important to point out that the SORM-cf requires the evaluation of the *Hessian matrix* at P^* used for the definition of $\widetilde{\mathbf{A}}$. This Hessian matrix $\nabla^2 G\left(\mathbf{u}^*\right)$ is computed numerically by finite differences in FERUM 4.x by means of $n + n\left(n+1\right)/2$ calls to the LSF. All these calls are independent and therefore run in parallel by FERUM.

5.2.2.2. *Point-fitting SORM*

An alternative method proposed by Der Kiureghian *et al.* (1987) and known as *point-fitting* SORM (SORM-pf) consists of approximating the LSS by a piece-wise paraboloid surface, which must be tangent to the LSS at the MPFP and interpolates the LSS at a set of fitting points. This method is advantageous for slightly noisy LSF, such as those resulting from calls to a FE code for which the computation of the Hessian matrix by finite differences cannot be achieved with sufficient accuracy, for high-dimensional problems, i.e. large n, or when the computation of curvatures fails.

The method works in a rotated standard normal space such that the last coordinate axis u'_n coincides with the direction (O, P^*), as defined in the SORM-cf method (note that this choice of axes is made regardless of the orientation of the principal directions of the LSS, whose determination would require us to calculate the Hessian of G at the MPFP). The orthogonal transformation to this new space is, as previously described, defined by $\mathbf{u}' = \mathbf{R}_1 \mathbf{u}$, where $\mathbf{u}'$ denotes the coordinate vector in the new space.

For each coordinate u'_i, $i = 1, \ldots, n-1$, the objective is to find two fitting points of the LSS with coordinates $(-k\beta, u'_{n,-i})$ and $(k\beta, u'_{n,+i})$, where β is the reliability index and k is a preselected parameter that controls the distance of these points from the (O, P^*) axis (see Figure 5.6). The ordinates $u'_{n,-i}$ and $u'_{n,+i}$ of these two points are solutions of the following equation:

$$G\left(\mathbf{R}_1^{-1}\mathbf{u}'\right) = 0, \qquad [5.48]$$

where $\mathbf{u}' = (0, \ldots, 0, u'_i, 0, \ldots, 0, u'_n)^{\mathrm{T}}$ and $u'_i = \pm k\beta$, $u'_n = u'_{n,\pm i}$. The corresponding optimization problem is solved in FERUM 4.x using a vectorized version of the Matlab® fzero function, which allows the $2(n-1)$ independent searches to be run in parallel.

Two semi-parabolas can then be defined in the (u'_i, u'_n) plane, which are tangent to the LSS at P^* and pass through the fitting points (see Figure 5.6). The corresponding curvatures at P^* are given by $\kappa_{\pm i} = 2(u'_{n,\pm i} - \beta)/(k\beta)^2$ and the Hohenbicher and Rackwitz approximation reads:

$$p_{\mathrm{f}}^{\text{SORM-pf}} \approx \Phi(-\beta) \prod_{i=1}^{n-1} \frac{1}{2} \left(\frac{1}{\sqrt{1 + \kappa_{-i}\psi(\beta)}} + \frac{1}{\sqrt{1 + \kappa_{+i}\psi(\beta)}} \right). \qquad [5.49]$$

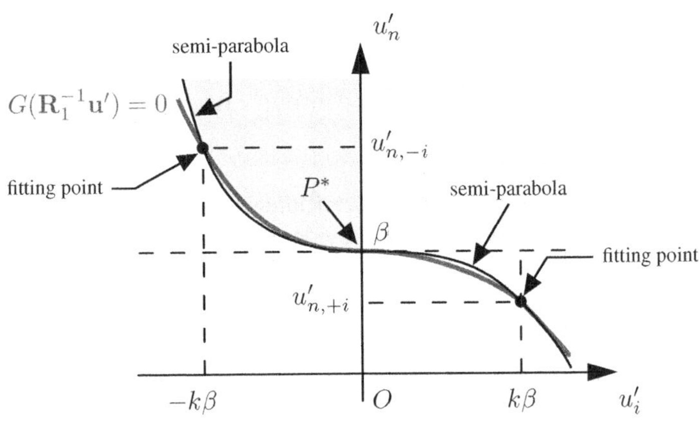

Figure 5.6. *Concept of SORM-pf approximation in the (u'_i, u'_n) plane. For a color version of this figure, see www.iste.co.uk/gogu/uncertainties.zip*

5.3. Simulation methods

This section presents the main basis of some popular *sampling methods*, which are often used for the reliability assessment of rare events. The common idea of all these methods is to generate samples in the outcome space, e.g. random realizations from the joint PDF $f_{\mathbf{X}}$ in a crude Monte Carlo (MC) simulation, in order to find an estimate of the failure probability p_{f}. The goal is to construct an *estimator* $\widehat{p}_{\mathrm{f}}$ that is close to the unknown failure probability p_{f} for any generated sample. This "proximity" to the true and unknown quantity p_{f} is defined in terms of the following important properties of the estimator $\widehat{p}_{\mathrm{f}}$: *bias* of $\widehat{p}_{\mathrm{f}}$ w.r.t. the true and unknown probability p_{f} (the objective is to have an unbiased estimator such that $\mathbb{E}\left[\widehat{p}_{\mathrm{f}}\right] = p_{\mathrm{f}}$), and *variance* $\mathrm{Var}\left[\widehat{p}_{\mathrm{f}}\right]$ (the objective is to construct an estimator with the lowest achievable variance in order to obtain an estimate close to the unknown value p_{f} for any sample set that is drawn). Note that other properties, such as consistency and efficiency, are also of interest for the characterization of $\widehat{p}_{\mathrm{f}}$.

The well-known MC method, applied in numerous fields of physics and engineering, is introduced in section 5.3.1. For improved efficiency in the context of rare events, two *variance reduction* techniques are then presented: subset simulation, proposed by Au and Beck (2001) in section 5.3.2, and importance sampling (IS), through its adaptive variant known as the cross-entropy (CE) method (Rubinstein 1999) in section 5.3.3. The objective of variance reduction techniques is to retain most of the advantages of the MC method (sampling in the outcome space, no assumption about the complexity of the LSF and independence w.r.t. the dimension n of the outcome space) with a lower variance of the estimator $\widehat{p}_{\mathrm{f}}$ for a fixed computational budget. For complementary details about the sampling methods addressed in this section and some of the related useful techniques (e.g. random number generation and Markov chains), the reader may refer to general textbooks such as those of Robert and Casella (2004), Rubinstein and Kroese (2007) and Kroese *et al.* (2011).

5.3.1. *Crude MC simulation*

In order to define the estimator of p_{f} in a *crude* MC simulation, equation [5.1] is rewritten as:

$$p_{\mathrm{f}} = \mathbb{P}(E) = \int_{\mathcal{X}} \mathbb{1}_{\mathcal{F}_{\mathbf{x}}}(\mathbf{x})\, f_{\mathbf{X}}(\mathbf{x})\, \mathrm{d}\mathbf{x} = \mathbb{E}_{f_{\mathbf{X}}}\left[\mathbb{1}_{\mathcal{F}_{\mathbf{x}}}(\mathbf{X})\right], \qquad [5.50]$$

where $\mathbb{E}_h$ denotes the mathematical expectation operator w.r.t. a given PDF h.

The crude MC estimator of p_{f} is simply derived from equation [5.50] as the sample mean of the failure indicator function:

$$\widehat{p}_{\mathrm{f}}^{\mathrm{MC}} = \frac{1}{N}\sum_{j=1}^{N} \mathbb{1}_{\mathcal{F}_{\mathbf{x}}}\left(\mathbf{X}^{(j)}\right), \qquad [5.51]$$

where $\{\mathbf{X}^{(j)}, 1 \leq j \leq N\}$ are N independent copies of the random vector $\mathbf{X}$, i.e. $\mathbf{X}^{(1)}, \ldots, \mathbf{X}^{(N)} \overset{\text{i.i.d.}}{\sim} f_{\mathbf{X}}$, and N denotes the sample size.

From simple calculations, we can find that the coefficient of variation (c.o.v.) of the crude MC estimator $\widehat{p}_{\mathrm{f}}^{\mathrm{MC}}$ is written as follows:

$$\delta_{\widehat{p}_{\mathrm{f}}^{\mathrm{MC}}} = \frac{\sqrt{\operatorname{Var}\left[\widehat{p}_{\mathrm{f}}^{\mathrm{MC}}\right]}}{\mathbb{E}\left[\widehat{p}_{\mathrm{f}}^{\mathrm{MC}}\right]} = \frac{\sqrt{\dfrac{p_{\mathrm{f}}\,(1-p_{\mathrm{f}})}{N}}}{p_{\mathrm{f}}} = \sqrt{\frac{1-p_{\mathrm{f}}}{N p_{\mathrm{f}}}}. \qquad [5.52]$$

From equation [5.52], it can be noticed that the convergence speed of a crude MC simulation is a function of N (the c.o.v. of $\widehat{p}_{\mathrm{f}}^{\mathrm{MC}}$ is proportional to $1/\sqrt{N}$). As an illustration, a 10% c.o.v. on $\widehat{p}_{\mathrm{f}}^{\mathrm{MC}}$ by crude MC requires a sample size N approximately equal to $100/p_{\mathrm{f}}$ in the case of a rare event ($p_{\mathrm{f}} \ll 1$). Crude MC hence appears inefficient for estimating probabilities of rare events. The efficiency can, however, be enhanced by means of variance reduction techniques, such as those presented in sections 5.3.2 and 5.3.3. It is also important to mention that the c.o.v. of $\widehat{p}_{\mathrm{f}}^{\mathrm{MC}}$ is independent of the dimension n of the problem. This independence w.r.t. the dimension is a key advantage of the crude MC approach over other techniques. Last but not least, it is of paramount importance to remember that the crude MC method works regardless of the complexity of the LSF, which is also an advantage over approximation methods presented in section 5.2 or approaches based on surrogate models.

The crude MC method is implemented in FERUM 4.x and takes full advantage of the independence of $\{\mathbf{X}^{(j)}, 1 \leq j \leq N\}$, which allows the respective calls to the LSF to be sent in parallel.

5.3.2. *Subset simulation*

This section describes the *subset simulation* (SS) method developed by Au and Beck (2001), which has become popular in the structural reliability community over the years. Methods with a very similar basis have been proposed in the statistical community, starting with the seminal work of Kahn and Harris (1951) in the setting of particle transmission. Such methods are also known as *adaptive multilevel splitting* (Cérou and Guyader 2007; Botev and Kroese 2012). These methods were developed in a framework that may differ from the specific one of structural reliability, such as, for example, the classical context of Markovian processes (Cérou and Guyader 2007).

5.3.2.1. *Conceptual idea of subset simulation*

The conceptual idea of SS is to represent the small probability of a rare failure event E as a product of larger and sequentially estimated probabilities of wisely selected *intermediate events* E_s, $s = 1, \ldots, m$. These m intermediate events are *nested* such that $E_m \subset E_{m-1} \subset \cdots \subset E_2 \subset E_1$ where $E_m = E$. From the inclusion rule and by successive conditioning we can write:

$$\begin{aligned} p_{\mathrm{f}} &= \mathbb{P}(E) \\ &= \mathbb{P}(E_m) \\ &= \mathbb{P}(E_m|E_{m-1})\,\mathbb{P}(E_{m-1}) \\ &= \ldots \\ &= \mathbb{P}(E_m|E_{m-1})\,\mathbb{P}(E_{m-1}|E_{m-2})\ldots\mathbb{P}(E_2|E_1)\,\mathbb{P}(E_1) \\ &= \prod_{s=1}^{m} p_s, \end{aligned} \quad [5.53]$$

where p_1 denotes the probability of E_1, and $p_s = \mathbb{P}(E_s|E_{s-1})$ the conditional probability of E_s given E_{s-1} for $s = 2, \ldots, m$. Hence, p_{f} appears as the product of m probabilities, each of which is necessarily larger than p_{f} and therefore easier to estimate than p_{f} by sampling.

We will here assume a component-wise independence of the input random vector for the subsequent application of the Metropolis algorithm modified by Au and Beck (2001) (m-M) (see section 5.3.2.3). This can be conveniently achieved by considering the reliability assessment problem in the standard normal space (see equation [5.5]), wherein $\varphi_n(\mathbf{u}) = \prod_{i=1}^{n} \varphi(u_i)$. All the components U_i for $i = 1, \ldots, n$ have the same univariate standard normal distribution, which, moreover, greatly simplifies the choice of the proposal multivariate PDF used for MC Markov chain sampling and the tuning of the algorithm parameters.

We define the following set of nested *intermediate failure domains* in the standard normal space, for $s = 1, \ldots, m$ (see Figure 5.7):

$$\mathcal{F}_{\mathbf{u},\,s} = \{\mathbf{u} \in \mathbb{R}^n : G(\mathbf{u}) \leq y_s\}, \quad [5.54]$$

where y_s represents intermediate levels of the LSF such that $y_m = 0$ and $y_m < y_{m-1} < \ldots < y_1$. The intermediate events E_s are such that $E_s = \{G(\mathbf{U}) \leq y_s\}$ for $s = 1, \ldots, m$. They correspond to the intermediate failure domains $\mathcal{F}_{\mathbf{u},\,s}$ and they meet the previously defined inclusion rule $E_m \subset E_{m-1} \subset \cdots \subset E_2 \subset E_1$.

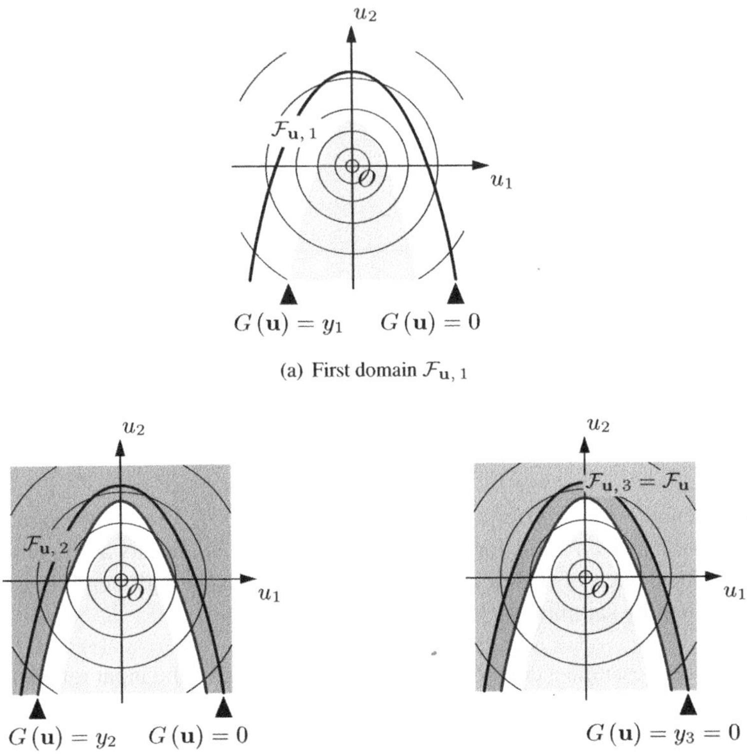

(a) First domain $\mathcal{F}_{\mathbf{u},1}$

(b) Second domain $\mathcal{F}_{\mathbf{u},2}$

(c) Last domain $\mathcal{F}_{\mathbf{u},3}$

Figure 5.7. *Intermediate failure domains $\mathcal{F}_{\mathbf{u},s}$ for $s = 1, \ldots, m$ (here $m = 3$). For a color version of this figure, see www.iste.co.uk/gogu/uncertainties.zip*

5.3.2.2. *Description of the SS algorithm*

The rare event estimation problem formulated in equation [5.5] is replaced by a sequence of m problems with larger probabilities to estimate w.r.t. the intermediate levels y_s, for $s = 1$:

$$p_1 = \mathbb{P}(E_1) = \mathbb{E}_{\varphi_n}\left[\mathbb{1}_{\mathcal{F}_{\mathbf{u},1}}(\mathbf{U})\right], \qquad [5.55]$$

and, for $s = 2, \ldots, m$:

$$p_s = \mathbb{P}(E_s|E_{s-1}) = \mathbb{E}_{\varphi_n(\cdot|E_{s-1})}\left[\mathbb{1}_{\mathcal{F}_{\mathbf{u},s}}(\mathbf{U})\right]. \qquad [5.56]$$

Estimators of these probabilities are given by the fraction of independent samples (for $s = 1$) or conditional independent samples (for $s > 1$) that cause failure w.r.t. the selected intermediate level y_s, respectively, for $s = 1$:

$$\widehat{p}_1 = \frac{1}{N} \sum_{j=1}^{N} \mathbb{1}_{\mathcal{F}_{\mathbf{u},1}} \left(\mathbf{U}_1^{(j)} \right) \text{ where } \mathbf{U}_1^{(1)}, \ldots, \mathbf{U}_1^{(N)} \overset{\text{i.i.d.}}{\sim} \varphi_n, \quad [5.57]$$

and for $s = 2, \ldots, m$:

$$\widehat{p}_s = \frac{1}{N} \sum_{j=1}^{N} \mathbb{1}_{\mathcal{F}_{\mathbf{u},s}} \left(\mathbf{U}_s^{(j)} \right) \text{ where } \mathbf{U}_s^{(1)}, \ldots, \mathbf{U}_s^{(N)} \overset{\text{i.i.d.}}{\sim} \varphi_n \left(\cdot | E_{s-1} \right), \quad [5.58]$$

where N denotes the sample size (the same value is used at each level y_s) and where the conditional PDF $\varphi_n \left(\cdot | E_{s-1} \right)$ is defined as follows, for $s = 2, \ldots, m$:

$$\varphi_n \left(\mathbf{u} | E_{s-1} \right) = \frac{\varphi_n \left(\mathbf{u} \right) \mathbb{1}_{\mathcal{F}_{\mathbf{u},s-1}} \left(\mathbf{u} \right)}{\mathbb{P} \left(E_{s-1} \right)}. \quad [5.59]$$

In SS, the intermediate levels y_s are selected such that, by construction, all the estimates $\widehat{p}_s$ for $s = 1, \ldots, (m-1)$ are set equal to p_0, where p_0 is a prescribed probability level (a usual choice is $p_0 = 0.1$). Note that targeting equal probability levels in terms of conditional probabilities corresponds to the optimal choice for a minimal asymptotic variance of $\widehat{p}_{\mathrm{f}}^{\mathrm{SS}} = \prod_{s=1}^{m} \widehat{p}_s$, as pointed out by Guyader (2011). This probability level p_0 governs how many intermediate failure domains $\mathcal{F}_{\mathbf{u},s}$ are required to reach the failure domain $\mathcal{F}_{\mathbf{u}}$. A small value for p_0 ensures a fast exploration but with the main disadvantage of poorly estimated probabilities w.r.t. the intermediate levels y_s. A large value for p_0 enables good accuracy on the estimated probabilities but at the cost of a lengthy exploration (large m). According to Zuev *et al.* (2012), the optimal range for p_0 is 0.1–0.3. Optimality here means that we are searching for a value of p_0 that minimizes the variance of $\widehat{p}_{\mathrm{f}}^{\mathrm{SS}}$ for a given total number of samples $N_{\mathrm{t}} = mN$. It is assumed here that both $1/p_0$ and $p_0 N$ are positive integers. If not, we can work with slightly different values for p_0 and N, such that $p_0 = 1/k_0$ and $N = k_1 k_0$, where $k_0 = \lfloor 1/p_0 \rceil$, $k_1 = \lfloor N/k_0 \rceil$ and $\lfloor x \rceil$ denotes the nearest integer to x.

The first intermediate level y_1 is straightforward to obtain. It is defined as the p_0-quantile of $\mathcal{Y}_1 = \{G(\mathbf{u}_1^{(j)}), 1 \leq j \leq N\}$, where $\mathcal{U}_1 = \{\mathbf{u}_1^{(j)}, 1 \leq j \leq N\}$ are the N i.i.d. samples drawn from φ_n, and we therefore have $\widehat{p}_1 = p_0$. At the following levels $s > 1$, we proceed similarly by defining y_s as the p_0-quantiles of $\mathcal{Y}_s = \{G(\mathbf{u}_s^{(j)}), 1 \leq j \leq N\}$, where $\mathcal{U}_s = \{\mathbf{u}_s^{(j)}, 1 \leq j \leq N\}$ are the i.i.d. samples drawn from $\varphi_n(\cdot|E_{s-1})$. The SS algorithm ends when the p_0-quantile of $\mathcal{Y}_s = \{G(\mathbf{u}_s^{(j)}), 1 \leq j \leq N\}$ becomes lower than zero. The last level y_m is set to zero, and the corresponding probability $\widehat{p}_m$ is given by $\widehat{p}_m = \#\mathcal{J}/N$, where $\mathcal{J} = \{j \in \{1, \ldots, N\} : G(\mathbf{u}_m^{(j)}) \leq 0\}$ and $\#\mathcal{J}$ denotes the cardinality of the set $\mathcal{J}$. The probability $\widehat{p}_m$ estimated at the last level $y_m = 0$ is therefore greater than or equal to p_0, and the estimation of p_f by SS is simply given by:

$$\widehat{p}_\mathrm{f}^{\mathrm{SS}} = p_0^{m-1}\widehat{p}_m. \qquad [5.60]$$

In SS, we need to generate independent samples from the n-dimensional standard normal PDF φ_n at the first level ($s = 1$) (see equation [5.57]), and from the conditional n-dimensional standard normal PDF $\varphi_n(\cdot|E_{s-1})$ at the next levels ($s > 1$) (see equation [5.58]). Sampling from φ_n is straightforward and corresponds to a crude MC simulation. Sampling from the conditional PDF $\varphi_n(\cdot|E_{s-1})$ for $s > 1$ is, however, not trivial. In SS, this is performed by means of *Markov chain Monte Carlo* (MCMC), which is a useful technique for approximate sampling from an arbitrary distribution (see, e.g. (Robert and Casella 2004; Kroese *et al.* 2011)). The specific algorithm used in SS is detailed in the next section.

5.3.2.3. *Approximate sampling from* $\varphi_n(\cdot|E_{s-1})$ *for* $s = 2, \ldots, m$

It should first be noted that an acceptance–rejection method would be inefficient to generate samples from $\varphi_n(\cdot|E_{s-1})$ for $s = 2, \ldots, m$ (draw samples from φ_n, accept those that belong to $\mathcal{F}_{\mathrm{u},s-1}$, reject otherwise). The acceptance probability is proportional to $\mathbb{P}(E_{s-1})$, which decreases with s to reach a value for $s = m$ in the order of the unknown and small probability p_f.

In order to simulate samples from $\varphi_n(\cdot|E_{s-1})$ for $s = 2, \ldots, m$, SS resorts to MCMC. This technique was invented soon after crude MC and dates back to 1953 with the original work of Metropolis *et al.* (1953) at Los Alamos. This section briefly presents the principle of Markov chains to generate samples from an arbitrary multivariate target distribution, which cannot be directly sampled (e.g. known up to a scaling factor). The *Metropolis–Hastings algorithm*, one of the most popular MCMC techniques, is first described. This algorithm was adapted by Au and Beck (2001) for the specific needs of SS and this important variant is then detailed. Note that the presentation of Markov chains and the above-mentioned algorithms is made in the standard normal space $\mathbb{R}^n$, in coherence with the formulation of the SS method. A presentation in similar settings can be found in Papaioannou *et al.* (2015).

Let us introduce a time-homogeneous discrete-time *Markov chain* $\{\mathbf{U}^k, k \in \mathbb{N}\}$ in the continuous state space $\mathbb{R}^n$ (here the standard normal space). It is assumed that $\{\mathbf{U}^k, k \in \mathbb{N}\}$ is a first-order Markov chain:

$$\mathbb{P}\left(\mathbf{U}^{k+1} \in A \,|\, \bigcap_{m=0}^{k}\{\mathbf{U}^m = \mathbf{u}^m\}\right) = \mathbb{P}\left(\mathbf{U}^{k+1} \in A \,|\, \{\mathbf{U}^k = \mathbf{u}^k\}\right) \stackrel{\text{def}}{=} K\left(A|\mathbf{u}^k\right), \quad [5.61]$$

for any $A \in \mathcal{B}$, $\mathcal{B}$ being the Borel σ-field of $\mathbb{R}^n$, and where K is the *transition kernel* of the Markov chain. $K\left(A|\mathbf{u}^k\right) = \int_A K\left(\mathrm{d}\mathbf{v}|\mathbf{u}^k\right)$ is a conditional distribution function that represents the probability of moving from $\mathbf{u}^k$ to a point in the set A. K is assumed to be independent of k, which makes the chain time independent (or stationary). The marginal distribution of $\mathbf{U}^0$ is called the *initial distribution* of the chain.

At any given level $s > 1$ of the SS algorithm, the objective is to sample a Markov chain $\{\mathbf{U}^k, k \in \mathbb{N}\}$ whose invariant (or stationary) distribution is the *target distribution* $\varphi_n\left(\cdot|E_{s-1}\right)$. MCMC algorithms start from a set of initial states of the chain, which are, in general, not distributed according to the target distribution. The samples generated from the Markov chain are expected to follow the target distribution for sufficiently large k after a transient stage known as the *burn-in* period. It is important to point out that the generated samples are expected to be identically distributed according to the target density after this burn-in period but they are not independent. Some level of dependence, in fact, appears in the chain due to the use of a proposal PDF q, introduced later in this section. Convergence to a unique invariant distribution occurs under some regularity conditions of the constructed Markov chain, known as aperiodicity and irreducibility (see, e.g. (Kroese *et al.* 2011)). These conditions are met in usual MCMC samplers. They are, in particular, satisfied for proposal PDFs q with a restricted support, such as uniform PDFs centered at the current state of the chain with a finite width (Chib and Greenberg 1995), as applied in SS.

The potential move in the standard normal space from a current state denoted $\mathbf{u}$ to a new state denoted $\mathbf{v}$ is characterized by a transition kernel K, generally expressed as follows:

$$K\left(\mathrm{d}\mathbf{v}|\mathbf{u}\right) = p\left(\mathbf{v}|\mathbf{u}\right)\mathrm{d}\mathbf{v} + r\left(\mathbf{u}\right)\delta_{\mathbf{u}}\left(\mathrm{d}\mathbf{v}\right), \quad [5.62]$$

where $p : \mathbb{R}^n \times \mathbb{R}^n \to \mathbb{R}_{\geq 0}, (\mathbf{u}, \mathbf{v}) \mapsto p(\mathbf{v}|\mathbf{u})$ is a given function, such that $p\left(\mathbf{u}|\mathbf{u}\right) = 0$, δ is the Dirac delta function, such that $\delta_{\mathbf{u}}\left(\mathrm{d}\mathbf{v}\right) = 1$ if $\mathbf{u} \in \mathrm{d}\mathbf{v}$ and 0 otherwise, and $r\left(\mathbf{u}\right) = 1 - \int_{\mathbb{R}^n} p\left(\mathbf{v}|\mathbf{u}\right)\mathrm{d}\mathbf{v}$. The first continuous term of the sum defines the move from $\mathbf{u}$ to $\mathbf{v}$ according to $p\left(\mathbf{v}|\mathbf{u}\right)$, and the second discrete term represents the stagnation at $\mathbf{u}$ with probability $r\left(\mathbf{u}\right)$.

The invariant distribution of the Markov chain must satisfy the following condition, which states that if $\mathbf{U}^k$ is distributed according to the target distribution $\varphi_n(\cdot|E_{s-1})$, then $\mathbf{U}^{k+1}$ and all subsequent elements of the chain are also distributed according to $\varphi_n(\cdot|E_{s-1})$:

$$\varphi_n(\mathbf{v}|E_{s-1})\,\mathrm{d}\mathbf{v} = \int_{\mathbb{R}^n} K(\mathrm{d}\mathbf{v}\,|\,\mathbf{u})\,\varphi_n(\mathbf{u}|E_{s-1})\,\mathrm{d}\mathbf{u}. \qquad [5.63]$$

A sufficient condition for the target distribution $\varphi_n(\cdot|E_{s-1})$ to be a stationary (or invariant) distribution for K, is to satisfy the so-called *reversibility* condition a.k.a. *detailed balance*:

$$\varphi_n(\mathbf{u}|E_{s-1})\,p(\mathbf{v}|\mathbf{u}) = \varphi_n(\mathbf{v}|E_{s-1})\,p(\mathbf{u}|\mathbf{v}). \qquad [5.64]$$

It is easy to show that equation [5.64] implies equation [5.63] (see, e.g. (Tierney 1994; Chib and Greenberg 1995)).

Here, we will be specifically interested in the Metropolis–Hastings kernel (Hastings 1970), defined such that:

$$p(\mathbf{v}|\mathbf{u}) = p_{\mathrm{MH}}(\mathbf{v}|\mathbf{u}) = \begin{cases} q(\mathbf{v}|\mathbf{u})\,\alpha(\mathbf{u},\mathbf{v}) & \text{if } \mathbf{u} \neq \mathbf{v}, \\ 0 & \text{if } \mathbf{u} = \mathbf{v}, \end{cases} \qquad [5.65]$$

where $q(\mathbf{v}|\mathbf{u})$ is a *proposal* PDF (note that we have $\int_{\mathbb{R}^n} q(\mathbf{v}|\mathbf{u})\,\mathrm{d}\mathbf{v} = 1$), and $\alpha(\mathbf{u},\mathbf{v})$ is the probability of moving from $\mathbf{u}$ to $\mathbf{v}$. The reversibility of the transition kernel w.r.t. $\varphi_n(\cdot|E_{s-1})$ is imposed by construction, by defining $\alpha(\mathbf{u},\mathbf{v})$ as follows:

$$\alpha(\mathbf{u},\mathbf{v}) = \begin{cases} \min\left\{ \dfrac{\varphi_n(\mathbf{v}|E_{s-1})\,q(\mathbf{u}|\mathbf{v})}{\varphi_n(\mathbf{u}|E_{s-1})\,q(\mathbf{v}|\mathbf{u})}, 1 \right\} & \text{if } \varphi_n(\mathbf{u}|E_{s-1})\,q(\mathbf{v}|\mathbf{u}) > 0, \\ 1 & \text{if } \varphi_n(\mathbf{u}|E_{s-1})\,q(\mathbf{v}|\mathbf{u}) = 0. \end{cases} \qquad [5.66]$$

The original Metropolis sampler (Metropolis *et al.* 1953) is obtained by assuming that the proposal PDF q is symmetric, i.e. such that $q(\mathbf{v}|\mathbf{u}) = q(\mathbf{u}|\mathbf{v})$. In such a case, the move probability (a.k.a. acceptance probability) is given by:

$$\alpha(\mathbf{u},\mathbf{v}) = \begin{cases} \min\left\{ \dfrac{\varphi_n(\mathbf{v}|E_{s-1})}{\varphi_n(\mathbf{u}|E_{s-1})}, 1 \right\} & \text{if } \varphi_n(\mathbf{u}|E_{s-1})\,q(\mathbf{v}|\mathbf{u}) > 0, \\ 1 & \text{if } \varphi_n(\mathbf{u}|E_{s-1})\,q(\mathbf{v}|\mathbf{u}) = 0. \end{cases} \qquad [5.67]$$

If $\varphi_n(\mathbf{v}|E_{s-1}) > \varphi_n(\mathbf{u}|E_{s-1})$, the move from $\mathbf{u}$ to $\mathbf{v}$ is always accepted. Otherwise, the move is accepted with a probability given by the ratio

$\varphi_n(\mathbf{v}|E_{s-1})/\varphi_n(\mathbf{u}|E_{s-1})$. A straightforward choice to ensure symmetry in the proposal PDF q, is to consider $\mathbf{v} = \mathbf{u} + \mathbf{z}$, where $\mathbf{z}$ is a realization of a random vector $\mathbf{Z}$ independent of $\mathbf{U}$ and whose n-variate PDF $f_{\mathbf{Z}}$ is symmetric about zero, i.e. such that $f_{\mathbf{Z}}(\mathbf{z}) = f_{\mathbf{Z}}(-\mathbf{z})$. The proposal PDF is then defined as $q(\mathbf{v}|\mathbf{u}) = f_{\mathbf{Z}}(\mathbf{v} - \mathbf{u})$. Straightforward choices for $f_{\mathbf{Z}}$ include the normal or uniform multi-variate distributions. Because the candidate state is equal to the current state of the chain plus noise, samplers of this form are usually termed *random-walk* Metropolis samplers. Note that if q does not satisfy the symmetry property $q(\mathbf{v}|\mathbf{u}) = q(\mathbf{u}|\mathbf{v})$, the constructed sampler is usually referred to as a Metropolis–Hastings sampler.

Assuming that $\varphi_n(\mathbf{u}|E_{s-1})\, q(\mathbf{v}|\mathbf{u}) > 0$ (i.e. $\mathbf{u} \in \mathcal{F}_{\mathbf{u},\, s-1}$ and $q(\mathbf{v}|\mathbf{u}) > 0$), we can express the move probability $\alpha(\mathbf{u}, \mathbf{v})$ as follows, using equation [5.59] (Papaioannou *et al.* 2015):

$$
\begin{aligned}
\alpha(\mathbf{u}, \mathbf{v}) &= \min\left\{ \frac{\varphi_n(\mathbf{v}|E_{s-1})\, q(\mathbf{u}|\mathbf{v})}{\varphi_n(\mathbf{u}|E_{s-1})\, q(\mathbf{v}|\mathbf{u})}, 1 \right\} \\
&= \min\left\{ \frac{\varphi_n(\mathbf{v})\, q(\mathbf{u}|\mathbf{v})}{\varphi_n(\mathbf{u})\, q(\mathbf{v}|\mathbf{u})} \mathbb{1}_{\mathcal{F}_{\mathbf{u},\, s-1}}(\mathbf{v}), 1 \right\} \\
&= \min\left\{ \frac{\varphi_n(\mathbf{v})\, q(\mathbf{u}|\mathbf{v})}{\varphi_n(\mathbf{u})\, q(\mathbf{v}|\mathbf{u})}, 1 \right\} \mathbb{1}_{\mathcal{F}_{\mathbf{u},\, s-1}}(\mathbf{v}) \\
&= \alpha_{\varphi_n}(\mathbf{u}, \mathbf{v})\, \mathbb{1}_{\mathcal{F}_{\mathbf{u},\, s-1}}(\mathbf{v}).
\end{aligned}
\qquad [5.68]
$$

From the above expression, it is found that the move probability $\alpha(\mathbf{u}, \mathbf{v})$ of the Metropolis–Hastings kernel can be expressed as the product of the move probability $\alpha_{\varphi_n}(\mathbf{u}, \mathbf{v})$, based on the unconditional n-dimensional standard normal PDF φ_n and the indicator function of the intermediate failure domain $\mathcal{F}_{\mathbf{u},\, s-1}$. This enables a two-step implementation of the Metropolis–Hastings algorithm for generating samples from the target distribution $\varphi_n(\cdot|E_{s-1})$:

1) First, a candidate sample $\mathbf{v}$ is drawn from the proposal PDF $q(\mathbf{v}|\mathbf{u})$ and the move from $\mathbf{u}$ to $\mathbf{v}$ is temporarily accepted with probability $\alpha_{\varphi_n}(\mathbf{u}, \mathbf{v})$. This corresponds to a single application of a Metropolis–Hastings kernel where $p_{\mathrm{MH}}(\mathbf{v}|\mathbf{u}) = q(\mathbf{v}|\mathbf{u})\, \alpha_{\varphi_n}(\mathbf{u}, \mathbf{v})$ for $\mathbf{u} \neq \mathbf{v}$ (see equation [5.65]).

2) Second, the move from $\mathbf{u}$ to $\mathbf{v}$ is finally accepted if $\mathbf{v} \in \mathcal{F}_{\mathbf{u},\, s-1}$. If not, the chain remains in its current state $\mathbf{u}$. This second step requires the evaluation of the LSF at $\mathbf{v}$ to know whether $\mathbf{v} \in \mathcal{F}_{\mathbf{u},\, s-1}$ or not.

An implementation of the standard Metropolis–Hastings sampler is given in Table 5.1. For a given intermediate level with $s \in \{2, \ldots, m\}$, the objective is to construct the sample set $\mathcal{U}_s = \{\mathbf{u}_s^{(j)}, 1 \leq j \leq N\}$, as defined in section 5.3.2.2.

These samples are obtained from $K = 1/p_0$ successive draws of $N_c = p_0 N$ separate Markov chains (see lines 4–14 in Table 5.1; note that we do not consider any burn-in period as explained later in this section, all the draws are gathered in $\mathcal{U}_s$ starting from the first element of the chain and for each chain[4]). The initial states of these N_c chains (called the seeds of the chains) are taken as the samples of $\mathcal{U}_{s-1}$, whose LSF evaluations are lower than y_{s-1} (see lines 1–3 in Table 5.1). The algorithm proposed in Table 5.1 presents some deficiencies in high-dimensional spaces (large n), which make it inapplicable in such a context. The probability that a given Markov chain will remain in its current state $\mathbf{u}_s^{(j),k-1}$ approaches 1 when the dimension n increases (see (Au and Beck 2001, Appendix); (Papaioannou *et al.* 2015, section 3.1.2)). The direct consequence is to obtain repeated samples in $\mathcal{U}_s$ sets and therefore, a high level of correlation between the samples in these sets. This results in a poor convergence of $\widehat{p}_s$ estimators given in equation [5.58] compared to the ideal case based on i.i.d. samples. A geometric interpretation of why the standard Metropolis–Hastings sampler fails in high dimensions can also be given (Katafygiotis and Zuev 2008).

1	$\mathcal{J}_{\text{seeds}} = \{j \in \{1, \ldots, N\} : G(\mathbf{u}_{s-1}^{(j)}) \leq y_{s-1}\}$	// index set of seeds in $\mathcal{U}_{s-1}$ (size $N_c = p_0 N$)
2	$\mathcal{J}_{\text{chains}} = \{1, \ldots, N_c\}$	// index set of Markov chains in $\mathcal{U}_s$ (size $N_c = p_0 N$)
3	$\mathbf{u}_s^{(\mathcal{J}_{\text{chains}}),0} = \mathbf{u}_{s-1}^{(\mathcal{J}_{\text{seeds}})}$	// set initial state of N_c Markov chains
4	$K = 1/p_0$	
5	**for** $k = 1$ **to** K	
6	**for** $j = 1$ **to** N_c	
7	// apply Metropolis-Hasting algorithm	
8	Sample $\mathbf{v}$ from proposal PDF $q(\cdot\|\mathbf{u}_s^{(j),k-1})$	
9	Compute acceptance ratio $\alpha_{\varphi_n}^k = \alpha_{\varphi_n}\left(\mathbf{u}_s^{(j),k-1}, \mathbf{v}\right) \dfrac{\varphi_n(\mathbf{v})\, q\left(\mathbf{u}_s^{(j),k-1} \mid \mathbf{v}\right)}{\varphi_n\left(\mathbf{u}_s^{(j),k-1}\right) q\left(\mathbf{v} \mid \mathbf{u}_s^{(j),k-1}\right)}$	
10	Accept or reject $\mathbf{v}$: $\widetilde{\mathbf{v}} = \begin{cases} \mathbf{v} & \text{with probability } \min\{1, \alpha_{\varphi_n}^k\} \\ \mathbf{u}_s^{(j),k-1} & \text{with probability } 1 - \min\{1, \alpha_{\varphi_n}^k\} \end{cases}$	
11	Accept or reject $\widetilde{\mathbf{v}}$: $\mathbf{u}_s^{(j),k} = \begin{cases} \widetilde{\mathbf{v}} & \text{if } \widetilde{\mathbf{v}} \in \mathcal{F}_{\mathbf{u},\, s-1} \\ \mathbf{u}_s^{(j),k-1} & \text{if } \widetilde{\mathbf{v}} \notin \mathcal{F}_{\mathbf{u},\, s-1} \end{cases}$	
12	**end for**	
13	**end for**	
14	$\mathcal{U}_s = \{\mathbf{u}_s^{(j),k}, 1 \leq j \leq N_c, 1 \leq k \leq K\}$	// define new sample set $\mathcal{U}_s$

Note: $\lceil x \rceil$ stands for the smallest integer not lower than x.

Table 5.1. *Application of the standard Metropolis-Hasting sampler, construction of the sample set* $\mathcal{U}_s = \{\mathbf{u}_s^{(j)}, 1 \leq j \leq N\}$ *for* $s = 2, \ldots, m$

4 The notation $\mathbf{u}_s^{(j),k}$ will be used if an explicit mention of the kth sample in the jth chain is required. If not, the shorter notation $\mathbf{u}_s^{(j)}$ will be used for reference to the jth sample in the set $\mathcal{U}_s$.

In order to address high-dimensional reliability problems, Au and Beck (2001) proposed a *componentwise* adaptation of the standard Metropolis sampler. This adaptation is known as the *modified Metropolis* (m-M) algorithm, which represents a key ingredient of SS. We again assume a two-step implementation of the m-M algorithm, as used for the standard Metropolis–Hastings sampler. The transition kernel applied in the first step is now expressed as a product of the transition kernels of each coordinate (Zuev and Katafygiotis 2011a; Zuev *et al.* 2012):

$$K(\mathrm{d}\mathbf{v}|\mathbf{u}) = \prod_{i=1}^{n} K_i(\mathrm{d}v_i|u_i), \qquad [5.69]$$

where $\mathrm{d}\mathbf{v} = (\mathrm{d}v_1, \ldots, \mathrm{d}v_n)^{\mathrm{T}}$, $\mathbf{u} = (u_1, \ldots, u_n)^{\mathrm{T}}$ and K_i represents the transition kernel of the ith coordinate. For $i = 1, \ldots, n$, K_i is taken as a standard Metropolis kernel defined as follows:

$$K_i(\mathrm{d}v_i|u_i) = p_i(v_i|u_i)\,\mathrm{d}v_i + r_i(u_i)\,\delta_{u_i}(\mathrm{d}v_i), \qquad [5.70]$$

where $p_i : \mathbb{R} \times \mathbb{R} \to \mathbb{R}_{\geq 0}$

$$(u_i, v_i) \mapsto p_i(v_i|u_i) = \begin{cases} q_i(v_i|u_i)\,\alpha_\varphi(u_i, v_i) & \text{if } u_i \neq v_i, \\ 0 & \text{if } u_i = v_i, \end{cases}$$

in which q_i is a univariate proposal PDF (assumed symmetric in SS) and $\alpha_\varphi(u_i, v_i) = \dfrac{\varphi(v_i)}{\varphi(u_i)}$, and where $r_i(u_i) = 1 - \int_{\mathbb{R}} p_i(v_i|u_i)\,\mathrm{d}v_i$.

The algorithm proposed in SS by Au and Beck (2001) is given in Table 5.2. Note that this algorithm is similar to that proposed in Table 5.1, except lines 7–11 which are replaced by lines 7–13. The initial states of the $N_c = p_0 N$ chains are again defined as the samples of $\mathcal{U}_{s-1}$, whose LSF evaluations are lower than y_{s-1} (see the two-dimensional example for $s = 2$ given in Figure 5.8(a)). The n proposal PDFs q_i are chosen as uniform PDFs of width $2w_i = 2w$, centered at the ith coordinate of the current state $\mathbf{u}_s^{(j),k-1}$, i.e. such that $q_i(v_i|u_i) = \frac{1}{2w}\mathbb{1}_{[u_i - w, u_i + w]}(v_i)$, see Figure 5.8(b) for $k = 1$ (centering at the initial state of the chains). Note that the w_i parameters are identically set to a unique value denoted w, since we are working in the standard normal space, with the same "scale" along each axis u_i.

$\mathcal{J}_{\text{seeds}} = \{j \in \{1, \dots, N\} : G(\mathbf{u}_{s-1}^{(j)}) \leq y_{s-1}\}$ // index set of seeds in $\mathcal{U}_{s-1}$ (size $N_c = p_0 N$)
$\mathcal{J}_{\text{chains}} = \{1, \dots, N_c\}$ // index set of Markov chains in $\mathcal{U}_s$ (size $N_c = p_0 N$)
$\mathbf{u}_s^{(\mathcal{J}_{\text{chains}}),0} = \mathbf{u}_{s-1}^{(\mathcal{J}_{\text{seeds}})}$ // set initial state of N_c Markov chains
$K = 1/p_0$
for $k = 1$ **to** K
  for $j = 1$ **to** N_c
    // apply componentwise Metropolis algorithm from Au and Beck (2001)
    for $i = 1$ **to** n
      Sample v_i from proposal PDF $q_i(\cdot | u_{s,i}^{(j),k-1})$
      Compute acceptance ratio $\alpha_{\varphi,i}^k = \alpha_\varphi\left(u_{s,i}^{(j),k-1}, v_i\right) = \dfrac{\varphi(v_i)}{\varphi\left(u_{s,i}^{(j),k-1}\right)}$
      Accept or reject v_i : $\widetilde{v}_i = \begin{cases} v_i & \text{with probability } \min\{1, \alpha_{\varphi,i}^k\} \\ u_{s,i}^{(j),k-1} & \text{with probability } 1 - \min\{1, \alpha_{\varphi,i}^k\} \end{cases}$
    end for
    Accept or reject $\widetilde{\mathbf{v}} = (\widetilde{v}_1, \dots, \widetilde{v}_n)^{\mathrm{T}}$: $\mathbf{u}_s^{(j),k} = \begin{cases} \widetilde{\mathbf{v}} & \text{if } \widetilde{\mathbf{v}} \in \mathcal{F}_{\mathbf{u},s-1} \\ \mathbf{u}_s^{(j),k-1} & \text{if } \widetilde{\mathbf{v}} \notin \mathcal{F}_{\mathbf{u},s-1} \end{cases}$
  end for
end for
$\mathcal{U}_s = \{\mathbf{u}_s^{(j),k}, 1 \leq j \leq N_c, 1 \leq k \leq K\}$ // define new sample set $\mathcal{U}_s$

Note: $\lceil x \rceil$ stands for the smallest integer not lower than x.

Table 5.2. *Application of the componentwise Metropolis sampler proposed by Au and Beck (2001), construction of the sample set* $\mathcal{U}_s = \{\mathbf{u}_s^{(j)}, 1 \leq j \leq N\}$ *for* $s = 2, \dots, m$

The m-M algorithm is characterized by the following important properties:

1) the m-M algorithm works componentwise with univariate proposal PDFs q_i, which allows independent movement along each axis. Such a technique reduces the number of repeated samples in $\mathcal{U}_s$ sets compared to the original Metropolis–Hastings algorithm, which makes SS well adapted to high-dimensional reliability problems (large n);

2) the target distribution of samples with MCMC is usually attained as the stationary distribution of the Markov chain, which generally imposes a burn-in period. In SS, the seeds already follow the target distribution $\varphi_n(\cdot | E_{s-1})$ which means that any further states of the chain are distributed according to this target distribution (see, e.g.(Au and Beck 2001; Zuev *et al.* 2012) for a demonstration). This property, known as *perfect sampling* in the MCMC literature, makes any burn-in period unnecessary. For this reason, the $\mathcal{U}_s$ sets are composed of all the states of the Markov chains from $k = 1$ to $K = 1/p_0$;

3) the samples generated with the m-M algorithm are not independent. The level of correlation in the chains (and therefore in the samples of the $\mathcal{U}_s$ sets) depends on the choice of the n proposal PDFs q_i. The idea of the m-M algorithm of SS is to use proposal PDFs q_i, which are sufficiently local, to ensure that the sample candidates $\widetilde{\mathbf{v}}$

mostly lie in $\mathcal{F}_{\mathbf{u},\,s-1}$. Practically, this is tuned by means of the unique parameter w of the uniform proposal PDFs q_i. If w is too small, the probability that $\widetilde{\mathbf{v}}$ lies in $\mathcal{F}_{\mathbf{u},\,s-1}$ is high, which thus avoids repeated samples in the $\mathcal{U}_s$ set. However, the candidate state $\widetilde{\mathbf{v}}$ is close to the current state $\mathbf{u}_s^{(j),k-1}$, which induces a high level of correlation in the chain due to the small movements and therefore reduces the efficiency of the conditional estimators $\widehat{p}_s$ for $s = 2, \ldots, m$. Moreover, proposal PDFs that are too local do not enable a sufficiently wide exploration of the intermediate failure domain $\mathcal{F}_{\mathbf{u},\,s-1}$, which may result in a biased estimate $\widehat{p_{\mathrm{f}}^{\mathrm{SS}}}$ if important failure zones are missed at intermediate levels. If w is too large, the proposal PDFs allow a sufficiently wide exploration of $\mathcal{F}_{\mathbf{u},\,s-1}$, but the probability that $\widetilde{\mathbf{v}}$ lies in $\mathcal{F}_{\mathbf{u},\,s-1}$ is low. The chain is expected to remain at the current state $\mathbf{u}_s^{(j),k-1}$, which results in an increase in the number of repeated samples in the $\mathcal{U}_s$ set. The consequence is again a high level of correlation in the chains, which therefore implies a drop in the efficiency of the conditional estimators $\widehat{p}_s$. According to Au and Beck (2001), the efficiency of the m-M algorithm is not sensitive to the type of proposal PDFs, but to their spreads. A tradeoff is made in SS with a parameter w set to 2 (Au and Beck 2001). It is worth pointing out that attempts to reduce the correlation in the chains have been made in a few works (Miao and Ghosn 2011; Santoso *et al.* 2011; Zuev and Katafygiotis 2011b; Zuev *et al.* 2012; Papaioannou *et al.* 2015). The reader may refer to the recent paper from Papaioannou *et al.* (2015) for a comparative review of these works.

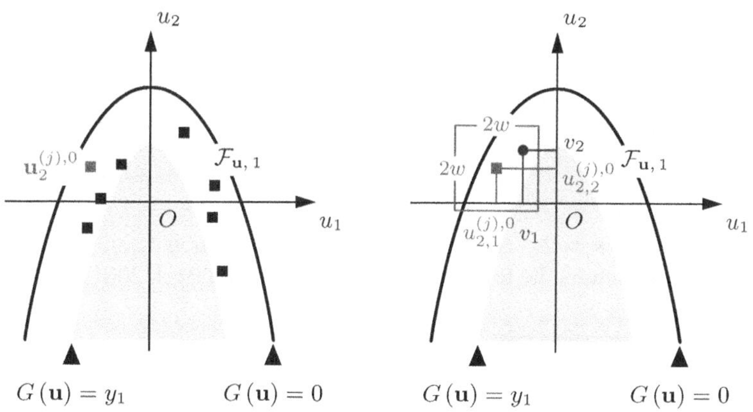

(a) Initial state of Markov chains $\mathbf{u}_s^{(j),0}$ for $s = 2$

(b) Sampling from univariate proposal PDFs $q_i(\cdot|u_{2,i}^{(j),0})$, $i = 1, 2$

Figure 5.8. *Modified Metropolis algorithm: seeds of Markov chains and proposal PDFs. For a color version of this figure, see www.iste.co.uk/gogu/uncertainties.zip*

5.3.2.4. *Statistical properties of the SS failure probability estimator*

5.3.2.4.1. Coefficient of variation of $\widehat{p}_{\mathrm{f}}^{\mathrm{SS}}$

The coefficient of variation of $\widehat{p}_{\mathrm{f}}^{\mathrm{SS}}$ can be estimated from a single run of the SS method. This is important for assessing the accuracy of the probability estimate obtained by this method. Only the main expressions required to calculate this quantity are recalled here, and the reader may refer to Au and Beck (2001) for further details and proofs of the given expressions. The coefficient of variation of $\widehat{p}_{\mathrm{f}}^{\mathrm{SS}}$ denoted $\delta_{\widehat{p}_{\mathrm{f}}^{\mathrm{SS}}}$ can be bounded above according to the following expression:

$$\delta_{\widehat{p}_{\mathrm{f}}^{\mathrm{SS}}}^{2} = \mathbb{E}\left[\left(\frac{\widehat{p}_{\mathrm{f}}^{\mathrm{SS}} - p_{\mathrm{f}}}{p_{\mathrm{f}}}\right)^{2}\right] \leq \sum_{r,s=1}^{m} \delta_r \delta_s + o\left(\frac{1}{N}\right) = \mathcal{O}\left(\frac{1}{N}\right), \qquad [5.71]$$

where δ_s represents the coefficient of variation of $\widehat{p}_s$ for $s = 1, \ldots, m$. This upper bound is established under the assumption of fully correlated probability estimators $\widehat{p}_s$. It is worth pointing out that the coefficient of variation $\delta_{\widehat{p}_{\mathrm{f}}^{\mathrm{SS}}}$ is expressed in terms of a deviation w.r.t. the unknown probability p_{f}, and not w.r.t. the expectation of the probability estimator as defined in equation [5.52] for the MC estimator. For this reason, $\delta_{\widehat{p}_{\mathrm{f}}^{\mathrm{SS}}}$ accounts for the bias effect.

The coefficients of variation δ_s appearing in equation [5.71] for $s = 1, \ldots, m$ are calculated as follows. The first level of SS is merely a crude MC simulation and therefore the coefficient of variation of $\widehat{p}_1$ reads:

$$\delta_1 = \sqrt{\frac{1 - p_1}{N p_1}}. \qquad [5.72]$$

At the next level $s = 2, \ldots, m$, the coefficient of variation δ_s, of the conditional estimators $\widehat{p}_s$, is given by the following expression (Au and Beck 2001):

$$\delta_s = \sqrt{\frac{1 - p_s}{N p_s}(1 + \gamma_s)}, \qquad [5.73]$$

where:

$$\gamma_s = 2 \sum_{\ell=1}^{K-1} \left(1 - \frac{\ell}{K}\right) \rho_s(\ell), \qquad [5.74]$$

in which $K = N/N_{\mathrm{c}} = 1/p_0$ and $\rho_s(\ell) = R_s(\ell)/R_s(0)$, and where $R_s(\ell)$ is the covariance between $\mathbb{1}_{\mathcal{F}_{\mathrm{u},s}}\left(\mathbf{u}_s^{(j),k}\right)$ and $\mathbb{1}_{\mathcal{F}_{\mathrm{u},s}}\left(\mathbf{u}_s^{(j),k+\ell}\right)$ for any lag $\ell = 0, \ldots, (K-1)$ (note that this covariance should be independent of k due to stationarity, and independent of j since all chains are probabilistically equivalent):

$$R_s(\ell) = \mathbb{E}\left[\mathbb{1}_{\mathcal{F}_{\mathrm{u},s}}\left(\mathbf{u}_s^{(j),k}\right)\mathbb{1}_{\mathcal{F}_{\mathrm{u},s}}\left(\mathbf{u}_s^{(j),k+\ell}\right)\right] - p_s^2. \quad [5.75]$$

The autocovariance sequence $\{R_s(\ell), \ell = 0, \ldots, K-1\}$ is practically estimated from the samples of the set $\mathcal{U}_s$, by averaging on both the N_c chains and their elements according to the following expression:

$$\widehat{R}_s(\ell) = \left(\frac{1}{N_c(K-\ell)}\sum_{j=1}^{N_c}\sum_{k=1}^{K-\ell}\mathbb{1}_{\mathcal{F}_{\mathrm{u},s}}\left(\mathbf{u}_s^{(j),k}\right)\mathbb{1}_{\mathcal{F}_{\mathrm{u},s}}\left(\mathbf{u}_s^{(j),k+\ell}\right)\right) - \widehat{p}_s^2. \quad [5.76]$$

The coefficient $\gamma_s > 0$ appearing in equation [5.73] makes the coefficient of variation δ_s of the conditional probabilities $\widehat{p}_s$ larger than that of the MC estimate with i.i.d. samples, which corresponds to the case $\gamma_s = 0$. Therefore, working with correlated MCMC samples appears less efficient than a simulation with i.i.d. samples. For a better efficiency in the SS method, one could search for optimal values of the parameter w that control the width of the n uniform proposal PDFs in order to minimize γ_s at each level $s = 2, \ldots, m$. This problem investigated by Zuev *et al.* (2012) unfortunately has no practically implementable solution. An alternative proposed by these authors is to adapt w for an average number of accepted candidate states between 30 and 50%.

Another expression, less pessimistic than the upper bound given in equation [5.71], is often used to estimate the coefficient of variation of $\widehat{p}_{\mathrm{f}}^{\mathrm{SS}}$. This expression, based on the assumption of independent probability estimators $\widehat{p}_s$, writes (Au and Beck 2001):

$$\delta_{\widehat{p}_{\mathrm{f}}^{\mathrm{SS}}}^2 \approx \sum_{s=1}^{m}\delta_s^2. \quad [5.77]$$

This second expression is reported to give fairly good approximations of the true coefficient of variation of $\widehat{p}_{\mathrm{f}}^{\mathrm{SS}}$ in Au and Beck (2001), although it underestimates it. From the author's experience, neither of these two expressions is more accurate than the other. The upper bound sometimes appears closer to the true coefficient of variation in some problems.

5.3.2.4.2. Bias of $\widehat{p}_{\mathrm{f}}^{\mathrm{SS}}$

The estimator $\widehat{p}_{\mathrm{f}}^{\mathrm{SS}} = \prod_{s=1}^{m}\widehat{p}_s$ is biased for any sample size N due to the correlation that exists between the estimators $\widehat{p}_s$ for $s = 1, \ldots, m$ (Au and Beck

2001). This correlation is due to the way the seeds of the chains are selected: the seeds at level s are samples generated at the previous level $(s-1)$ (see lines 1–3 of Table 5.2). This results in a correlation between $\widehat{p}_s$ and $\widehat{p}_{s-1}$ for $s=2,\ldots,m$. The estimator $\widehat{p}_\mathrm{f}^\mathrm{SS}$ is asymptotically unbiased, and its relative bias can be bounded from above by the following expression (Au and Beck 2001):

$$\left|\mathbb{E}\left[\frac{\widehat{p}_\mathrm{f}^\mathrm{SS}-p_\mathrm{f}}{p_\mathrm{f}}\right]\right| \leq \sum_{r>s} \delta_r \delta_s + o\left(\frac{1}{N}\right) = \mathcal{O}\left(\frac{1}{N}\right). \quad [5.78]$$

This bias of order $1/N$ thus becomes negligible compared to the standard deviation when N is increased. Moreover, it can be shown that the bias is always positive (Guyader 2011).

5.3.2.5. *Remark about the computer implementation of SS*

The SS method is available in FERUM 4.x as proposed by Au and Beck (2001). The implemented algorithm takes full advantage of the independence of:

– the first N that calls to the LSF evaluated at the initial level $s=1$ of SS (crude MC), which are sent as independent tasks;

– the N_c chains that are sampled in parallel for a given k (see lines 6–12 of Table 5.2).

In SS, the LSF evaluation tasks are sent in the following sequence: N independent calls to the LSS for $s=1$, then p_0N independent calls times K (length of the chains) times $(m-1)$ (for each intermediate level $s=2,\ldots,m$). SS therefore appears less suited to distributed computations than crude MC. Improving the parallelization efficiency of the method may be of interest, as recently investigated by Walter (2015) in the framework of moving particles. Walter proposes a parallel multilevel splitting algorithm that allows the simultaneous and independent movement of several particles.

5.3.3. *IS and CE methods*

IS is one of the most well-known variance reduction techniques used for assessing small failure probabilities. The idea of IS is to draw samples following another distribution than the original, in order to populate the failure domain more frequently. The failure probability estimate assessed by IS is then obtained as a weighted average of these draws. The basic ideas of IS were outlined in the early fifties and this technique, used in conjunction with FORM, was later introduced in structural reliability (Schuëller and Stix 1987; Melchers 1989a). Several IS methods are available and the reader is invited to refer to general textbooks for an exhaustive list (see, e.g. (Kroese *et al.* 2011)). IS is also used in adaptive versions, one of which, known as the CE method, is presented in this chapter.

5.3.3.1. *Importance sampling*

Let us consider another PDF $h_{\mathbf{X}}$ in the problem formulated in equation [5.50], called *importance sampling PDF* or *instrumental PDF*, such that $f_{\mathbf{X}}$, from now on referred to as *nominal PDF*, is dominated by $h_{\mathbf{X}}$ in the absolutely continuous sense:

$$h_{\mathbf{X}}(\mathbf{x}) = 0 \Rightarrow \mathbb{1}_{\mathcal{F}_{\mathbf{x}}}(\mathbf{x})\, f_{\mathbf{X}}(\mathbf{x}) = 0 \qquad [5.79]$$

or equivalently

$$\mathbb{1}_{\mathcal{F}_{\mathbf{x}}}(\mathbf{x})\, f_{\mathbf{X}}(\mathbf{x}) \neq 0 \Rightarrow h_{\mathbf{X}}(\mathbf{x}) \neq 0. \qquad [5.80]$$

Note that this condition simply means that the support of $\mathbb{1}_{\mathcal{F}_{\mathbf{x}}}(\cdot)\, f_{\mathbf{X}}(\cdot)$ must be included in the support $\mathcal{D}_{h_{\mathbf{X}}}$ of $h_{\mathbf{X}}$.

Using the IS PDF $h_{\mathbf{X}}$, the unknown failure probability p_{f} is now rewritten as follows:

$$\begin{aligned} p_{\mathrm{f}} &= \int_{\mathcal{D}_{f_{\mathbf{X}}}} \mathbb{1}_{\mathcal{F}_{\mathbf{x}}}(\mathbf{x})\, f_{\mathbf{X}}(\mathbf{x})\, \mathrm{d}\mathbf{x} \\ &= \int_{\mathcal{D}_{h_{\mathbf{X}}}} \mathbb{1}_{\mathcal{F}_{\mathbf{x}}}(\mathbf{x})\, \frac{f_{\mathbf{X}}(\mathbf{x})}{h_{\mathbf{X}}(\mathbf{x})}\, h_{\mathbf{X}}(\mathbf{x})\, \mathrm{d}\mathbf{x} \\ &= \mathbb{E}_{h_{\mathbf{X}}}\left[\mathbb{1}_{\mathcal{F}_{\mathbf{x}}}(\mathbf{X})\, \frac{f_{\mathbf{X}}(\mathbf{X})}{h_{\mathbf{X}}(\mathbf{X})}\right] \\ &= \mathbb{E}_{h_{\mathbf{X}}}\left[\mathbb{1}_{\mathcal{F}_{\mathbf{x}}}(\mathbf{X})\, W(\mathbf{X})\right], \end{aligned} \qquad [5.81]$$

where the ratio of densities $W(\mathbf{x}) = f_{\mathbf{X}}(\mathbf{x})/h_{\mathbf{X}}(\mathbf{x})$ is called the *likelihood ratio* or *IS quotient*.

The IS estimator of p_{f} therefore reads:

$$\widehat{p_{\mathrm{f}}}^{\mathrm{IS}} = \frac{1}{N}\sum_{j=1}^{N} \mathbb{1}_{\mathcal{F}_{\mathbf{x}}}\left(\mathbf{X}^{(j)}\right) \frac{f_{\mathbf{X}}\left(\mathbf{X}^{(j)}\right)}{h_{\mathbf{X}}\left(\mathbf{X}^{(j)}\right)}, \qquad [5.82]$$

where $\{\mathbf{X}^{(j)}, 1 \leq j \leq N\}$ are N independent copies of the random vector $\mathbf{X}$ following the IS PDF $h_{\mathbf{X}}$, i.e. $\mathbf{X}^{(1)}, \ldots, \mathbf{X}^{(N)} \overset{\text{i.i.d.}}{\sim} h_{\mathbf{X}}$, and N denotes the sample size.

The IS estimator is unbiased (i.e. $\mathbb{E}_{h_{\mathbf{X}}}\left[\widehat{p}_{\mathrm{f}}^{\mathrm{IS}}\right] = p_{\mathrm{f}}$) and its variance is given by:

$$\mathrm{Var}_{h_{\mathbf{X}}}\left[\widehat{p}_{\mathrm{f}}^{\mathrm{IS}}\right] = \frac{1}{N}\left(\mathbb{E}_{h_{\mathbf{X}}}\left[\left(\mathbb{1}_{\mathcal{F}_{\mathbf{x}}}(\mathbf{X})\,\frac{f_{\mathbf{X}}(\mathbf{X})}{h_{\mathbf{X}}(\mathbf{X})}\right)^{2}\right] - p_{\mathrm{f}}^{2}\right). \qquad [5.83]$$

The accuracy of the approximation given by IS critically depends on the choice of the IS PDF $h_{\mathbf{X}}$. The optimal choice for this PDF is obtained by minimizing the variance of $\widehat{p}_{\mathrm{f}}^{\mathrm{IS}}$ defined in equation [5.83]. The optimal solution $h_{\mathbf{X}}^{*}$ is given in equation [5.84] (see, e.g. (Rubinstein and Kroese 2007)). It is easy to show that this choice leads to a zero variance of $\widehat{p}_{\mathrm{f}}^{\mathrm{IS}}$ from equation [5.83].

$$h_{\mathbf{X}}^{*}(\mathbf{x}) = \frac{\mathbb{1}_{\mathcal{F}_{\mathbf{x}}}(\mathbf{x})\, f_{\mathbf{X}}(\mathbf{x})}{p_{\mathrm{f}}}. \qquad [5.84]$$

The use of this optimal PDF $h_{\mathbf{X}}^{*}$ is unfortunately not possible because it depends on the failure probability p_{f}, which is unknown. Nevertheless, it is expected that the "good" candidates for IS sampling correspond to PDFs, which are not too far from this optimal solution. One important condition that needs to be fulfilled is that the IS estimator $\mathrm{Var}_{h_{\mathbf{X}}}\left[\widehat{p}_{\mathrm{f}}^{\mathrm{IS}}\right]$ should have a finite variance, as expressed in equation [5.85]. Such a condition is met if $h_{\mathbf{X}}$ does not have lighter tails than $f_{\mathbf{X}}$, or if the likelihood ratio W is bounded (Kroese *et al.* 2011).

$$\mathbb{E}_{h_{\mathbf{X}}}\left[\mathbb{1}_{\mathcal{F}_{\mathbf{x}}}^{2}(\mathbf{X})\,\frac{f_{\mathbf{X}}^{2}(\mathbf{X})}{h_{\mathbf{X}}^{2}(\mathbf{X})}\right] = \mathbb{E}_{f_{\mathbf{X}}}\left[\mathbb{1}_{\mathcal{F}_{\mathbf{x}}}^{2}(\mathbf{X})\,\frac{f_{\mathbf{X}}(\mathbf{X})}{h_{\mathbf{X}}(\mathbf{X})}\right] < \infty. \qquad [5.85]$$

In structural reliability, several types of IS PDF have been defined in the context of FORM (see a review of some of these strategies in Engelund and Rackwitz 1993). The most common choice that has emerged is to perform IS in the standard normal space, by means of a multivariate standard normal PDF centered on the supposedly unique MPFP identified by FORM (Schuëller and Stix 1987; Melchers 1989a):

$$h_{\mathbf{U}}(\mathbf{u}) = \varphi_{n}(\mathbf{u} - \mathbf{u}^{*}). \qquad [5.86]$$

In the case of multiple MPFPs, solutions based on mixtures of multivariate standard normal PDFs need to be considered. However such approaches assume that these MPFPs have been determined, which is often hardly possible. If some MPFPs corresponding to high probability content failure subdomains are missed, then the failure probability estimate may be seriously biased and therefore, inaccurate.

In order to fit the optimal IS PDF, some works consider adaptively adjusting the IS PDF by iteratively using wisely selected samples in order to populate the subdomains of higher probability densities within the failure domain. Such methods are known as *adaptive IS* and examples of such a strategy can be found in Bucher (1988); Karamchandani *et al.* (1989); Melchers (1989b, 1990); Au and Beck (1999); Zou *et al.* (2003); Morio (2012). It is worth pointing out that these methods also relate to particle filters (Chopin 2002) and sequential MC methods (Doucet *et al.* 2001; Johansen *et al.* 2005; Del Moral *et al.* 2006; Cérou *et al.* 2012).

A convenient solution for the IS PDF is to consider that it belongs to some *parametric family* of distributions, e.g. the same as that of the nominal PDF $f_{\mathbf{X}}$, as assumed later. Let us consider that $h_{\mathbf{X}}$ belongs to the following family:

$$h_{\mathbf{X}}(\cdot) \in \{h(\cdot;\mathbf{q}), \mathbf{q} \in \mathcal{Q}\}, \qquad [5.87]$$

where $\mathbf{q} = (q_1, \ldots, q_{n_\mathbf{q}})$ is a finite-dimensional vector of $n_\mathbf{q}$ real parameters. The objective is therefore to select $\mathbf{q}$ such that $h(\cdot;\mathbf{q})$ is as "close" as possible to the optimal IS PDF $h^*_{\mathbf{X}}$.

The first and natural solution consists of finding the value of $\mathbf{q}$ that minimizes the variance of the IS estimator $\widehat{p}_{\mathrm{f}}^{\mathrm{IS}}$. The optimal solution $\mathbf{q}^*$ is given by:

$$\mathbf{q}^* = \arg\min_{\mathbf{q}} \int \mathbb{1}^2_{\mathcal{F}_\mathbf{x}}(\mathbf{x}) \frac{f_{\mathbf{X}}(\mathbf{x})}{h(\mathbf{x};\mathbf{q})} f_{\mathbf{X}}(\mathbf{x})\, d\mathbf{x}, \qquad [5.88]$$

or equivalently by:

$$\mathbf{q}^* = \arg\min_{\mathbf{q}} \int \frac{h^{*2}_{\mathbf{X}}(\mathbf{x})}{h(\mathbf{x};\mathbf{q})}\, d\mathbf{x}. \qquad [5.89]$$

Solving this first type of optimization problem is known as the *variance minimization* (VM) *method* (Kroese *et al.* 2011).

A second solution is obtained by choosing the IS PDF $h(\cdot;\mathbf{q})$ that is the closest to $h^*_{\mathbf{X}}$ w.r.t. a given measure, here the *Kullback–Leibler* (KL) *divergence* or *KL distance*[5] (Kullback and Leibler 1951). The KL distance between $h^*_{\mathbf{X}}$ and $h(\cdot;\mathbf{q})$ is given by:

$$\begin{aligned} \mathcal{D}_{\mathrm{KL}}(h^*_{\mathbf{X}}, h(\cdot;\mathbf{q})) &= \int h^*_{\mathbf{X}}(\mathbf{x}) \ln\left(\frac{h^*_{\mathbf{X}}(\mathbf{x})}{h(\mathbf{x};\mathbf{q})}\right) d\mathbf{x} \\ &= \int h^*_{\mathbf{X}}(\mathbf{x}) \ln(h^*_{\mathbf{X}}(\mathbf{x}))\, d\mathbf{x} - \int h^*_{\mathbf{X}}(\mathbf{x}) \ln(h(\mathbf{x};\mathbf{q}))\, d\mathbf{x}. \end{aligned} \qquad [5.90]$$

5 The KL divergence is in fact, not a distance, since it is not symmetric: $\mathcal{D}_{\mathrm{KL}}(f,g) \neq \mathcal{D}_{\mathrm{KL}}(g,f)$. However, we have $\mathcal{D}_{\mathrm{KL}}(f,g) \geq 0$ with equality if and only if $f = g$.

The optimal solution $\mathbf{q}^*$ that leads to the minimal distance is therefore given by:

$$\mathbf{q}^* = \arg\min_{\mathbf{q}} \int h^*_{\mathbf{X}}(\mathbf{x}) \ln\left(\frac{h^*_{\mathbf{X}}(\mathbf{x})}{h(\mathbf{x};\mathbf{q})}\right) \mathrm{d}\mathbf{x}, \quad [5.91]$$

or equivalently by:

$$\mathbf{q}^* = \arg\max_{\mathbf{q}} \int h^*_{\mathbf{X}}(\mathbf{x}) \ln\left(h(\mathbf{x};\mathbf{q})\right) \mathrm{d}\mathbf{x}, \quad [5.92]$$

which can also be expressed as follows:

$$\mathbf{q}^* = \arg\max_{\mathbf{q}} \int \mathbb{1}_{\mathcal{F}_{\mathbf{x}}}(\mathbf{x}) \ln\left(h(\mathbf{x};\mathbf{q})\right) f_{\mathbf{X}}(\mathbf{x}) \mathrm{d}\mathbf{x}. \quad [5.93]$$

Solving this second type of optimization problem is known as the CE *method* (Kroese *et al.* 2011).

Although the solutions of these two optimization problems are different, it is shown by Chan *et al.* (2011) that the CE and VM methods provide parameters $\mathbf{q}^*$ that are asymptotically identical from the treated examples. The optimization problem that needs to be solved using the VM method is known to be highly nonlinear and noisy (see, e.g. Botev and L'Ecuyer 2011). In contrast, the optimization problem of the CE method appears easier to solve than its VM counterpart, especially when $h(\cdot;\mathbf{q})$ belongs to a natural exponential family (NEF).

The CE method is often used with a parametric IS PDF $h_{\mathbf{X}}(\cdot) \in \{h(\cdot;\mathbf{q}), \mathbf{q} \in \mathcal{Q}\}$[6]. As described in the next section, the CE method consists of an adaptive scheme that updates the vector of parameters $\mathbf{q}$ iteratively, until the IS PDF $h(\cdot;\mathbf{q})$ becomes sufficiently close to the optimal IS PDF $h^*_{\mathbf{X}}$. It is important to mention that the performances of both the CE and VM methods used with a parametric IS PDF are closely related to how well this parametric PDF can approximate the optimal IS PDF. For highly nonlinear LSS with possibly disconnected failure subdomains, such "rigid" parametric IS PDFs may not be able to approximate $h^*_{\mathbf{X}}$ sufficiently well. Such a problem was recently addressed by Kurtz and Song (2013), who used the CE method with a multimodal IS PDF based on a Gaussian mixture. Moreover, for high-dimensional problems, the number of

6 Note that the CE method can also be used in a non-parametric framework, i.e. without assuming any specific family of distributions. Examples of such approaches are found in Botev *et al.* (2007), in which a χ^2-divergence and kernel mixture densities are used, and in Homem-de-Mello (2007), in which a product form on the IS distribution is assumed.

parameters in the vector $\mathbf{q}$ necessarily becomes large and the CE and VM optimization problems could become too hard and costly to solve.

As an alternative to IS based on parametric PDFs, the recourse to surrogate-based (and often adaptive) IS has been explored in several works. In such approaches, the first step consists of drawing samples that adaptively populate the subdomains of the failure domain with high probability content. This may be carried out by using Markov chains (see, e.g. Au and Beck 1999; Dai *et al.* 2012). From these samples, a non-parametric PDF is constructed as an approximation of the optimal IS PDF and is then used for sampling. Several surrogate models are available to construct such approximate IS PDFs. The first type of surrogate model is known as *kernel density*, which is most often based on the choice of a Gaussian kernel. Such kernel PDFs were used by Au and Beck (1999); Morio (2011); Dai *et al.* (2012) among others. Kriging surrogate models constitute a second type of surrogate model, which was used for adaptive IS by Balesdent *et al.* (2013); Echard *et al.* (2013). Kriging was also used by Dubourg *et al.* (2013) to define a kriging-based probabilistic classification function as a surrogate for the true failure indicator function, which appears in the optimal IS PDF. A modification of the AK-IS algorithm of Echard *et al.* (2013), also based on kriging, is proposed in the work of Cadini *et al.* (2014), in which the assumption of a unique MPFP found by FORM is relaxed. This work makes use of the metamodel-based IS of Dubourg *et al.* (2013). A third and less common solution is to use support vector machines to build an approximate IS PDF, as in the work of Dai *et al.* (2012).

As a final and important remark about IS, it is worth pointing out that finding a suitable IS PDF in high dimensions is a challenging task, especially for highly non-linear LSS and possibly disconnected failure subdomains of similar probability weights. IS may yield erroneous failure probability estimates that are even worse than those obtained by crude MC, as pointed out by Au and Beck (2003); Katafygiotis and Zuev (2008). Regarding surrogate-based IS, it is also hard to find an approximate PDF sufficiently close to the optimal one. If the surrogate-based IS PDF is too crude of an approximation of the optimal IS PDF, then the surrogate-based IS method may require too many samples or, even worse, may fail to give a correct failure probability approximation even if a large number of samples is used, as pointed out by Dubourg *et al.* (2013).

5.3.3.2. *CE method*

The CE method, whose basis was briefly introduced in section 5.3.3.1, is a generic approach that addresses a variety of problems, such as the probability estimation of rare events (Rubinstein 1997) and the optimization of discrete or continuous problems (Rubinstein 1999). Here, the presentation will be limited to rare-event probability estimation for the so-called static problems introduced in section 5.1. For a broader view and understanding of the CE method, the reader may refer to the general textbook of Kroese *et al.* (2011) and the tutorial of de Boer *et al.* (2005) in addition to the seminal papers of Rubinstein cited above.

5.3.3.2.1. CE method for estimation

Let us now assume that the IS PDF $h_{\mathbf{X}}$ we use belongs to the same family as that of $f_{\mathbf{X}}$:

$$f_{\mathbf{X}}(\cdot), h_{\mathbf{X}}(\cdot) \in \{f(\cdot;\mathbf{q}), \mathbf{q} \in \mathcal{Q}\} \qquad [5.94]$$

for some reference vector of parameters $\mathbf{q}$. Let us moreover consider that the nominal PDF $f_{\mathbf{X}}$ is parameterized by the vector of parameters $\mathbf{p}$, such that $f_{\mathbf{X}}(\cdot) = f(\cdot;\mathbf{p})$.

The optimization problem defined in equation [5.93] thus rewrites as:

$$\mathbf{q}^* = \arg\max_{\mathbf{q}} \int \mathbb{1}_{\mathcal{F}_{\mathbf{x}}}(\mathbf{x}) \ln(f(\mathbf{x};\mathbf{q}))\, f(\mathbf{x};\mathbf{p})\, \mathrm{d}\mathbf{x}, \qquad [5.95]$$

or equivalently:

$$\mathbf{q}^* = \arg\max_{\mathbf{q}} \mathbb{E}_{f(\cdot;\mathbf{p})}[\mathbb{1}_{\mathcal{F}_{\mathbf{x}}}(\mathbf{X}) \ln(f(\mathbf{X};\mathbf{q}))]. \qquad [5.96]$$

By IS with a change of measure $f(\cdot;\mathbf{r})$ corresponding to *any* vector of parameters $\mathbf{r}$ (we will define how to choose $\mathbf{r}$ in the following), we can write:

$$\mathbf{q}^* = \arg\max_{\mathbf{q}} \mathbb{E}_{f(\cdot;\mathbf{r})}[\mathbb{1}_{\mathcal{F}_{\mathbf{x}}}(\mathbf{X})\, W(\mathbf{X};\mathbf{p},\mathbf{r}) \ln(f(\mathbf{X};\mathbf{q}))], \qquad [5.97]$$

where $W(\mathbf{x};\mathbf{p},\mathbf{r}) = f(\mathbf{x};\mathbf{p})/f(\mathbf{x};\mathbf{r})$ is the likelihood ratio at $\mathbf{x}$ between $f(\cdot;\mathbf{p})$ and $f(\cdot;\mathbf{r})$.

Practically, this problem is solved by simulation using the so-called *stochastic counterpart* of the CE program:

$$\max_{\mathbf{q}} \frac{1}{N} \sum_{j=1}^{N} \mathbb{1}_{\mathcal{F}_{\mathbf{x}}}\left(\mathbf{X}^{(j)}\right) W\left(\mathbf{X}^{(j)};\mathbf{p},\mathbf{r}\right) \ln\left(f\left(\mathbf{X}^{(j)};\mathbf{q}\right)\right), \qquad [5.98]$$

where $\mathbf{X}^{(1)}, \ldots, \mathbf{X}^{(N)} \overset{\text{i.i.d.}}{\sim} f(\cdot;\mathbf{r})$.

If we define $\mathcal{X}_{\text{el}} = \left\{\mathbf{X}^{(j)} : \mathbb{1}_{\mathcal{F}_{\mathbf{x}}}\left(\mathbf{X}^{(j)}\right) = 1\right\}$ as the set of samples that belongs to the failure domain $\mathcal{F}_{\mathbf{x}}$, known as the *elite* set of samples, equation [5.98] becomes:

$$\max_{\mathbf{q}} \widehat{D}(\mathbf{q}) = \max_{\mathbf{q}} \frac{1}{N} \sum_{\mathbf{X}^{(j)} \in \mathcal{X}_{\text{el}}} W\left(\mathbf{X}^{(j)};\mathbf{p},\mathbf{r}\right) \ln\left(f\left(\mathbf{X}^{(j)};\mathbf{q}\right)\right). \qquad [5.99]$$

The above-defined function $\widehat{D}(\mathbf{q})$ is often convex and differentiable, and the solution of equation [5.99] can be obtained from the following system of equations:

$$\frac{1}{N} \sum_{\mathbf{X}^{(j)} \in \mathcal{X}_{\text{el}}} W\left(\mathbf{X}^{(j)}; \mathbf{p}, \mathbf{r}\right) \nabla_{\mathbf{q}} \ln\left(f\left(\mathbf{X}^{(j)}; \mathbf{q}\right)\right) = \mathbf{0}. \qquad [5.100]$$

For distributions that belong to an NEF (for a definition see Kroese *et al.* 2011, Appendix D), the system of equations defined in equation [5.100] have an analytical solution, which is recognized as being a key advantage of the CE method.

If we take $\mathbf{r} = \mathbf{p}$ in the case of rare events (i.e. $p_{\text{f}} \ll 1$), most of the evaluations of the indicator function $\mathbb{1}_{\mathcal{F}_{\mathbf{x}}}$ in equation [5.98] are equal to zero for a moderate sample size N. This results in an excessively small or, even worse, an empty elite set of samples, and it is therefore difficult or impossible to solve equation [5.99] or equation [5.100]. This motivates the introduction of the multilevel version of the CE method, which will be presented next.

5.3.3.2.2. Multilevel CE method

In order to avoid an elite set of samples that is too small, and therefore too few terms in the sums appearing in equation [5.99] and equation [5.100], the idea is to build a sequence of reference parameters $\{\widehat{\mathbf{q}}_s,\ s \in \mathbb{N}\}$ with $\widehat{\mathbf{q}}_0 = \mathbf{p}$ and a sequence of LSF levels $\{\widehat{y}_s,\ s \in \mathbb{N}_{>0}\}$, and iterate on both $\widehat{y}_s$ and $\widehat{\mathbf{q}}_s$. The algorithm first proceeds with an update of $\widehat{y}_s$ before updating $\widehat{\mathbf{q}}_s$.

The algorithm of the *multilevel CE method* is given in Table 5.3. It requires the specification of the so-called *rarity parameter* ρ (typically between 0.01 and 0.1), the sample size N used at each level s and the sample size N_1 used for the final estimation of the failure probability.

5.4. Sensitivity measures

5.4.1. *Introduction*

In uncertainty propagation, it is of interest to quantify, in a probabilistic sense, how each random input contributes to a given output of interest. The first type of approach, known as a global sensitivity analysis (Saltelli *et al.* 2008), consists of assessing the sensitivity to inputs over the entire support of the joint PDF $f_{\mathbf{X}}$. Several methods can be used to perform such a global analysis, among which are screening methods, variance-based methods, such as those based on Sobol's indices (Sobol' 1993), non-parametric methods and density-based methods. Note that global sensitivity analysis is not considered within the scope of this chapter.

Define ρ, N and N_1 // define the parameters of the multilevel CE algorithm
Set $s = 0, \widehat{\mathbf{q}}_0 = \mathbf{p}$ and $\widehat{y}_0 = +\infty$ // initialize the algorithm
Do
   $s = s + 1$
   Generate a sample $\mathbf{X}^{(1)}, \ldots, \mathbf{X}^{(N)} \overset{\text{i.i.d.}}{\sim} f\left(\cdot; \widehat{\mathbf{q}}_{s-1}\right)$
   Assess $g\left(\mathbf{X}^{(1)}\right), \ldots, g\left(\mathbf{X}^{(N)}\right)$
   Define the sample ρ-quantile $\widehat{y}_s$ of $\mathcal{Y} = \left\{g(\mathbf{X}^{(j)}), 1 \leq j \leq N\right\}$
   if $\widehat{y}_s < 0$, **then** $\widehat{y}_s = 0$, **end if** // reset $\widehat{y}_s$ if lower than zero
   Define the elite set of samples $\mathcal{X}_{\text{el}} = \left\{\mathbf{X}^{(j)} : g\left(\mathbf{X}^{(j)}\right) \leq \widehat{y}_s\right\}$
   Solve the stochastic program defined in equation [5.99] with $\mathbf{r} = \widehat{\mathbf{q}}_{s-1}$ and denote the solution by $\widetilde{\mathbf{q}}_s$:
   $$\widetilde{\mathbf{q}}_s = \arg\max_{\mathbf{q}} \frac{1}{N} \sum_{\mathbf{X}^{(j)} \in \mathcal{X}_{\text{el}}} W\left(\mathbf{X}^{(j)}; \mathbf{p}, \widehat{\mathbf{q}}_{s-1}\right) \ln\left(f\left(\mathbf{X}^{(j)}; \mathbf{q}\right)\right)$$
   Update the reference parameter: $\widehat{\mathbf{q}}_s = \widetilde{\mathbf{q}}_s$
while $\widehat{y}_s > 0$
Set $m = s$ // define the total number of levels used
Generate a sample $\mathbf{X}^{(1)}, \ldots, \mathbf{X}^{(N_1)} \overset{\text{i.i.d.}}{\sim} f\left(\cdot; \widehat{\mathbf{q}}_m\right)$
Estimate p_{f} by IS: $\widehat{p}_{\text{f}}^{\text{CE}} = \frac{1}{N_1} \sum_{j=1}^{N_1} \mathbb{1}_{\mathcal{F}_{\mathbf{X}}}\left(\mathbf{X}^{(j)}\right) W\left(\mathbf{X}^{(j)}; \mathbf{p}, \widehat{\mathbf{q}}_m\right)$

Table 5.3. *Multilevel CE algorithm for rare-event probability estimation*

If reliability assessment is the objective pursued, then the analyst is usually interested in the sensitivities to random inputs at failure. These sensitivities are therefore of a local nature and need to be evaluated at failure, e.g. at the MPFP in a FORM analysis. Such local importance measures introduced in the context of FORM are briefly reviewed in section 5.4.2.1, either with independent or Nataf-correlated random inputs.

In engineering problems of practical interest, the statistical data available are often limited, and assessing how the results of an uncertainty propagation analysis are sensitive to the statistical model used as an input, turns out to be of paramount importance. Several uncertainties may be considered, such as the types of marginal distributions, the dependence structure between the random components or any related model parameters that are used in the analysis. In this chapter, the scope will be limited to the sensitivity of the failure probability p_{f} w.r.t. distribution parameters, assuming that choices have been made regarding the types of marginal distributions and the dependence structure that is used (here linear correlation, accounted for by means of the Nataf model). We assume that all the distribution parameters are gathered in a vector denoted $\boldsymbol{\theta} = (\theta_1, \ldots, \theta_{n_{\theta}})$ in which θ_k designates any given parameter of this vector. These parameters include the statistical moments of the components X_i of the random vector $\mathbf{X}$ for $i = 1, \ldots, n$, but also, and more generally, any parameters used to define the corresponding marginal distributions. The appropriate notation for the joint PDF of $\mathbf{X}$ would be $f_{\mathbf{X}}\left(\cdot; \boldsymbol{\theta}\right)$ but we will simply use $f_{\mathbf{X}}(\cdot)$ in the following, without making the dependence on $\boldsymbol{\theta}$ for shorter notations explicit .

The sensitivities of p_{f} w.r.t. distribution parameters in the FORM context and under the assumption of a Nataf model are presented first. Section 5.4.2 addresses FORM sensitivities w.r.t. parameters of the marginal distributions of $\mathbf{X}$ and also the w.r.t. correlations between the components of $\mathbf{X}$. Note that these sensitivity measures are available in FERUM 4.x (Bourinet *et al.* 2009).

Sensitivity w.r.t. distribution parameters are then placed in the broader context of MC methods. The sensitivities of p_{f} w.r.t. distribution parameters can be obtained from a crude MC analysis, by means of the score function introduced by Rubinstein (1976), as described in section 5.4.3.1. Such sensitivities were later derived in the SS framework by Song *et al.* (2009), as presented in section 5.4.3.2. Note that it is also possible to assess such sensitivities by means of surrogate models, as recently proposed by Dubourg and Sudret (2014), in which the sensitivities of p_{f} w.r.t. distribution parameters are assessed by means of the meta-IS adaptive refinement technique initially proposed for reliability assessment (Dubourg *et al.* 2013) .

5.4.2. *FORM*

5.4.2.1. *Importance factors*

In a FORM analysis, the variance of $G_1(\mathbf{U})$ defined in equation [5.36] can be expressed as follows (see, e.g. (Der Kiureghian 1999; Ditlevsen and Madsen 2007)):

$$\begin{aligned}\mathrm{Var}\left[G_1(\mathbf{U})\right] &= \mathrm{Var}\left[-\left\|\nabla G\left(\mathbf{u}^*\right)\right\|\left(\boldsymbol{\alpha}^{\mathrm{T}}\mathbf{U}-\beta\right)\right] \\ &= \left\|\nabla G\left(\mathbf{u}^*\right)\right\|^2 \mathrm{Var}\left[\boldsymbol{\alpha}^{\mathrm{T}}\mathbf{U}\right] \\ &= \left\|\nabla G\left(\mathbf{u}^*\right)\right\|^2 \left(\alpha_1^2+\alpha_2^2+\ldots+\alpha_n^2\right),\end{aligned} \qquad [5.101]$$

where α_i for $i = 1, \ldots, n$ are the components of the unit vector $\boldsymbol{\alpha}$.

From this expression, it appears that the squared values of the components of the $\boldsymbol{\alpha}$-vector are indicative of the relative importance of the corresponding random inputs in the standard normal space in the neighborhood of the MPFP. These α_i^2-values are known as *importance factors*. If the random inputs are independent, these importance factors clearly express the relative importance of each random component X_i. This is no longer true when correlated inputs are used in reliability analysis. In such a context, Der Kiureghian introduced a new vector γ of importance factors by considering an "equivalent" normal vector in the x-space at the MPFP $\mathbf{x}^*$, denoted $\widehat{\mathbf{X}}$, such that $\mathbf{u} = \mathbf{u}^* + \mathbf{J}_T(\widehat{\mathbf{x}} - \mathbf{x}^*)$ where $\mathbf{J}_T = \mathbf{J}_{\mathbf{u},\mathbf{x}}$ is the Jacobian matrix of the diffeomorphism T evaluated at $\mathbf{x}^*$. This vector of importance factors is expressed as follows (Der Kiureghian 1999):

$$\boldsymbol{\gamma}^{\mathrm{T}} = \frac{\boldsymbol{\alpha}^{\mathrm{T}} \mathbf{J}_T \widehat{\mathbf{D}}}{\|\boldsymbol{\alpha}^{\mathrm{T}} \mathbf{J}_T \widehat{\mathbf{D}}\|}, \qquad [5.102]$$

where $\widehat{\mathbf{D}} = [\widehat{D}_{ii}]_{1\leq i\leq n}$ is a $(n \times n)$-diagonal matrix such that $\widehat{D}_{ii} = \sqrt{J_{2,ii}}$ for $i = 1, \ldots, n$ and $J_{2,ii}$ is the ith diagonal term of the $(n \times n)$-matrix $\mathbf{J}_2 = [J_{2,ij}]_{1\leq i,j\leq n} = \mathbf{J}_{T^{-1}} \mathbf{J}_{T^{-1}}^{\mathrm{T}}$ where $\mathbf{J}_{T^{-1}} = \mathbf{J}_{\mathbf{x},\mathbf{u}}$ is the Jacobian matrix of the inverse T^{-1} of T evaluated at $\mathbf{u}^*$. It should be noted that $\boldsymbol{\gamma} = \boldsymbol{\alpha}$ for the specific case of independent components X_i.

5.4.2.2. *Sensitivities of* p_{f} w.r.t. *distribution parameters*

In the context of a FORM analysis, the gradient of the failure probability approximation $p_{\mathrm{f}}^{\mathrm{FORM}}$ w.r.t. the distribution parameters $\boldsymbol{\theta}$ is given by:

$$\nabla_{\boldsymbol{\theta}} p_{\mathrm{f}}^{\mathrm{FORM}} = -\varphi(\beta)\, \nabla_{\boldsymbol{\theta}} \beta. \qquad [5.103]$$

The gradient of the reliability index β w.r.t. $\boldsymbol{\theta}$ reads as follows, as shown by Hohenbichler and Rackwitz (1986):

$$\nabla_{\boldsymbol{\theta}} \beta = \mathbf{J}_{\mathbf{u},\boldsymbol{\theta}} (\mathbf{u}^*; \boldsymbol{\theta})^{\mathrm{T}} \boldsymbol{\alpha}, \qquad [5.104]$$

where $\mathbf{J}_{\mathbf{u},\boldsymbol{\theta}}(\mathbf{u}; \boldsymbol{\theta}) = [\partial u_i / \partial \theta_j]_{1\leq i\leq n\,,\, 1\leq j\leq n_{\boldsymbol{\theta}}}$ represents the Jacobian of the transformation T w.r.t. the distribution parameters $\boldsymbol{\theta}$. Note that this Jacobian is expressed at the MPFP $\mathbf{u}^*$.

The kth column of this Jacobian matrix for $k = 1, \ldots, n_{\boldsymbol{\theta}}$ can be obtained by differentiating $\mathbf{u}$ defined in equation [5.25], w.r.t. the kth element θ_k of the vector of distribution parameters $\boldsymbol{\theta}$:

$$\frac{\partial \mathbf{u}}{\partial \theta_k} = \mathbf{L}_0^{-1} \frac{\partial \mathbf{z}}{\partial \theta_k} + \frac{\partial \mathbf{L}_0^{-1}}{\partial \theta_k} \mathbf{z}. \qquad [5.105]$$

Two main cases deserve attention, as detailed in section 5.4.2.2.1 and section 5.4.2.2.2 respectively:

1) the first case appears when θ_k is a correlation coefficient of $\mathbf{R}$. The components $z_i = \Phi^{-1}(F_{X_i}(x_i))$ of $\mathbf{z}$ for $i = 1, \ldots, n$ are expressed in terms of the parameters of the marginal distributions of $\mathbf{X}$, and therefore do not depend on θ_k. For this reason, the first term of equation [5.105] vanishes;

2) the second case appears when θ_k is a parameter of a given marginal distribution (e.g. its mean or standard deviation or, more generally, any parameter of this marginal distribution). The dependence of $\mathbf{z}$ on θ_k is clear from the transformation $(T_{vz} \circ T_{xv})$ defined in equation [5.105]. The calculation of the first term of equation [5.105] is required to assess the sensitivity to θ_k. The contribution of the second term is less clear and often neglected. It is however non-zero, due to μ_i, σ_i, μ_j and σ_j that appear in equation [5.30], which may imply a dependence of $\mathbf{R}_0$, and therefore of $\mathbf{L}_0^{-1}$, on θ_k.

5.4.2.2.1. Sensitivities w.r.t. correlation

Sensitivity to correlation in the framework of the Nataf transformation has been studied in very few works (Žanić and Žiha 1998, 2001; Bourinet and Lemaire 2008; Bourinet 2017). The methods proposed by Žanić and Žiha are based on the approximate formulas $F = \rho_{0,ij}/\rho_{ij}$ of Liu and Der Kiureghian (1986) derived for pairs of commonly used marginal distributions. Two methods are proposed in this work, one based on the Cholesky decomposition, as later used by Bourinet and Lemaire (2008) and presented in the following, and the other based on a spectral decomposition of $\mathbf{R}_0$. The objective here is to assess these sensitivities by numerical integration from the integral problem defined in equation [5.30], in order to obtain highly accurate sensitivities.

The sensitivity to a given correlation coefficient ρ_{ij} of $\mathbf{R}$ requires evaluating the following expression at $\mathbf{u}^*$:

$$\frac{\partial \mathbf{u}}{\partial \rho_{ij}} = \frac{\partial \mathbf{L}_0^{-1}}{\partial \rho_{ij}} \mathbf{z}. \qquad [5.106]$$

From $\mathbf{L}_0^{-1}\mathbf{L}_0 = \mathbf{I}$, where $\mathbf{I}$ is the $n \times n$ identity matrix, we can straightforwardly obtain:

$$\frac{\partial \mathbf{L}_0^{-1}}{\partial \rho_{ij}} = -\mathbf{L}_0^{-1} \frac{\partial \mathbf{L}_0}{\partial \rho_{ij}} \mathbf{L}_0^{-1}. \qquad [5.107]$$

This expression shows that the sensitivity of $\mathbf{L}_0^{-1}$ w.r.t. ρ_{ij} can be obtained from that of $\mathbf{L}_0$. The evaluation of $\partial \mathbf{L}_0 / \partial \rho_{ij}$ is carried out in two steps, as described in the following.

In the first step, the sensitivity of $\mathbf{R}_0$ is derived from equation [5.30]. All the elements of the matrix $\partial \mathbf{R}_0 / \partial \rho_{ij}$ are zeros except the one in the ith row and jth column, which can be expressed as follows:

$$\frac{\partial \rho_{0,ij}}{\partial \rho_{ij}} = \left(\frac{\partial \rho_{ij}}{\partial \rho_{0,ij}} \right)^{-1}. \qquad [5.108]$$

The sensitivity $\partial \rho_{ij} / \partial \rho_{0,ij}$ is assessed from the following equation by numerical integration with the same rule that was used to solve equation [5.30] (see equation [5.31]):

$$\frac{\partial \rho_{ij}}{\partial \rho_{0,ij}} = \int_{\mathbb{R}} \int_{\mathbb{R}} h\left(z_i, z_j, \mu_i, \mu_j, \sigma_i, \sigma_j\right) \frac{\partial \varphi_2\left(z_i, z_j, \rho_{0,ij}\right)}{\partial \rho_{0,ij}} \mathrm{d}z_i \mathrm{d}z_j. \quad [5.109]$$

In the second step, the sensitivity of $\mathbf{L}_0$ w.r.t. ρ_{ij} is obtained from that of $\mathbf{R}_0$ through a step-by-step differentiation of the Cholesky algorithm. The algorithm used to calculate both $\mathbf{L}_0$ and $\partial \mathbf{L}_0 / \partial \rho_{ij}$ can be found in Bourinet (2017).

5.4.2.2.2. Sensitivities w.r.t. other distribution parameters

For $k = 1, \ldots, n_{\boldsymbol{\theta}}$, we want to assess $\partial \mathbf{u} / \partial \theta_k$ in equation [5.105], where θ_k represents any given marginal distribution parameter. Evaluating the first term $\mathbf{L}_0^{-1} \partial \mathbf{z} / \partial \theta_k$ is quite standard from the following expression obtained from equation [5.22], for $i = 1, \ldots, n$:

$$\frac{\partial z_i}{\partial \theta_k} = \frac{1}{\varphi\left(F_{X_i}\left(x_i\right)\right)} \frac{\partial F_{X_i}\left(x_i\right)}{\partial \theta_k}, \quad [5.110]$$

in which $\partial F_{X_i} / \partial \theta_k$ has to be derived for any usual statistical distribution.

Evaluating the second term $(\partial \mathbf{L}_0^{-1} / \partial \theta_k) \mathbf{z}$ is less easy. The sensitivity of $\mathbf{L}_0^{-1}$ to θ_k is obtained in three steps, similarly to the process for the sensitivity of $\mathbf{L}_0^{-1}$ to correlation in the previous section:

1) we first express the sensitivity $\dfrac{\partial \mathbf{R}_0}{\partial \theta_k}$, as described in the following;

2) $\dfrac{\partial \mathbf{L}_0}{\partial \theta_k}$ is obtained from $\mathbf{R}_0$ and $\dfrac{\partial \mathbf{R}_0}{\partial \theta_k}$ by means of the algorithm given in Bourinet (2017);

3) $\dfrac{\partial \mathbf{L}_0^{-1}}{\partial \theta_k}$ is obtained from $\dfrac{\partial \mathbf{L}_0}{\partial \theta_k}$ (see equation [5.107]).

The sensitivity of $\mathbf{R}_0$ to θ_k is obtained from the following equation. For any $i, j = 1, \ldots, n$, we have:

$$\frac{\partial \rho_{ij}}{\partial \theta_k} = \int_{\mathbb{R}} \int_{\mathbb{R}} \left[\frac{\partial h\left(z_i, z_j, \mu_i, \mu_j, \sigma_i, \sigma_j\right)}{\partial \theta_k} \varphi_2\left(z_i, z_j, \rho_{0,ij}\right) + \right. \\ \left. h\left(z_i, z_j, \mu_i, \mu_j, \sigma_i, \sigma_j\right) \frac{\partial \varphi_2\left(z_i, z_j, \rho_{0,ij}\right)}{\partial \theta_k} \right] \mathrm{d}z_i \mathrm{d}z_j, \quad [5.111]$$

where:

$$\frac{\partial h}{\partial \theta_k} = \frac{\partial h}{\partial z_i}\frac{\partial z_i}{\partial \theta_k} + \frac{\partial h}{\partial z_j}\frac{\partial z_j}{\partial \theta_k} + \frac{\partial h}{\partial \mu_i}\frac{\partial \mu_i}{\partial \theta_k} + \frac{\partial h}{\partial \mu_j}\frac{\partial \mu_j}{\partial \theta_k} + \frac{\partial h}{\partial \sigma_i}\frac{\partial \sigma_i}{\partial \theta_k} + \frac{\partial h}{\partial \sigma_j}\frac{\partial \sigma_j}{\partial \theta_k}, \quad [5.112]$$

and:

$$\frac{\partial \varphi_2}{\partial \theta_k} = \frac{\partial \varphi_2}{\partial z_i}\frac{\partial z_i}{\partial \theta_k} + \frac{\partial \varphi_2}{\partial z_j}\frac{\partial z_j}{\partial \theta_k} + \frac{\partial \varphi_2}{\partial \rho_{0,ij}}\frac{\partial \rho_{0,ij}}{\partial \theta_k}. \quad [5.113]$$

It is important to note that the unknown sensitivity $\partial \rho_{0,ij}/\partial \theta_k$ (i.e. the element of $\partial \mathbf{R}_0/\partial \theta_k$ in the ith row and jth column) appears in the integrand of equation [5.111] (see last term of equation [5.113]). This sensitivity is obtained by solving equation [5.111] by numerical integration, in which we have $\partial \rho_{ij}/\partial \theta_k = 0$. We again use the same rule for numerical integration as defined for the calculation of $\mathbf{R}_0$. Note that the partial derivatives of μ_i, μ_j, σ_i and σ_j w.r.t. θ_k in equation [5.112] are all zeros if θ_k is a distribution parameter of a component of $\mathbf{X}$, other than X_i or X_j.

From the examples investigated in Bourinet (2017), it is found that the contribution of the second term of equation [5.105] is only important when correlated components with sufficiently large variances are present in $\mathbf{X}$. Accounting for this second term is only required when at least one component of $\mathbf{X}$ has a distribution belonging to Group 2, as defined by Liu and Der Kiureghian (1986). As a reminder, this group includes the lognormal, gamma, type-II largest value and type-III smallest value distributions. When X_i belongs to Group 2, $\rho_{0,ij}$ (therefore $\mathbf{R}_0$) depends on the coefficient of variation δ_i of X_i, see approximate expressions of $F = \rho_{0,ij}/\rho_{ij}$ given by Liu and Der Kiureghian. This implies a dependence of $\mathbf{L}_0^{-1}$ on the marginal distribution parameters of X_i and, as a consequence, a non-zero sensitivity of $\mathbf{L}_0^{-1}$ w.r.t. those parameters.

5.4.3. *Crude MC simulation and subset simulation*

5.4.3.1. *Crude MC sensitivities of p_f w.r.t. distribution parameters: the score function*

The sensitivities of p_f w.r.t. distribution parameters can be assessed as a simple post-processing of a crude MC simulation, by means of the score function introduced by Rubinstein (1976, 1986). Such an approach was later brought to the structural reliability community by Wu (1994).

We wish to assess the sensitivity of p_{f} w.r.t. a given distribution parameter θ_k of $\boldsymbol{\theta}$:

$$\frac{\partial p_{\mathrm{f}}}{\partial \theta_k} = \frac{\partial}{\partial \theta_k} \int_{\mathcal{X}} \mathbb{1}_{\mathcal{F}_{\mathbf{x}}}(\mathbf{x})\, f_{\mathbf{X}}(\mathbf{x})\, \mathrm{d}\mathbf{x}. \tag{5.114}$$

Assuming that (i) the joint PDF of $\mathbf{X}$ is continuously differentiable w.r.t. θ_k, and that (ii) the integration domain $\mathcal{X}$ does not depend on θ_k, we can write:

$$\frac{\partial p_{\mathrm{f}}}{\partial \theta_k} = \int_{\mathcal{X}} \mathbb{1}_{\mathcal{F}_{\mathbf{x}}}(\mathbf{x})\, \frac{\partial f_{\mathbf{X}}(\mathbf{x})}{\partial \theta_k}\, \mathrm{d}\mathbf{x}. \tag{5.115}$$

It is important to point out that assumption (ii) is not met for truncated distributions. An example of such distributions is the uniform distribution for which the bounds of the integration domain depend on its statistical moments (see, e.g. (Lee *et al.* 2011a,b)).

This integral can be assessed by IS with the joint PDF $f_{\mathbf{X}}$, as detailed in the first line of equation [5.116]. It is worth pointing out that this IS PDF is the same as that used in a crude MC simulation, which therefore implies that the sample drawn for the estimation of p_{f} by crude MC can also be used to estimate the sensitivity of p_{f} w.r.t. any $\theta_k \in \boldsymbol{\theta}$. The function $\kappa_{\theta_k}(\mathbf{x}) = \partial \ln(f_{\mathbf{X}}(\mathbf{x}))/\partial \theta_k$, which is introduced in line 3 of this equation, is known as the *score function*.

$$\begin{aligned}
\frac{\partial p_{\mathrm{f}}}{\partial \theta_k} &= \int_{\mathcal{X}} \mathbb{1}_{\mathcal{F}_{\mathbf{x}}}(\mathbf{x})\, \frac{\dfrac{\partial f_{\mathbf{X}}(\mathbf{x})}{\partial \theta_k}}{f_{\mathbf{X}}(\mathbf{x})}\, f_{\mathbf{X}}(\mathbf{x})\, \mathrm{d}\mathbf{x} \\
&= \int_{\mathcal{X}} \mathbb{1}_{\mathcal{F}_{\mathbf{x}}}(\mathbf{x})\, \frac{\partial \ln(f_{\mathbf{X}}(\mathbf{x}))}{\partial \theta_k}\, f_{\mathbf{X}}(\mathbf{x})\, \mathrm{d}\mathbf{x} \\
&= \mathbb{E}_{f_{\mathbf{X}}}\left[\mathbb{1}_{\mathcal{F}_{\mathbf{x}}}(\mathbf{X})\, \kappa_{\theta_k}(\mathbf{X})\right].
\end{aligned} \tag{5.116}$$

The corresponding MC estimator of $\partial p_{\mathrm{f}}/\partial \theta_k$ is given by:

$$\widehat{\frac{\partial p_{\mathrm{f}}}{\partial \theta_k}} = \frac{1}{N} \sum_{j=1}^{N} \mathbb{1}_{\mathcal{F}_{\mathbf{x}}}\left(\mathbf{X}^{(j)}\right) \kappa_{\theta_k}\left(\mathbf{X}^{(j)}\right), \tag{5.117}$$

where $\{\mathbf{X}^{(j)}, 1 \leq j \leq N\}$ are N independent copies of the random vector $\mathbf{X}$, i.e. $\mathbf{X}^{(1)}, \ldots, \mathbf{X}^{(N)} \overset{\text{i.i.d.}}{\sim} f_{\mathbf{X}}$, and N denotes the sample size.

5.4.3.2. *SS sensitivities of* p_{f} w.r.t. *distribution parameters*

A similar approach can be applied to SS to derive the sensitivities of p_{f} w.r.t. distribution parameters, as proposed by Song *et al.* (2009), also see the presentation by Dubourg (2011, pp. 161–163). From equation [5.53], the sensitivity of the failure probability p_{f} w.r.t. a given distribution parameter $\theta_k \in \boldsymbol{\theta}$ can be expressed as follows:

$$\frac{\partial p_{\mathrm{f}}}{\partial \theta_k} = \sum_{s=1}^{m} \frac{p_{\mathrm{f}}}{p_s} \frac{\partial p_s}{\partial \theta_k}, \qquad [5.118]$$

where $p_1 = \mathbb{P}(E_1)$ and $p_s = \mathbb{P}(E_s|E_{s-1})$ for $s = 2, \ldots, m$, and where the partial derivatives of the intermediate probabilities p_s for $s = 1, \ldots, m$ are defined by recurrence (Song *et al.* 2009):

$$\frac{\partial p_1}{\partial \theta_k} = \mathbb{E}_{\varphi_n}\left[\mathbb{1}_{\mathcal{F}_{\mathrm{u},1}}(\mathbf{U})\, K_{\theta_k}(\mathbf{U})\right], \qquad [5.119]$$

and, for $s = 2, \ldots, m$:

$$\frac{\partial p_s}{\partial \theta_k} = \mathbb{E}_{\varphi_n(\cdot|E_{s-1})}\left[\mathbb{1}_{\mathcal{F}_{\mathrm{u},s}}(\mathbf{U})\left(K_{\theta_k}(\mathbf{U}) - \sum_{r=1}^{s-1} \frac{1}{p_r} \frac{\partial p_r}{\partial \theta_k}\right)\right], \qquad [5.120]$$

where $K_{\theta_k}(\mathbf{u}) = (\kappa_{\theta_k} \circ T^{-1})(\mathbf{u})$ denotes the score function expressed in the standard normal space.

The corresponding sensitivity estimator at the first level of SS is:

$$\widehat{\frac{\partial p_1}{\partial \theta_k}} = \frac{1}{N} \sum_{j=1}^{N} \left[\mathbb{1}_{\mathcal{F}_{\mathrm{u}}}\left(\mathbf{U}^{(j)}\right) K_{\theta_k}\left(\mathbf{U}^{(j)}\right)\right], \qquad [5.121]$$

where $\{\mathbf{U}^{(j)}, 1 \leq j \leq N\}$ are N independent copies of the random vector $\mathbf{U}$ at the first SS level, i.e. such that $\mathbf{U}^{(1)}, \ldots, \mathbf{U}^{(N)} \overset{\text{i.i.d.}}{\sim} \varphi_n$, and N denote the sample size of the SS levels.

At the next levels $s = 2, \ldots, m$, the sensitivity estimator writes:

$$\widehat{\frac{\partial p_s}{\partial \theta_k}} = \frac{1}{N} \sum_{j=1}^{N} \left[\mathbb{1}_{\mathcal{F}_{\mathrm{u},s}}\left(\mathbf{U}^{(j)}\right)\left(K_{\theta_k}\left(\mathbf{U}^{(j)}\right) - \sum_{r=1}^{s-1} \frac{1}{p_r} \widehat{\frac{\partial p_r}{\partial \theta_k}}\right)\right], \qquad [5.122]$$

where $\{\mathbf{U}^{(j)}, 1 \leq j \leq N\}$ are N independent copies of the random vector $\mathbf{U}$ at level s, i.e. such that $\mathbf{U}^{(1)}, \ldots, \mathbf{U}^{(N)} \overset{\text{i.i.d.}}{\sim} \varphi_n(\cdot|E_{s-1})$.

5.5. References

Au, S.-K. and Beck, J.L. (1999). A new adaptive importance sampling scheme for reliability calculations. *Structural Safety*, 21(2), 135–158.

Au, S.-K. and Beck, J.L. (2001). Estimation of small failure probabilities in high dimensions by subset simulation [Online]. *Probabilistic Engineering Mechanics*, 16(4), 263–277. Available at: http://dx.doi.org/10.1016/S0266-8920(01)00019-4.

Au, S.-K. and Beck, J.L. (2003). Important sampling in high dimensions [Online]. *Structural Safety*, 25(2), 139–163. Available at: http://dx.doi.org/10.1016/S0167-4730(02)00047-4.

Balesdent, M., Morio, J., Marzat, J. (2013). Kriging-based adaptive importance sampling algorithms for rare event estimation [Online]. *Structural Safety*, 44, 1–10. Available at: http://dx.doi.org/10.1016/j.strusafe.2013.04.001.

de Boer, P.-T., Kroese, D.P., Mannor, S., Rubinstein, R.Y. (2005). A tutorial on the cross-entropy method [Online]. *Annals of Operations Research*, 134(1), 19–67. Available at: http://dx.doi.org/10.1007/s10479-005-5724-z.

Botev, Z.I. and Kroese, D.P. (2012). Efficient Monte Carlo simulation via the generalized splitting method [Online]. *Statistics and Computing*, 22(1), 1–16. Available at: http://dx.doi.org/10.1007/s11222-010-9201-4.

Botev, Z.I. and L'Ecuyer, P. (2011). Semi-parametric importance sampling for rare-event probability estimation [Online]. In *Proceedings of 8th IMACS Seminar on Monte Carlo Methods*. Sabelfeld, K., Dimov, I. (eds). De Gruyter, Borovets, Bulgaria, August 29–September 2. Available at: https://www.degruyter.com/view/journals/mcma/mcma-overview.xml.

Botev, Z.I., Kroese, D.P., Taimre, T. (2007). Generalized cross-entropy methods with applications to rare-event simulation and optimization [Online]. *Simulation*, 83(11), 785–806. Available at: http://dx.doi.org/10.1177/0037549707087067.

Bourinet, J.-M. (2017). FORM sensitivities to distribution parameters with the Nataf transformation [Online]. In *Risk and Reliability Analysis: Theory and Applications, Edited Volume in Honor of Prof. Armen Der Kiureghian,* Gardoni, P. (ed.). Springer International Publishing, Cham, Switzerland, 277–302. Available at: https://link.springer.com/chapter/10.1007%2F978-3-319-52425-2_12.

Bourinet, J.-M. and Lemaire, M. (2008). FORM sensitivities to correlation: Application to fatigue crack propagation based on Virkler data. In *Proceedings of 4th International ASRANet Colloquium*, Das, P.K. (ed.). Athens, Greece, June 25–27.

Bourinet, J.-M., Mattrand, C., Dubourg, V. (2009). A review of recent features and improvements added to FERUM software [Online]. In *Proceedings of 10th International Conference on Structural Safety and Reliability (ICOSSAR 2009)*, Furuta, H., Frangopol, D.M., Shinozuka, M. (eds). CRC Press, Osaka, Japan, September 13–17. Available at: https://www.routledge.com/Safety-Reliability-and-Risk-of-Structures-Infrastructures-and-Engineering/Furuta-Frangopol-Shinozuka/p/book/9780415475570.

Breitung, K. (1984). Asymptotic approximations for multinormal integrals [Online]. *Journal of Engineering Mechanics*, 110(3), 357–366. Available at: http://dx.doi.org/10.1061/(ASCE)0733-9399(1984)110:3(357).

Bucher, C.G. (1988). Adaptive sampling – An iterative fast Monte Carlo procedure [Online]. *Structural Safety*, 5(2), 119–126. Available at: http://dx.doi.org/10.1016/0167-4730(88)90020-3.

Cadini, F., Santos, F., Zio, E. (2014). An improved adaptive Kriging-based importance technique for sampling multiple failure regions of low probability [Online]. *Reliability Engineering & System Safety*, 131, 109–117. Available at: http://dx.doi.org/10.1016/j.ress.2014.06.023.

Cancela, H., El Khadiri, M., Rubino, G. (2009). *Rare Event Analysis by Monte Carlo Techniques in Static Models* [Online]. John Wiley & Sons. Available at: http://dx.doi.org/10.1002/9780470745403.ch7.

Cérou, F. and Guyader, A. (2007). Adaptive multilevel splitting for rare event analysis [Online]. *Stochastic Analysis and Applications*, 25(2), 417–443. Available at: http://dx.doi.org/10.1080/07362990601139628.

Cérou, F., Del Moral, P., Furon, T., Guyader, A. (2012). Sequential Monte Carlo for rare event estimation [Online]. *Statistics and Computing*, 22(3), 795–808. Available at: http://dx.doi.org/10.1007/s11222-011-9231-6.

Chan, J.C.C., Glynn, P.W., Kroese, D.P. (2011). A comparison of cross-entropy and variance minimization strategies [Online]. *Journal of Applied Probability*, 48A, 183–194. Available at: http://dx.doi.org/10.1239/jap/1318940464.

Chib, S. and Greenberg, E. (1995). Understanding the Metropolis-Hastings algorithm [Online]. *The American Statistician*, 49(4), 327–335. Available at: http://dx.doi.org/10.1080/00031305.1995.10476177.

Chopin, N. (2002). A sequential particle filter method for static models. *Biometrika*, 89(3), 539–552. Available at: http://dx.doi.org/10.1093/biomet/89.3.539

Dai, H., Zhang, H., Wang, W. (2012). A support vector density-based importance sampling for reliability assessment [Online]. *Reliability Engineering & System Safety*, 106, 86–93. Available at: http://dx.doi.org/10.1016/j.ress.2012.04.011.

Del Moral, P., Doucet, A., Jasra, A. (2006). Sequential Monte Carlo samplers [Online]. *Journal of the Royal Statistical Society: Series B (Statistical Methodology)*, 68(3), 411–436. Available at: http://dx.doi.org/10.1111/j.1467-9868.2006.00553.x.

Der Kiureghian, A. (1999). Introduction to structural reliability. Ce229 class notes, Dept. of Civil and Environmental Engineering, University of California, Berkeley, USA.

Der Kiureghian, A. and Dakessian, T. (1998). Multiple design points in first and second-order reliability [Online]. *Structural Safety*, 20(1), 37–49. Available at: http://dx.doi.org/ 10.1016/S0167-4730(97)00026-X.

Der Kiureghian, A., Lin, H.-Z. and Hwang, S.-J. (1987). Second-order reliability approximations [Online]. *Journal of Engineering Mechanics*, 113(8), 1208–1225. Available at: http://dx.doi.org/10.1061/(ASCE)0733-9399(1987)113:8(1208).

Ditlevsen, O. and Madsen, H.O. (2007). Structural reliability methods, Internet Edition 2.3.7 [Online]. Available at: http://od-website.dk/books/OD-HOM-StrucRelMeth-Ed2.3.7.pdf.

Doucet, A., de Freitas, N., Gordon, N. (eds) (2001). *Sequential Monte Carlo Methods in Practice* [Online]. Springer, New York, USA. Available at: http://dx.doi.org/10.1007/978-1-4757-3437-9.

Dubourg, V. (2011). Adaptive surrogate models for reliability analysis and reliability-based design optimization [Online]. PhD Thesis, Blaise Pascal University, Clermont-Ferrand, France. Available at: https://tel.archives-ouvertes.fr/tel-00697026v2.

Dubourg, V. and Sudret, B. (2014). Meta-model-based importance sampling for reliability sensitivity analysis [Online]. *Structural Safety*, Special Issue In Honor of Prof. Wilson H. Tang, 49, 27–36. Available at: http://dx.doi.org/10.1016/j.strusafe.2013.08.010.

Dubourg, V., Sudret, B., Deheeger, F. (2013). Metamodel-based importance sampling for structural reliability analysis [Online]. *Probabilistic Engineering Mechanics*, 33, 47–57. Available at: http://dx.doi.org/10.1016/j.probengmech.2013.02.002.

Echard, B., Gayton, N., Lemaire, M. and Relun, N. (2013). A combined importance sampling and Kriging reliability method for small failure probabilities with time-demanding numerical models [Online]. *Reliability Engineering & System Safety*, 111, 232–240. Available at: http://dx.doi.org/10.1016/j.ress.2012.10.008.

Engelund, S. and Rackwitz, R. (1993). A benchmark study on importance sampling techniques in structural reliability [Online]. *Structural Safety*, 12(4), 255–276. Available at: http://dx.doi.org/10.1016/0167-4730(93)90056-7.

Guyader, A. (2011). Contributions à l'estimation non paramétrique et à la simulation d'évènements rares [Online]. HDR report (in French), Rennes 2 University, Rennes, France. Available at: http://www.lsta.upmc.fr/guyader/files/hdr.pdf.

Hasofer, A.M. and Lind, N.C. (1974). Exact and invariant second-moment code format. *Journal of the Engineering Mechanics Division*, 100(1), 111–121.

Hastings, W.K. (1970). Monte Carlo sampling methods using Markov chains and their applications [Online]. *Biometrika*, 57(1), 97–109. Available at: http://dx.doi.org/10.1093/biomet/57.1.97.

Haukaas, T. and Der Kiureghian, A. (2004). Finite element reliability and sensitivity methods for performance-based earthquake engineering. Peer report 2003/14, Pacific Earthquake Engineering Research Center, California, USA.

Hohenbichler, M. and Rackwitz, R. (1981). Non-normal dependent vectors in structural safety. *Journal of the Engineering Mechanics Division*, 107(6), 1227–1238.

Hohenbichler, M. and Rackwitz, R. (1986). Sensitivity and importance measures in structural reliability [Online]. *Civil Engineering Systems*, 3(4), 203–209. Available at: http://dx.doi.org/10.1080/02630258608970445.

Hohenbichler, M. and Rackwitz, R. (1988). Improvement of second-order reliability estimates by importance sampling [Online]. *Journal of Engineering Mechanics*, 114(12), 2195–2199. Available at: http://dx.doi.org/10.1061/(ASCE)0733-9399(1988)114:12(2195).

Homem-de-Mello, T. (2007). A study on the cross-entropy method for rare-event probability estimation [Online]. *INFORMS Journal on Computing*, 19(3), 381–394. Available at: http://dx.doi.org/10.1287/ijoc.1060.0176.

Homem-de-Mello, T. and Rubinstein, R.Y. (2002). Rare event estimation for static models via cross-entropy and importance sampling [Online]. Available at: http://citeseerx.ist.psu.edu/viewdoc/summary?doi=10.1.1.15.6923.

Johansen, A.M., Del Moral, P. and Doucet, A. (2005). Sequential Monte Carlo samplers for rare event estimation [Online]. Technical report cued/f-infeng/543, Department of Engineering, University of Cambridge, Trumpington, UK. Available at: http://citeseerx.ist.psu.edu/viewdoc/summary?doi=10.1.1.175.8309.

Kahn, H. and Harris, T.E. (1951). Estimation of particle transmission by random sampling. *National Bureau of Standards Applied Mathematics Series*, 12, 27–30.

Karamchandani, A., Bjerager, P., Cornell, C.A. (1989). Adaptive importance sampling. In *Proceedings of 5th International Conference on Structural Safety and Reliability (ICOSSAR '89), San Francisco, CA, USA, August 7–11, 1989,* Ang, A.H.-S., Shinozuka, M., and Schuëller, G. I. (eds). ASCE, New York, USA.

Katafygiotis, L.S. and Zuev, K.M. (2008). Geometric insight into the challenges of solving high-dimensional reliability problems. *Probabilistic Engineering Mechanics*, 23(2–3), 208–218.

Kroese, D.P., Taimre, T., Botev, Z.I. (2011). *Handbook of Monte Carlo Methods*, [Online]. Wiley, New York, USA. Available at: http://dx.doi.org/10.1002/9781118014967.

Kullback, S. and Leibler, R.A. (1951). On information and sufficiency [Online]. *The Annals of Mathematical Statistics*, 22(1), 79–86. Available at: http://dx.doi.org/10.1214/aoms/1177729694.

Kurtz, N. and Song, J. (2013). Cross-entropy-based adaptive importance sampling using Gaussian mixture [Online]. *Structural Safety*, 42, 35–44. Available at: http://dx.doi.org/10.1016/j.strusafe.2013.01.006.

Lebrun, R. (2013). Contributions à la modélisation de la dépendance stochastique [Online]. PhD Thesis (in english), Paris-Diderot University – Paris VII, France. Available at: https://tel.archives-ouvertes.fr/tel-00913510/.

Lebrun, R. and Dutfoy, A. (2009a). Do Rosenblatt and Nataf isoprobabilistic transformations really differ? [Online]. *Probabilistic Engineering Mechanics*, 24(4), 577–584. Available at: http://dx.doi.org/10.1016/j.probengmech.2009.04.006.

Lebrun, R. and Dutfoy, A. (2009b). A generalization of the Nataf transformation to distributions with elliptical copula [Online]. *Probabilistic Engineering Mechanics*, 24(2), 172–178. Available at: http://dx.doi.org/10.1016/j.probengmech.2008.05.001.

Lee, I., Choi, K.K., Noh, Y., Zhao, L., Gorsich, D. (2011a). Sampling-based stochastic sensitivity analysis using score functions for RBDO problems with correlated random variables [Online]. *Journal of Mechanical Design*, 133(2), 021003. Available at: http://dx.doi.org/10.1115/1.4003186.

Lee, I., Choi, K.K., Zhao, L. (2011b). Sampling-based RBDO using the stochastic sensitivity analysis and dynamic Kriging method [Online]. *Structural and Multidisciplinary Optimization*, 44(3), 299–317. Available at: http://dx.doi.org/10.1007/s00158-011-0659-2.

Lemaire, M., Chateauneuf, A., Mitteau, J.-C. (2010). *Structural Reliability* [Online]. ISTE Ltd London, UK, and John Wiley & Sons, New York, USA. Available at: https://onlinelibrary.wiley.com/doi/book/10.1002/9780470611708.

Liu, P.-L. and Der Kiureghian, A. (1986). Multivariate distribution models with prescribed marginals and covariance [Online]. *Probabilistic Engineering Mechanics*, 1(2), 105–112. Available at: http://dx.doi.org/10.1016/0266-8920(86)90033-0.

Liu, P.-L. and Der Kiureghian, A. (1991). Optimization algorithms for structural reliability [Online]. *Structural Safety*, 9(3), 161–177. http://dx.doi.org/10.1016/0167-4730(91)90041-7.

Melchers, R.E. (1989a). Importance sampling in structural systems [Online]. *Structural Safety*, 6(1), 3–10. Available at: http://dx.doi.org/10.1016/0167-4730(89)90003-9.

Melchers, R.E. (1989b). Improved importance sampling methods for structural system reliability calculation. In *Proceedings of 5th International Conference on Structural Safety and Reliability (ICOSSAR '89)*, San Francisco, CA, USA, August 7–11, Ang, A.H.-S., Shinozuka, M., and Schuëller, G.I. (eds). ASCE, New York, USA, 1185–1192.

Melchers, R.E. (1990). Search-based importance sampling [Online]. *Structural Safety*, 9(2), 117–128. Available at: http://dx.doi.org/10.1016/0167-4730(90)90003-8.

Metropolis, N., Rosenbluth, A.W., Rosenbluth, M.N., Teller, A.H., Teller, E. (1953). Equation of state calculations by fast computing machines [Online]. *The Journal of Chemical Physics*, 21(6), 1087–1092. Available at: http://dx.doi.org/10.1063/1.1699114.

Miao, F. and Ghosn, M. (2011). Modified subset simulation method for reliability analysis of structural systems [Online]. *Structural Safety*, 33(4–5), 251–260. Available at: http://dx.doi.org/10.1016/j.strusafe.2011.02.004.

Morio, J. (2011). Non-parametric adaptive importance sampling for the probability estimation of a launcher impact position [Online]. *Reliability Engineering & System Safety*, Special Issue on Safecomp 2008, 96(1), 178–183. Available at: http://dx.doi.org/10.1016/j.ress.2010.08.006.

Morio, J. (2012). Extreme quantile estimation with nonparametric adaptive importance sampling [Online]. *Simulation Modelling Practice and Theory*, 27, 76–89. Available at: http://dx.doi.org/10.1016/j.simpat.2012.05.008.

Nataf, A. (1962). Détermination des distributions dont les marges sont données. *Comptes rendus de l'Académie des Sciences*, 225, 42–43.

Nelsen, R.B. (1999). *An Introduction to Copulas*, Springer, New York, USA. Available at: http://dx.doi.org/10.1007/0-387-28678-0.

Papaioannou, I., Betz, W., Zwirglmaier, K., Straub, D. (2015). MCMC algorithms for subset simulation [Online]. *Probabilistic Engineering Mechanics* 41, 89–103. Available at: http://dx.doi.org/10.1016/j.probengmech.2015.06.006.

Rackwitz, R. and Fiessler, B. (1978). Structural reliability under combined random load sequences [Online]. *Computers & Structures*, 9(5), 489–494. Available at: http://dx.doi.org/10.1016/0045-7949(78)90046-9.

Robert, C.P. and Casella, G. (2004). *Monte Carlo Statistical Methods*, 2nd edition. Springer, New York, USA. Available at: http://dx.doi.org/10.1007/978-1-4757-4145-2.

Rosenblatt, M. (1952). Remarks on a multivariate transformation. *The Annals of Mathematical Statistics*, 23(3), 470–472.

Rubinstein, R.Y. (1976). A Monte Carlo method for estimating the gradient in a stochastic network. Unpublished manuscript, Technion, Haifa, Israel.

Rubinstein, R.Y. (1986). The score function approach for sensitivity analysis of computer simulation models [Online]. *Mathematics and Computers in Simulation*, 28(5), 351–379. Available at: http://dx.doi.org/10.1016/0378-4754(86)90072-8.

Rubinstein, R.Y. (1997). Optimization of computer simulation models with rare events [Onine]. *European Journal of Operational Research*, 99(1), 89–112. Available at: http://dx.doi.org/10.1016/S0377-2217(96)00385-2.

Rubinstein, R.Y. (1999). The cross-entropy method for combinatorial and continuous optimization. *Methodology and Computing in Applied Probability*, 1(2), 127–190.

Rubinstein, R.Y. and Kroese, D.P. (2007). *Simulation and the Monte Carlo Method*, 2nd edition John Wiley & Sons, New York, USA. Available at: http://dx.doi.org/10.1002/9780470230381.

Saltelli, A., Chan, K., Scott, E.M. (2008). *Sensitivity Analysis*. John Wiley & Sons, New York, USA.

Santoso, A.M., Phoon, K.K., Quek, S.T. (2011). Modified Metropolis-Hastings algorithm with reduced chain correlation for efficient subset simulation [Online]. *Probabilistic Engineering Mechanics*, 26(2), 331–341. Available at: http://dx.doi.org/10.1016/j.probengmech.2010.08.007.

Schuëller, G.I. and Stix, R. (1987). A critical appraisal of methods to determine failure probabilities [Online]. *Structural Safety*, 4(4), 293–309. Available at: http://dx.doi.org/10.1016/0167-4730(87)90004-X.

Sklar, A. (1959). Fonctions de répartition à n dimensions et leurs marges. *Publications de l'Institut de Statistique de l'Université de Paris*, Paris, France, 8, 229–231.

Sobol'. I.M. (1993). Sensitivity estimates for non-linear mathematical models. *Mathematical Modelling and Computational Experiments*, 1(4), 407–414.

Song, S., Lu, Z., Qiao, H. (2009). Subset simulation for structural reliability sensitivity analysis [Online]. *Reliability Engineering & System Safety*, 94(2), 658–665. Available at: http://dx.doi.org/10.1016/j.ress.2008.07.006.

Tierney, L. (1994). Markov chains for exploring posterior distributions [Online]. *The Annals of Statistics*, 22(4), 1701–1728. Available at: http://dx.doi.org/10.1214/aos/1176325750.

Tvedt, L. (1988). Second order reliability by an exact integral. In *Proceedings of 2nd IFIP WG 7.5 Working Conference on Reliability and Optimization of Structural Systems*, London, UK, September 26–28, Thoft-Christensen, P. (ed.). Springer, Berlin, Heidelberg, 377–384.

Walter, C. (2015). Moving particles: A parallel optimal multilevel splitting method with application in quantiles estimation and meta-model based algorithms [Online]. *Structural Safety*, 55, 10–25. Available at: http://dx.doi.org/10.1016/j.strusafe.2015.02.002.

Wu, Y.-T. (1994). Computational methods for efficient structural reliability and reliability sensitivity analysis [Online]. *AIAA Journal*, 32(8), 1717–1723. Available at: http://dx.doi.org/10.2514/3.12164.

Žanić, V. and Žiha, K. (1998). Sensitivity to correlation in multivariate models. *Computer Assisted Mechanics and Engineering Sciences*, 5(1), 75–84.

Žanić, V. and Žiha, K. (2001). Sensitivity to correlations in structural problems [Online]. *Transactions of FAMENA*, 25(2), 1–26. Available at: https://bib.irb.hr/datoteka/87824.87824-FAMENA2001V25N2.pdf.

Zhang, Y. and Der Kiureghian, A. (1993). Dynamic response sensitivity of inelastic structures [Online]. *Computer Methods in Applied Mechanics and Engineering*, 108(1), 23–36. Available at: https://doi.org/10.1016/0045-7825(93)90151-M.

Zhang, Y. and Der Kiureghian, A. (1994). Two improved algorithms for reliability analysis [Online]. In *Proceedings of 6th IFIP WG 7.5 Working Conference on Reliability and Optimization of Structural Systems*, Assisi, Italy, September 7–9, Rackwitz, R., Augusti, G., and Borri, A. (eds). Springer, USA. Available at: https://www.springer.com/gp/book/9780412636301.

Zou, T., Mahadevan, S., Mourelatos, Z.P. (2003). Reliability-based evaluation of automotive wind noise quality [Online]. *Reliability Engineering & System Safety* 82(2), 217–224. Available at: http://dx.doi.org/10.1016/S0951-8320(03)00170-4.

Zuev, K.M. and Katafygiotis, L.S. (2011a). The horseracing simulation algorithm for evaluation of small failure probabilities [Online]. *Probabilistic Engineering Mechanics*, 26(2), 157–164. Available at: http://dx.doi.org/10.1016/j.probengmech.2010.11.004.

Zuev, K.M. and Katafygiotis, L.S. (2011b). Modified Metropolis-Hastings algorithm with delayed rejection [Online]. *Probabilistic Engineering Mechanics*, 26(3), 405–412. Available at: http://dx.doi.org/10.1016/j.probengmech.2010.11.008.

Zuev, K.M., Beck, J.L., Au, S.-K., Katafygiotis, L.S. (2012). Bayesian post-processor and other enhancements of subset simulation for estimating failure probabilities in high dimensions [Online]. *Computers & Structures* 92–93, 283–296. Available at: http://dx.doi.org/10.1016/j.compstruc.2011.10.017.

6

Adaptive Kriging-based Methods for Failure Probability Evaluation: Focus on AK Methods

Cécile MATTRAND, Pierre BEAUREPAIRE and Nicolas GAYTON
SIGMA Clermont, Clermont-Ferrand, France

This chapter provides the reader with a fairly brief overview of recent adaptive methods for calculating failure probabilities based on the use of Kriging (so-called *Gaussian processes*) with a focus on so-called AK methods (active learning and Kriging-based method). For more information, the reader may refer to the PhD thesis of Lelièvre (2018), who is gratefully acknowledged for his work that contributed to this chapter.

6.1. Introduction

To evaluate the probability of the failure of a structure (or more broadly, the probability of the occurrence of a feared scenario), simulation methods remain essential methods that deal specially with problems with a complex topology (strong nonlinearity, several most probable points of failure, system reliability, etc.). The Monte Carlo simulation (MCS) method is the reference method that can theoretically address any type of problem. Its main disadvantage is the computational time required, especially for low probabilities. To reduce the computational time of the MCS method, a very large number of alternatives have

Mechanical Engineering under Uncertainties,
coordinated by Christian GOGU.

been proposed in recent years. The monotonous reliability method (MRM) proposed in De Rocquigny (2009) mainly allows for a reduction in the computational time of the MCS method when the monotony of the performance function is established (often the case for structural mechanics problems). Importance sampling (IS) methods (see, for example, Lemaire 2009) can be used to considerably reduce the computational time of the MCS method under the assumption of a particular topology of the performance function (single most probable point of failure, etc.). More recently, the subset simulations (SS) method (see Au and Beck 2001) seems to be the most successful method to reduce computational times without assumptions about the nature of the limit state.

However, all these simulation methods can hardly be considered for addressing problems, industrial or not, involving performance functions that are both complex and very costly in terms of computational time (finite element-based models). For such problems, the use of substitute models (also called response surface, metamodels, surrogate models, etc.) seems to have been gaining consensus in the scientific community for several years. The objective is to replace the costly model that is to be evaluated, by a model without computational costs and that is representative of the true model by means of a learning step.

The difficulty with these techniques lies in controlling the substitution error. The idea of substitution is not new. FORM/SORM methods make use of this principle since they aim to replace the performance function with a local, linear or quadratic model around the most probable failure point, a model from which the probability of failure finds an explicit form ($P_f = \Phi(-\beta)$ for FORM). The widespread use of polynomial response surfaces was also very fashionable in the 2000s (Bucher and Bourgund 1990). These methods quickly suffered from the "lack of flexibility" of polynomials, requiring one to increase their degree and thus leading to overfitting problems.

Subsequently, polynomial chaos techniques (Blatman and Sudret 2008) have also emerged with great efficiency, especially for global sensitivity analysis (Sudret 2008). After the work on optimization by Jones *et al.* (1998), the work by Romero *et al.* (2004) and Kaymaz (2005) were the first to highlight the full potential of using Kriging methods (so-called Gaussian processes) as surrogate models for the uncertainty propagation. The advantage of Kriging is the availability of the variance of estimation that constitutes a local indicator without calculating the quality of the substitution model. This estimation variance offers very interesting possibilities to define multiple adaptive enrichment criteria to identify enrichment points in an area close to the limit state.

The common feature of simulation methods (MCS, MRM, IS, SS) is the need to classify points according to the sign of the performance function. Based on this observation, the evaluation of a probability of failure therefore consists of classifying the points in the most economical possible way. This is the purpose of AK methods, presented in this chapter.

These methods form the AK family (see Figure 6.1), all based on the same classification principle, and which, because of their simplicity of implementation and their effectiveness, seem to be of interest to the scientific community, which regularly proposes new improvements to this family. These methods are primarily intended for the calculation of failure probabilities, but also for use in other fields (verification of geometric conformity, optimization under constraints, management of manufacturing deviations, etc.).

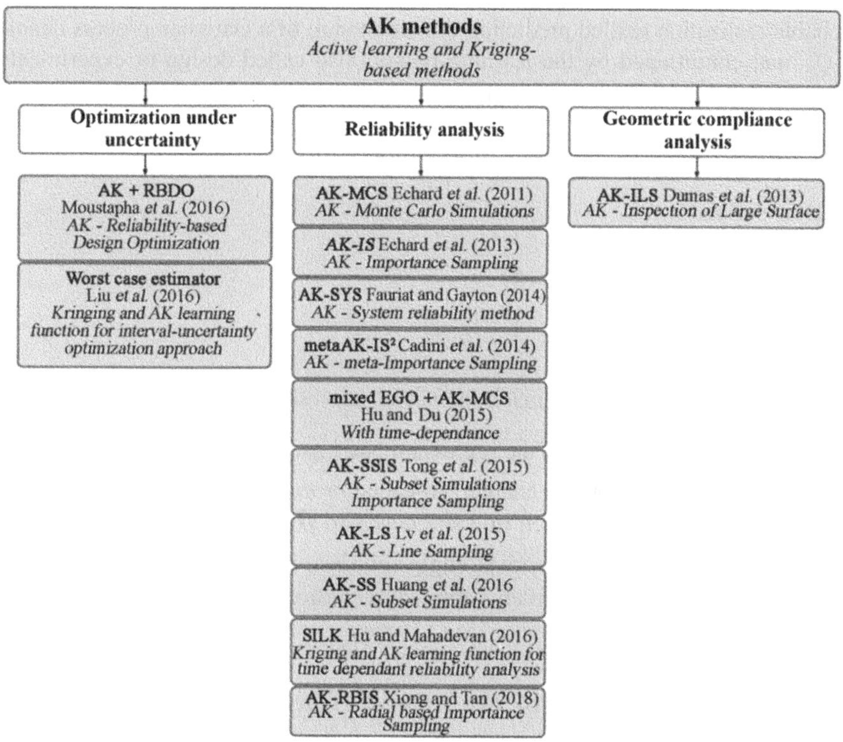

Figure 6.1. *Presentation of the AK family of methods (active learning and Kriging-based methods)*

This chapter is organized into several sections to provide a good understanding and illustration of AK methods. First of all, it will comprise a few elements to present the principle of Kriging. Then the various enrichment criteria used in the literature, the principle of generic classification of AK methods and the various basic methods (AK-MCS, AK-IS, AK-SYS and AK-HDMR1) will be presented.

6.2. Presentation of Kriging

6.2.1. *Principle*

Kriging, developed by Krige (1951) and theorized by Matheron (1973), allows the approximation of the performance function $G(\mathbf{X}(\omega))$ of the reliability problem, ω designating the uncertain and $\mathbf{X}(\omega)$ a vector of random variables, by the most probable realization (called prediction or metamodel) of a Gaussian process denoted by $\mathcal{G}$, and conditioned by the learning points (also called design of experiments). The objective of such a metamodel is to predict the value of the performance function at an unobserved point $\mathbf{x} \in \mathbb{R}^n$, n being the number of variables of the problem, from a set of m training points $\mathbf{x}_a^{(i)}, i = 1,...,m$. The Gaussian process model is defined by:

$$\mathcal{G}(\mathbf{x},\omega) = \mathbf{F}(\mathbf{x})^T \boldsymbol{\beta} + Z(\mathbf{x},\omega) \qquad [6.1]$$

where:

– $\mathbf{F}(\mathbf{x})^T \boldsymbol{\beta}$ is the deterministic part and represents the mean trend of the process, $\mathbf{F}(\mathbf{x})^T$ is the vector of the regression basis and $\boldsymbol{\beta}$ is the vector of regression coefficients. In the rest of this chapter, so-called *ordinary* Kriging is used, in which the deterministic part is reduced to a single scalar $\boldsymbol{\beta} \in \mathbb{R}$ independent of $\mathbf{x}$. This model is often sufficient even with strongly nonlinear functions (Bichon *et al.* 2011), and nonlinearity is then exclusively captured by the autocorrelation function;

– $Z(\mathbf{x},\omega)$ is a stationary scalar Gaussian process of zero mean and covariance function of the $C_{ZZ}(\mathbf{x}^{(1)},\mathbf{x}^{(2)}) = \sigma^2 R(\mathbf{x}^{(1)} - \mathbf{x}^{(2)},\boldsymbol{\theta})$ type where σ^2 corresponds to the process variance. Given that the process is stationary, the auto-correlation function R depends only on the difference between the points $\mathbf{x}^{(1)} - \mathbf{x}^{(2)}$ and on a set of hyperparameters $\theta \in \mathbb{R}_+^n$ for so-called *anisotropic* Kriging. There are many existing auto-correlation functions (called *kernels*) (exponential, Matérn, Gaussian, etc.).

The Gaussian kernel is recommended for fairly regular functions to be approximated. Its expression in the anisotropic case is given as follows:

$$R(\mathbf{x}^{(1)},\mathbf{x}^{(2)},\boldsymbol{\theta}) = \exp(-\sum_{i=1}^{n}\theta_i(x_i^{(1)} - x_i^{(2)})^2) \quad [6.2]$$

6.2.2. *Identification of Kriging hyperparameters*

Let $\mathbf{x}_a^{(i)}$ denote a set of m training points, associated with the response $G_a^{(i)} = G(\mathbf{x}_a^{(i)}), i = 1,...,m$. The identification of the hyperparameters β, σ and $\boldsymbol{\theta} = \{\theta_i, i = 1,...,m\}$ can be carried out either using the maximum likelihood method, described below and recommended in order to minimize postprocessing times and the size of the training set, or using cross-validation techniques (leave-one-out, cross-validation, etc.). The likelihood of the observations $\{\mathbf{x}_a, \mathbf{G}_a\}$ is written as:

$$\mathcal{L}(\beta,\sigma,\boldsymbol{\theta}) = \frac{1}{((2\pi\sigma^2)^m[\det\mathbf{R}(\boldsymbol{\theta})])^{\frac{1}{2}}}\exp\left[-\frac{1}{2\sigma^2}(\mathbf{G}_a - \beta)^T\mathbf{R}(\boldsymbol{\theta})^{-1}(\mathbf{G}_a - \beta)\right] \quad [6.3]$$

The matrix $\mathbf{R}(\theta)$ contains the correlations between the learning points, two by two. The hyperparameters β, σ^2 and $\boldsymbol{\theta}$ can be estimated by maximizing $\mathcal{L}$. The first-order optimality conditions give the following relations based only on the parameters $\boldsymbol{\theta}$:

$$\hat{\beta}(\boldsymbol{\theta}) = \left(\mathbf{1}^T\mathbf{R}(\boldsymbol{\theta})^{-1}\mathbf{1}\right)^{-1}\mathbf{1}^T\mathbf{R}(\boldsymbol{\theta})^{-1}\mathbf{G}_a \quad [6.4]$$

$$\hat{\sigma}^2(\boldsymbol{\theta}) = \frac{1}{m}(\mathbf{G}_a - \hat{\beta})^T\mathbf{R}(\boldsymbol{\theta})^{-1}(\mathbf{G}_a - \hat{\beta}) \quad [6.5]$$

Vector $\mathbf{1}$ is the unit column vector of size m. The identification of the hyperparameters $\boldsymbol{\theta}$ is performed by the likelihood maximization in which the two previous equations have been previously injected. The problem of hyperparameters identification is thus summarized by numerically minimizing the following function in dimension n:

$$\hat{\boldsymbol{\theta}} = \mathrm{argmin}_{\boldsymbol{\theta}}\hat{\sigma}^2(\boldsymbol{\theta})[\det\mathbf{R}(\boldsymbol{\theta})]^{\frac{1}{m}} \quad [6.6]$$

This optimization problem is complex especially for high-dimensional problems (that is for high n). Therefore, numerical solving methods must be used, such as the Hooke and Jeeves method (Lophaven *et al.* 2002), which is used by the DACE toolbox or the CMA-ES method (Hansenet *et al.* 2003).

6.2.3. *Kriging-based prediction*

The main objective of Kriging metamodeling is to predict the performance function G at one or more unobserved points $\mathbf{x} \in \mathbb{R}^n$. Hyperparameters are assumed to be identified. The Kriging estimator $\hat{G}(\mathbf{x})$ or an unobserved point $\mathbf{x}$, called BLUP (best linear unbiased predictor), is the Gaussian random variable of mean $\mu_{\hat{G}}(\mathbf{x})$ and variance $\sigma^2_{\hat{G}}(\mathbf{x})$ defined by:

$$\mu_{\hat{G}}(\mathbf{x}) = \hat{\beta} + \mathbf{r}(\mathbf{x})^T \mathbf{R}^{-1} \left(\mathbf{G}_a - \hat{\beta}\right) \qquad [6.7]$$

$$\sigma^2_{\hat{G}}(\mathbf{x}) = \hat{\sigma}^2 \left(1 - \mathbf{r}(\mathbf{x})^T \mathbf{R}^{-1} \mathbf{r}(\mathbf{x}) + \mathbf{u}(\mathbf{x})^T \left(\mathbf{1}^T \mathbf{R}^{-1} \mathbf{1}\right)^{-1} \mathbf{u}(\mathbf{x})\right) \qquad [6.8]$$

where $\mathbf{r}(\mathbf{x})$ is the vector of correlations between the point to be predicted and the points of the training set and $\mathbf{u}(\mathbf{x}) = \mathbf{1}^T \mathbf{R}^{-1} \mathbf{r}(\mathbf{x}) - 1$. The Kriging mean $\mu_{\hat{G}}(\mathbf{x})$ is utilized as a (metamodel) estimate of Kriging. The estimate is interpolating, that is, exact at the learning points $\mu_{\hat{G}}(\mathbf{x}_a^{(i)}) = G(\mathbf{x}_a^{(i)})$. The variance provides an *a priori* estimate of the quality of the prediction without computation costs. The variance is zero at the learning points $\sigma^2_{\hat{G}}(\mathbf{x}_a) = 0$.

6.2.4. *Illustration of Kriging-based prediction*

To illustrate Kriging-based prediction, let us consider the function $y = x\sin(x)$ as a function to be approximated over the interval [0;8] from a set of training points. Figure 6.2 illustrates the prediction using a Gaussian kernel and an exponential kernel. The 95% confidence interval is shown in yellow. It allows that an enrichment strategy be efficiently defined based on the desired information on the function $y = x\sin(x)$ (global approximation, finding the minimum, computation of probabilities, etc.).

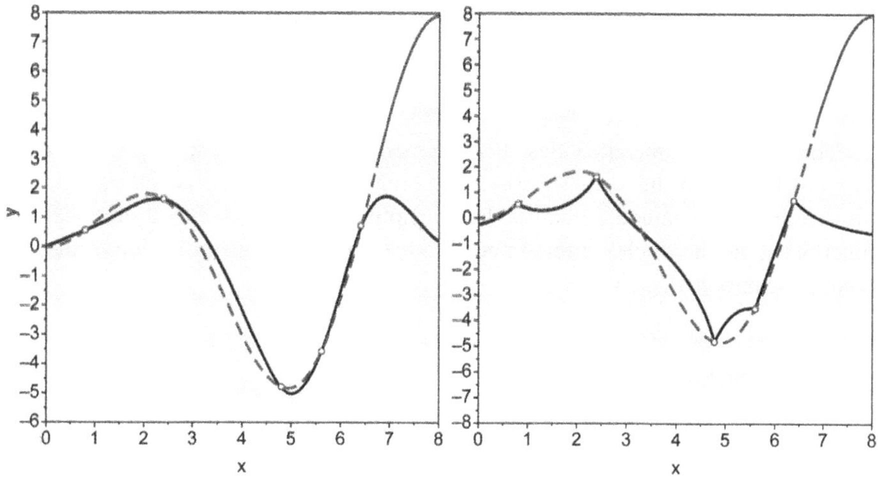

Figure 6.2. *Illustration of Kriging prediction with the function* y = xsin(x) *with (a) a Gaussian kernel and with (b) an exponential kernel. Red circles: training points; red dashed lines: function* xsin(x)*; black solid line: prediction; yellow area: 95% confidence interval. For a color version of this figure, see www.iste.co.uk/gogu/uncertainties.zip*

6.3. Employing Kriging to calculate failure probabilities

Most adaptive methods based on Kriging (but not exclusively) for optimization or reliability analysis are organized as follows:

1) definition of an *initial training basis* $\left\{\mathbf{x}_a^{(i)}, G_a^{(i)}\right\}, i = 1,\ldots,m$;

2) approximation of the *behavior model* by a Kriging-based metamodel, identification of the metamodel hyperparameters (learning phase);

3) calculation of the *enrichment criterion* and search for the best point to add to the training set;

4) calculation of the stopping criterion: if the stopping criterion is reached, the probability of failure is calculated from the metamodel (in reliability analysis) and the algorithm stopped. If the criterion is not reached, the best point is added to the training set and one returns to step 1.

Adaptive methods differ mainly according to their enrichment criterion, based on a learning function that identifies the best point to enrich, and according to their stopping criterion. A significant number of adaptive schemes based on Kriging have been proposed in the literature (see, for example, Bichon *et al.* 2008; Dubourg 2011). A non-exhaustive list of different learning functions is presented in the rest of this section.

These methods are called classification methods (this is true for the AK methods as well) for which the search for the best point to be enriched is carried out on a discrete basis of candidate points. These adaptive schemes are based on a learning function, some of which are presented below, allowing the identification of the best point to be enriched.

6.3.1. *The EFF function*

The efficient global reliability analysis (EGRA) method (Bichon *et al.* 2008) is based on learning managed by the expected feasibility function (EFF). This function expresses the expectation that the performance function $G(x)$ at a non-evaluated point $\mathbf{x}$ belongs to the interval $[-\varepsilon;+\varepsilon]$. This expectation is expressed as follows:

$$EFF(\mathbf{x}) = \mu_{\hat{G}}(\mathbf{x})\left[2\Phi\left(\frac{-\mu_{\hat{G}}(\mathbf{x})}{\sigma_{\hat{G}}(\mathbf{x})}\right)-\Phi\left(\frac{-\varepsilon-\mu_{\hat{G}}(\mathbf{x})}{\sigma_{\hat{G}}(\mathbf{x})}\right)-\Phi\left(\frac{\varepsilon-\mu_{\hat{G}}(\mathbf{x})}{\sigma_{\hat{G}}(\mathbf{x})}\right)\right]$$

$$-\sigma_{\hat{G}}(\mathbf{x})\left[2\varphi\left(\frac{-\mu_{\hat{G}}(\mathbf{x})}{\sigma_{\hat{G}}(\mathbf{x})}\right)-\varphi\left(\frac{-\varepsilon-\mu_{\hat{G}}(\mathbf{x})}{\sigma_{\hat{G}}(\mathbf{x})}\right)-\varphi\left(\frac{\varepsilon-\mu_{\hat{G}}(\mathbf{x})}{\sigma_{\hat{G}}(\mathbf{x})}\right)\right]$$

$$+\varepsilon(\mathbf{x})\left[\Phi\left(\frac{\varepsilon-\mu_{\hat{G}}(\mathbf{x})}{\sigma_{\hat{G}}(\mathbf{x})}\right)-\Phi\left(\frac{-\varepsilon-\mu_{\hat{G}}(\mathbf{x})}{\sigma_{\hat{G}}(\mathbf{x})}\right)\right]$$

When a point has a Kriging prediction close to 0 or a large Kriging variance, a high EFF value is observed. The learning process must therefore be done by searching over a continuous domain, which is of dimension n, the point that maximizes EFF($\mathbf{x}$).

6.3.2. The *U* function

The U function (Echard *et al.* 2011) is the common learning function of the AK methods that will be presented in the following sections. This criterion indicates the distance, in the number of Kriging standard deviations, at a point x from the predicted limit state. It is defined by:

$$U(\mathbf{x}) = \frac{|\mu_{\hat{G}}(\mathbf{x})|}{\sigma_{\hat{G}}(\mathbf{x})} \qquad [6.9]$$

The new enrichment point is the one that minimizes the U function making it possible to find a point that has a Kriging prediction close to 0 and/or a large variance.

6.3.3. *The IMSE$_T$ function*

The targeter integrated mean-squared error function proposed by Picheny *et al.* (2010) is defined in equation [6.10]. It proposes to observe the Kriging variance, particularly in the area of interest characterized by a mean level T and a band parameter ε. Therefore, the area of interest is defined by $[T-\varepsilon, T+\varepsilon]$. The point minimizing this function is the new point to be enriched:

$$IMSE_T(\mathbf{x}) = \int_D \sigma_{\hat{G}}^2(\mathbf{x}) W(\mathbf{x}) d\mathbf{x} \qquad [6.10]$$

where $W(\mathbf{x}) = \text{Prob}\{G(\mathbf{x}) \in [T-\varepsilon, T+\varepsilon]\}$ is a weighting function. The numerical procedure for estimating this criterion is presented by Picheny *et al.* (2010).

6.3.4. *The SUR function*

The SUR (stepwise uncertainty reduction) function considers the enrichment problem from a Bayesian perspective. As a matter of fact, the Bayesian framework provides here a means to obtain a measure of the estimation error. From a sequence of failure probability estimators $(\hat{\alpha}_n)_{n\leq 1}$ and the use of the quadratic loss function, equation [6.11] is defined and must be minimized:

$$J_n(\mathbf{x}) = \mathbf{E}_n((\alpha - \hat{\alpha}_{n+1})^2 \,|\, \mathbf{X}_{n+1} = \mathbf{x}) \qquad [6.11]$$

Since no effective numerical procedure was found to assess this criterion, four strategies are proposed in Bect *et al.* (2012) to estimate them.

6.3.5. *The H function*

Lv *et al.* (2015) propose another learning function based on entropy information. This function defined in equation [6.12] provides a measure of the uncertainty of the estimator $\hat{G}(\mathbf{x})$ at a point $\mathbf{x}$ and makes it possible to guarantee good accuracy in the vicinity of the limit state. The learning is carried out by selecting the point with the value of the largest H:

$$H(\mathbf{x}) = \left| \begin{array}{l} \ln\left(\sqrt{2\pi}\sigma_{\hat{G}}(\mathbf{x}) + \frac{1}{2}\right)\left[\Phi\left(\frac{2\sigma_{\hat{G}}(\mathbf{x}) - \mu_{\hat{G}}(\mathbf{x})}{\sigma_{\hat{G}}(\mathbf{x})}\right) - \Phi\left(\frac{-2\sigma_{\hat{G}}(\mathbf{x}) - \mu_{\hat{G}}(\mathbf{x})}{\sigma_{\hat{G}}(\mathbf{x})}\right)\right] \\ -\left[\frac{2\sigma_{\hat{G}}(\mathbf{x}) - \mu_{\hat{G}}(\mathbf{x})}{2}\varphi\left(\frac{2\sigma_{\hat{G}}(\mathbf{x}) - \mu_{\hat{G}}(\mathbf{x})}{\sigma_{\hat{G}}(\mathbf{x})}\right) + \frac{2\sigma_{\hat{G}}(\mathbf{x}) + \mu_{\hat{G}}(\mathbf{x})}{2}\varphi\left(\frac{-2\sigma_{\hat{G}}(\mathbf{x}) - \mu_{\hat{G}}(\mathbf{x})}{\sigma_{\hat{G}}(\mathbf{x})}\right)\right] \end{array} \right| \quad [6.12]$$

6.3.6. *The OBJ function*

The representative scheme proposed by Wen *et al.* (2016) requires the following optimization:

$$\text{maximize: } OBJ(\mathbf{x}) = \begin{cases} \mathbf{E}[F(\mathbf{x})], & \rho_{\mathbf{X}}(\mathbf{x}) > d_{thr} \\ 0, & \rho_{\mathbf{X}}(\mathbf{x}) < d_{thr} \end{cases} \quad [6.13]$$

where $\rho_{\mathbf{X}}(\mathbf{x})$ represents the joint probability density of the input variables, d_{thr} is a threshold and $\mathbf{E}[F(\mathbf{x})]$ is the *EFF* function.

6.3.7. The *L* function

Ma *et al.* (2018) propose a learning scheme based on the function L defined in equation [6.14]. This scheme identifies several enrichment points at each iteration. All the points not respecting $L(\mathbf{x}) \leq 10^{-t}$, where t is the number of the current iteration, are the new enrichment points:

$$L(\mathbf{x}) = \frac{\sigma_{\hat{G}}^2(\mathbf{x})}{\mu_{\hat{G}}(\mathbf{x})} \quad [6.14]$$

6.3.8. *Discussion*

The choice of a learning function depends on its efficiency, measured by the number of computations to the G function, and the postprocessing time needed to find the best point to enrich. To reduce postprocessing time, we recommend sampling the variables beforehand and to search again for the best point from these sampled points. It is then necessary to choose a simple learning function to be evaluated by a simple postprocessing of $\mu_{\hat{G}}(\mathbf{x})$ and/or $\sigma_{\hat{G}}(\mathbf{x})$.

The limit state function $G(\mathbf{x}) = 0$ must be precisely approximated by the metamodel to expect to obtain an accurate estimate of the probability of failure. Enrichment points should therefore be selected close to the limit state function. The *EFF*, *U* and *OBJ* functions seem to be more effective in approximating the limit state function. The function *L* proposes to enrich several points at each iteration of the method. This learning function may therefore cause too many unnecessary calls to the performance function.

The AK methods presented in the rest of this chapter are based on the *U* function for its simplicity (no additional computation is required to estimate its value, only the Kriging mean and standard deviation are needed), for its efficiency and for its speed associated with a preliminary sampling of random variables.

6.4. The AK-MCS method: presentation and generic principle

6.4.1. *Presentation of the AK-MCS method*

The AK methods form a family based on the principle of AK-MCS (Echard *et al.* 2011), which was chronologically the first method developed and the easiest to implement.

The first step of the method consists of sampling the random variables using the MCS method. The sample size n_{MC} must be large enough in order to be able to estimate the probability of failure accurately, however, no call to the performance function is carried out at this stage of the analysis.

A set of initial samples are then generated based on a design of experiments. The performance function is evaluated for each sample, the number of which must be sufficiently reduced to keep the numerical efforts reasonable. This design of experiments can be extracted from the initial Monte Carlo population, or can alternatively come from a grid or from numbers generated using one of the quasi-Monte Carlo methods (Sobol' sequences, Latin hypercube sampling, etc.), the latter methods covering the space more efficiently.

This design of experiments is then iteratively enriched using the procedure described as follows:

1) the Kriging metamodel is calibrated using all points of the design of experiments (and the corresponding values of the performance function);

2) the metamodel is then evaluated for every point of the Monte Carlo population for which the performance function is not known;

3) the U learning function defined in equation [6.9] is used by the AK-MCS method. It is therefore evaluated for every sample of the Monte Carlo population to be classified. This function is used to quantify the probability that each point in the initial population is misclassified. A point is misclassified if the metamodel predicts the wrong sign of the performance function. Under the assumption of Gaussianity, the probability that the point $\mathbf{x}^{(i)}$ be wrongly classified is equal to $\Phi\left(-U\left(\mathbf{x}^{(i)}\right)\right)$, Φ being the cumulative distribution function associated with the standard normal distribution. The metamodel is assumed to be sufficiently accurate if the value of the learning function U is greater than 2 for all points to be classified ($\min_{i=1,\dots,n_{MC}} U(\mathbf{x}^{(i)}) \geq 2$), which is equivalent to a misclassification probability of less than 2.3% for each of the unobserved points. If this criterion is met, the metamodel is considered to be sufficiently accurate and iterative enrichment is stopped, otherwise step 4 of the process is applied;

4) the point $\mathbf{x}_a^{(*)}$ associated with the lowest value of the learning function is identified and the performance function is evaluated for that point. The point is thus enriched and the method returns to step 1 described above.

Once the iterative enrichment process has converged and the probability of misclassification is sufficiently low for all of the points of the Monte Carlo population, the probability of failure is obtained using the well-known formula of the Monte Carlo–estimated probability of failure, but calculated from the metamodel:

$$\hat{P}_f = \frac{1}{n_{MC}} \sum_{i=1}^{n_{MC}} I_{\hat{G}}\left(\mathbf{x}^{(i)}\right) \quad [6.15]$$

with $\mathbf{x}^{(i)}$, $i = 1 \dots n_{MC}$, the samples from the initial population and $I_{\hat{G}}$ the indicator function given by:

$$I_{\hat{G}}\left(\mathbf{x}^{(i)}\right) = \begin{cases} 1 & \text{if } \mu_{\hat{G}}\left(\mathbf{x}^{(i)}\right) \leq 0 \\ 0 & \text{otherwise} \end{cases} \quad [6.16]$$

6.4.2. *Illustration of the AK-MCS method*

Figure 6.3 illustrates the AK-MCS method with an example involving a random variable. The first line represents the method before the first enrichment. The initial set of points contains five samples shown in red and green. The sample minimizing the learning function is identified (represented in black) and results in enriching the design of experiments. The performance function is evaluated for this sample and a new metamodel is calibrated. The operation is iteratively repeated until convergence of the method.

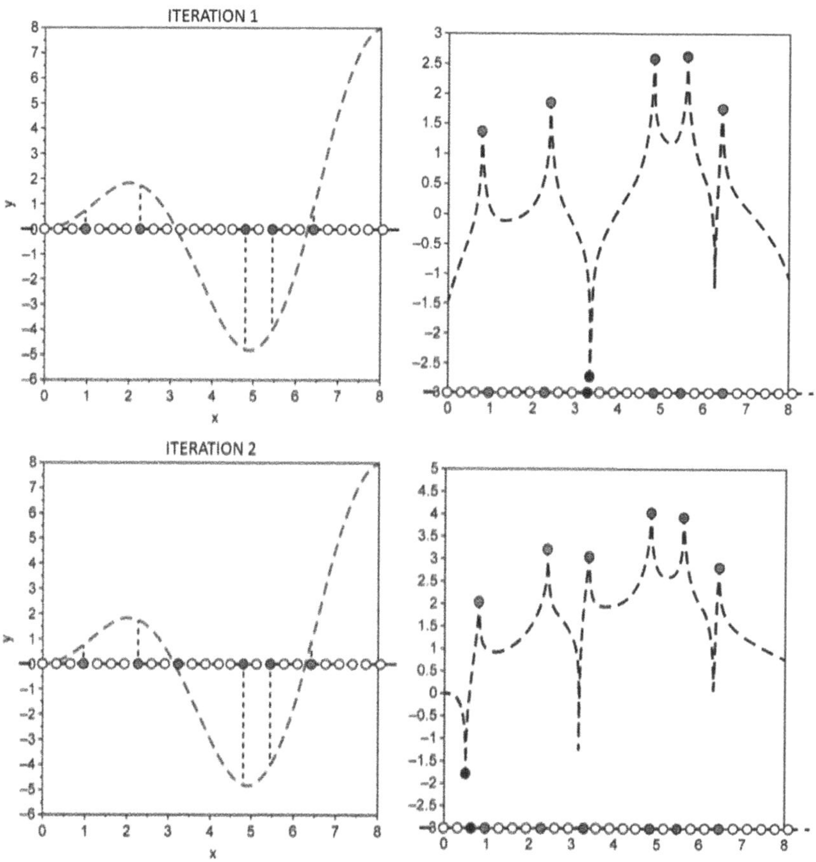

Figure 6.3. *Illustration of an example of the AK-MCS method with a single random variable. The left-hand column represents the performance function and the right-hand one represents the learning function with a logarithmic scale. For a color version of this figure, see www.iste.co.uk/gogu/uncertainties.zip*

Figure 6.4 shows the evolution of the method with a problem involving two random variables. The black line represents the limit state, the fine dots represent the Monte Carlo population (the red and green colors represent the samples that the metamodel classifies in the failure and safety domain, respectively), the larger black circles represent the enriched design of experiments, and the blue circle represents the next point to be enriched, which minimizes the learning function. In the first iteration, the failure domain is very roughly approximated. The quality of this approximation is progressively improved when adding points.

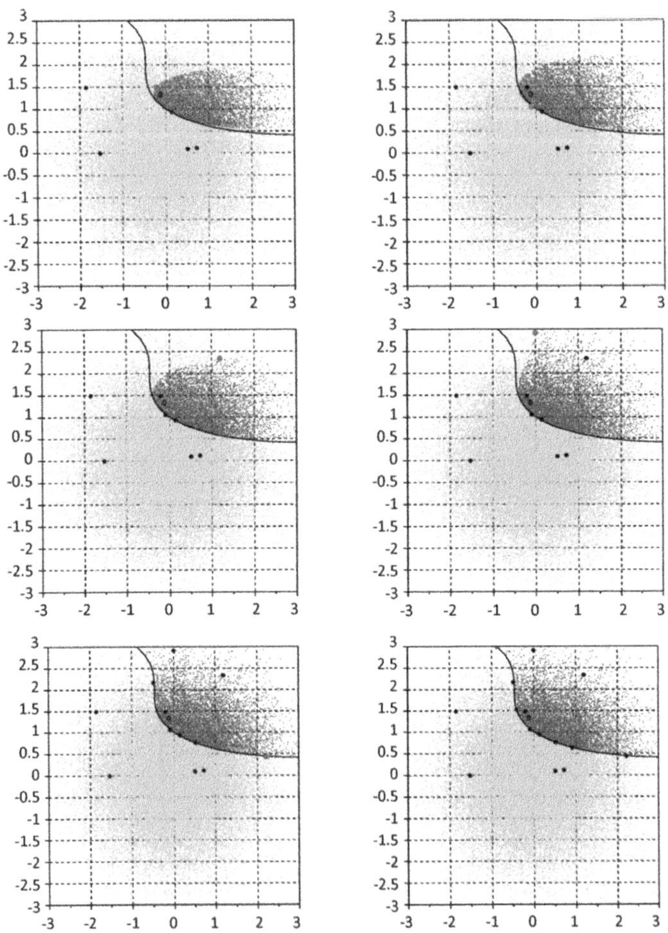

Figure 6.4. *Illustration of the AK-MCS method with an example with two random variables. These graphs respectively represent the first, second, third, fourth, seventh and 10th (and last) enrichment iteration. For a color version of this figure, see www.iste.co.uk/gogu/uncertainties.zip*

6.4.3. *Discussion*

The size of the initial Monte Carlo population has an impact on the accuracy of the estimation of the failure probability, which can be quantified, for example, by using the coefficient of variation. If the size of this population is insufficient at the end of the analysis with AK-MCS, it is possible to generate additional samples and repeat the analysis in point 1 described in section 6.4.1, keeping the enriched experimental design previously obtained. The performance function is then evaluated with selected samples from the new population until the method converges.

The methods based on adaptively enriched designs of experiments contain two phases. The initial set of points is associated with the *exploration* phase, which aims to cover the variable definition domain, and the enrichment iterations are associated with the *exploitation* phase, which aims to use the metamodel to generate samples in the area that contributes the most to the problem (namely near the limit state in the case of reliability). A good compromise between these two phases must be found to ensure that the method converges effectively. If exploration is insufficient, important areas may be overlooked; if exploitation is excessive, the metamodel is reliable everywhere, including in areas that contribute little to the problem. The search for the right compromise between these two phases is non-trivial; this is discussed in more detail, for example, by Forrester and Keane (2009).

6.5. The AK-IS method for estimating probabilities of rare events

6.5.1. *Presentation of the AK-IS method*

It is sometimes stated that it is necessary to use 10^{N+2} samples to estimate a probability of the order of 10^{-N} with the MCS method. The AK-MCS method is not suitable for estimating low probabilities (of the order of 10^{-6}–10^{-8}) because it is difficult to store all the points in the RAM of a computer. The calculation of the Kriging estimator for all these points represents prohibitive computational efforts, etc. The AK-IS method (Echard *et al.* 2013) has been proposed to circumvent such problems; it consists of coupling the principles of the active learning and Kriging-based method with IS.

The first step consists of identifying the most probable failure point, conventionally defined as the point in the failure domain that maximizes the joint probability density function of the random variables. A population is then generated using the instrumental density, combined with a multivariate normal distribution centered at the most probable point of failure with unit variance and independent variables.

The iterative enrichment procedure is identical to that of the AK-MCS method. An initial set of points is carried out and the performance function is evaluated. The Kriging metamodel is then calibrated and the U learning function is evaluated at every population point. As long as certain points have a learning function value of less than 2, the design of experiments is enriched in the same way as the AK-MCS method.

When the method has converged, the probability of failure is assessed using the metamodel as follows:

$$\hat{P}_f = \frac{1}{n_{MC}} \sum_{i=1}^{n_{MC}} I_{\hat{G}}\left(\mathbf{x}^{(i)}\right) \frac{\varphi\left(\mathbf{x}^{(i)}\right)}{f\left(\mathbf{x}^{(i)}\right)} \qquad [6.17]$$

φ being the joint probability density of the random variables expressed in the standard normal space and f being the instrumental density.

6.5.2. *Illustration of the AK-IS method*

Figure 6.5 shows the evolution of the method with a problem involving two random variables. The Monte Carlo population is centered at the most probable point of failure and therefore contains more samples in the vicinity of the limit state than for the AK-MCS method. In the first iteration, the failure domain is not correctly estimated and the precision is progressively improved when the set of points is iteratively enriched.

6.5.3. *Discussion*

The most probable point of failure is identified using an optimization algorithm, which requires additional calls to the performance function. However, it is not necessary to identify this point accurately, especially if a moderate reduction in accuracy results in a decrease in computational effort. This is because the results obtained from large runs converge toward the solution (when the population size tends toward infinity). The quality of the estimate of the probability of failure remains acceptable as long as the instrument density is centered at a point in the vicinity of the most likely point of failure.

Several so-called advanced simulation methods are available in the literature such as IS, but also SS, line samplings, etc. It is possible to combine the AK method with such advanced simulation algorithms, using the latter to generate the initial population with which the learning function is then evaluated. Huang *et al.* (2016) propose an AK method coupled with SS; Xiong and Tan (2018) describe a method coupled with radial IS. Alternatively, Lelièvre *et al.* (2018) propose to use several populations of reduced sizes for the estimation of probabilities of occurrence of rare events.

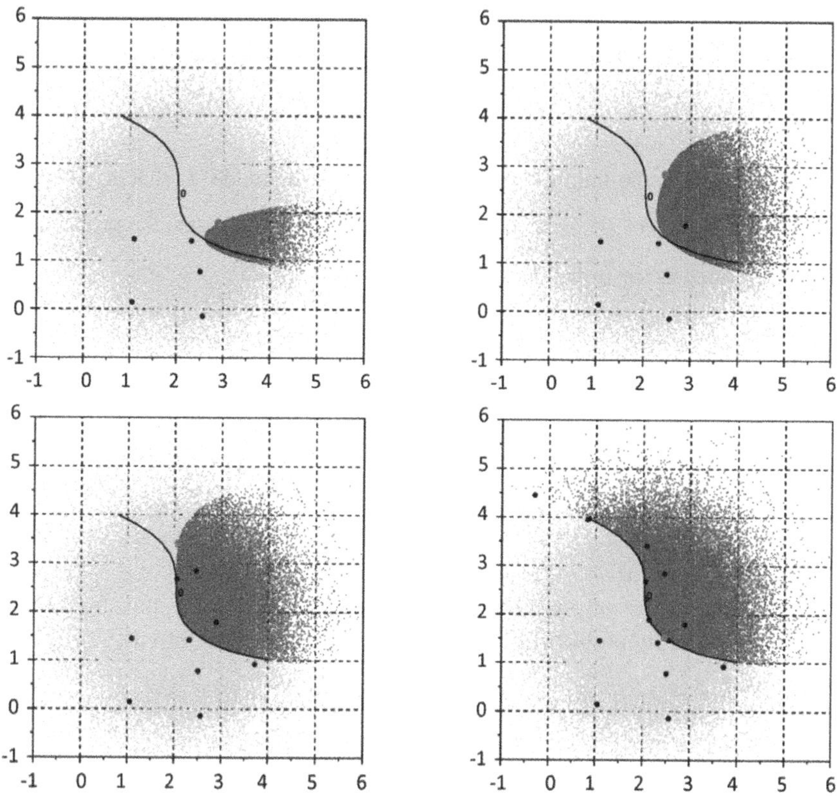

Figure 6.5. *Illustration of the AK-IS method with an example using two random variables. These graphs, respectively, represent the first, second, fifth and 13th (and last) enrichment iteration. For a color version of this figure, see www.iste.co.uk/gogu/uncertainties.zip*

6.6. The AK-SYS method for system reliability problems

6.6.1. *Some generalities about system reliability analysis*

The framework of the so-called component reliability analysis, in other words, the one described as employing a single G performance function, may no longer be sufficient when the system under study is likely to involve several modes of failure. Within this context, the so-called system reliability methods must be considered. Their objective is to evaluate the probability of failure when several performance functions $G_j, j = 1,\ldots,p$, characterizing the different failure scenarios are involved. The number of components p involved and the possible complexity of the relationship between the components are additional difficulties to be taken into account when calculating the failure probability, compared to a component reliability analysis.

We note here E_j the failure event associated with the mode j, that is, defined as: $E_j = \{\mathbf{x} \in \mathbb{R}^n : G_j(\mathbf{x}) \leq 0\}$ [1]. For series system reliability problems, the probability of failure is defined as the probability of the union of events E_j:

$$P_f = \mathrm{Prob}(\cup_{j=1}^{p} E_j) = \mathrm{Prob}(\cup_{j=1}^{p} G_j(\mathbf{x}) \leq 0) \qquad [6.18]$$

For parallel system reliability problems, the probability of failure is given by the probability of the intersection of events E_j:

$$P_f = \mathrm{Prob}(\cap_{j=1}^{p} E_j) = \mathrm{Prob}(\cap_{j=1}^{p} G_j(\mathbf{x}) \leq 0) \qquad [6.19]$$

These two types of system reliability problems can be transformed into component reliability problems by denoting the following equations as:

$$\{\mathbf{x} \in \mathbb{R}^n : \cup_{j=1}^{p} G_j(\mathbf{x}) \leq 0\} = \{\mathbf{x} \in \mathbb{R}^n : \min_{j=1,\ldots,p} G_j(\mathbf{x}) \leq 0\} \qquad [6.20]$$

$$\{\mathbf{x} \in \mathbb{R}^n : \cap_{j=1}^{p} G_j(\mathbf{x}) \leq 0\} = \{\mathbf{x} \in \mathbb{R}^n : \max_{j=1,\ldots,p} G_j(\mathbf{x}) \leq 0\} \qquad [6.21]$$

The nesting of parallel and serial subsets leads to more complex configurations and the state of the system can only be described from specific functions.

---

1 For the sake of brevity, the definition of a failed event coincides here with that of a failure domain.

6.6.2. Presentation of the AK-SYS method

The active learning and Kriging-based system reliability method (AK-SYS) was proposed by Fauriat and Gayton (2014) to deal with system reliability problems (mainly series and parallel) when the evaluation of performance functions is costly in computing time. It consists of an adaptation of the AK-MCS method to the management of multiple performance functions. Kriging-based metamodeling is thus used to minimize the total number of calls to system performance functions.

Each system performance function $G_j, j=1,...,p,$ is substituted by a Kriging metamodel, $\hat{G}_j, j=1,...,p,$ but the latter is enriched solely in areas that contribute significantly to the failure. If a component j is never activated during the AK-SYS enrichment procedure, the metamodel $\widehat{G}_j$ built from the initial design of experiments is not recalibrated, which reduces the computational efforts. The AK-SYS procedure can be summarized as follows:

1) generate a Monte Carlo population $\{\mathbf{x}^{(i)}, i=1,...,n_{MC}\}$ of size n_{MC}. This makes it possible to restrict the search for points to be enriched by only addressing the zones of the variable definition space having a non-negligible probability;

2) define a first design of experiments by drawing m points $\mathbf{x}_a^{(i)}$ in the Monte Carlo population and evaluate all the performance functions $G_j, j=1,...,p,$ on these points m;

3) calibrate each Kriging metamodel $\hat{G}_j, j=1,...,p,$ from the experience plan $\{\mathbf{x}_a^{(i)}; G_j(\mathbf{x}_a^{(i)})\}, i=1,...,m$;

4) select a single component $s^{(i)} \in \{1,...p\}$ activated by the system, for each point $\mathbf{x}^{(i)}, i=1,...,n_{MC}$, of the Monte Carlo population. This step depends on the configuration of the system. In the case of a series system reliability problem, the component $s^{(i)}$ is selected according to equation [6.20] for each point such that:

$$s^{(i)} = \text{argmin}_{j=1,...,p} \hat{G}_j(\mathbf{x}^{(i)}) \qquad [6.22]$$

In the case of a parallel system reliability problem we have, according to equation [6.21]:

$$s^{(i)} = \text{argmax}_{j=1,...,p} \hat{G}_j(\mathbf{x}^{(i)}) \qquad [6.23]$$

5) calculate the enrichment criterion $U(\mathbf{x}^{(i)})$ for every couple $\{\mathbf{x}^{(i)};s^{(i)}\}$, $i=1,\ldots,n_{MC}$:

$$U(\mathbf{x}^{(i)})=\frac{|\mu_{\hat{G}_{s^{(i)}}}(\mathbf{x}^{(i)})|}{\sigma_{\hat{G}_{s^{(i)}}}(\mathbf{x}^{(i)})} \qquad [6.24]$$

where $\mu_{\hat{G}_{s^{(i)}}}(\mathbf{x}^{(i)})$ and $\sigma_{\hat{G}_{s^{(i)}}}(\mathbf{x}^{(i)})$, respectively, denote the mean and the standard deviation of the variable $\hat{G}_{s^{(i)}}(\mathbf{x}^{(i)})$ conditioned by the point of the experimental design;

6) select the point $\mathbf{x}_a^{(*)}$ and the associated component s that minimizes the function U, that is, the point associated with the component s having the greatest prediction uncertainty:

$$\mathbf{x}_a^{(*)}=\mathrm{argmin}_{i=1,\ldots,n_{MC}}U(\mathbf{x}^{(i)}) \qquad [6.25]$$

The performance function G_s is only evaluated at the point of interest, the couple $\{\mathbf{x}_a^{(*)};G_s(\mathbf{x}_a^{(*)})\}$ is thus added to the design of experiments. The algorithm then returns to step 3 where only the metamodel $\hat{G}_s$ is recalibrated. If the other performance functions are obtained at the same time as the performance function G_s, namely at the same cost, then all the components $G_j, j=1,\ldots,p$, are evaluated and the metamodels $\hat{G}_j$ are recalibrated;

7) stop the algorithm when every point in the Monte Carlo population has a U value greater than 2.

An important remark concerns system reliability applications, where the performance functions involved are of different orders of magnitude (for example, in MPa, mm and N). This raises problems for the choice of s which then compares different quantities. A performance function may be considered to be important when it actually has no impact on reliability, which consequently results in slowing down the algorithm. To avoid this problem, it is suggested to express all performance functions with the same unit if possible or to normalize them. One possibility is, for example, to divide the performance functions by the values of a reference calculation (for instance, at the midpoint).

6.6.3. *Illustration of the AK-SYS method*

The AK-SYS method is illustrated based on a serial system consisting of two components, G_1 and G_2, both of which are a function of two independent Gaussian random variables $X_1(\omega)$ and $X_2(\omega)$, with zero means and unit variances. The initial design of experiments consists of five points (black circles in Figure 6.6). The evolution of the limit state during enrichments is illustrated in Figure 6.6.

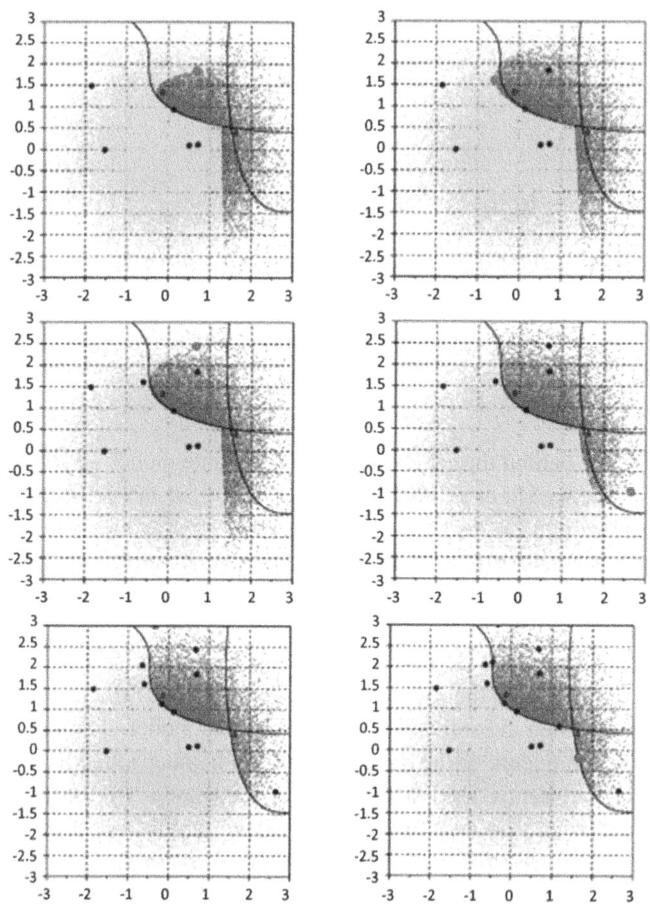

Figure 6.6. *Illustration of the AK-SYS method with a serial system using two components and two random variables. These graphs, respectively, represent the first, second, third, fourth, seventh and 13th (and last) enrichment iteration. For a color version of this figure, see www.iste.co.uk/gogu/uncertainties.zip*

From the initial design of experiments, it can be seen that many points in the Monte Carlo population are misclassified (the Monte Carlo population points are marked in red when predicted to fail by the metamodels $\hat{G}_1$ and $\hat{G}_2$ and in green when they belong to the safety domain, see Figure 6.6 in the upper left corner). The limit states $G_1(x_1,x_2)=0$ and $G_2(x_1,x_2)=0$ are plotted in black. The point to be learned (blue circle) minimizes the function U and activates the component $s=1$ of the system. Both metamodels $\hat{G}_1$ and $\hat{G}_2$ are here enriched at the same time at every iteration. It is assumed that the evaluation of G_1 and G_2 is simultaneous. A new point and the system component to be enriched are iteratively identified until the stop criterion $\min U_{i=1,\dots,n_{MC}} \geq 2$ is reached.

6.6.4. *Alternatives to the AK-SYS method*

Alternative approaches to the AK-SYS method using Kriging metamodeling have since been proposed. In the context of AK methods, which are based on the classification of a Monte Carlo population and whose enrichment strategies are based on using the U function, the works of Hu *et al.* (2017) and Yang *et al.* (2018) in particular can be cited.

The main idea behind the efficient Kriging surrogate modeling approach (EKSA) method, proposed by Hu *et al.* (2017), is to consider that the performance functions composing the system can be linked together, analogously to the vector output from a multiphysics simulation. Two approaches are deployed concomitantly to predict the response of the vector of performance functions at a point of the Monte Carlo population. The first approach considers the link between the system components. It is based on an eigenvalue decomposition of the matrix of performance functions evaluated at the points of the training set. Each function of the truncated set of the decomposition is then substituted by a Kriging metamodel. The second approach considers the components of the performance vector independently of one another. The same training set is used to calibrate a metamodel by performance vector component. At every point in the Monte Carlo population, the approach that maximizes the U criterion is then chosen, leading to a composite response over the space $\mathbf{x}$ of the variables. Metamodel enrichment is defined from the probability of poor classification of each point in the Monte Carlo population, which is an idea adjacent to the criterion U. For its part, the stopping criterion is defined on the basis of the estimation error of the probability of failure, which differs from the strategy initially proposed in the AK-MCS method.

The active learning Kriging with truncated candidate regions (ALK-TCR) method, proposed by Yang *et al.* (2018), makes it possible to address series or parallel system reliability issues. As mentioned by the authors, the main ideas of the AK-MCS and AK-SYS methods, namely the enrichment and stopping criteria based on the U function and the calibration of one metamodel per system component $j = 1,...,p$, are retained. The difference between the ALK-TCR and AK-SYS methods resides in the selection of the learning point and the metamodel j of the system to be recalibrated. The original idea of the ALK-TCR method is to reduce the search area of the candidate points for enrichment of each metamodel $j = 1,..., p$, by eliminating the points of the Monte Carlo population for which the state of the system is known with a certain level of certainty based on the other metamodels. The so-called truncated search areas are updated during the algorithm.

6.6.5. *Application to problems indexed by a subset*

The AK-SYS method is suitable for solving certain component reliability problems that are indexed by a subset T. This is the case, for example, when temporal phenomena, such as wear, are considered or when the performance function depends on space variables. T can thus refer to time and designate an interval of the real space $T \subset \mathbb{R}+$ or refer to the space and thus designate a subset of a multidimensional space $T \subset \mathbb{R}^d$. More generally, indexing can be done with respect to any variable (other than time or space) or by considering space and time simultaneously. It is assumed that the performance function G now depends on n random variables and an indexing variable t, which can be a vector if we consider the spatial coordinates of a structure $t = (x, y, z)$, for example. One therefore considers $G(\mathrm{x},t)$ where $\mathbf{x}$ denotes the vector of n random variables possibly derived from the representation of a stochastic process and t the index variable.

When examining the failed event defined by $E = \{\mathbf{x} \in \mathbb{R}^n : \exists t \in T, G(\mathbf{x},t) \leq 0\}$, the probability of failure is given by equation [6.26]:

$$\begin{aligned} P_f &= \text{Prob}(\{\mathbf{x} \in \mathbb{R}^n : \exists t \in T, G(\mathbf{x},t) \leq 0\}) \\ &= \text{Prob}(\{\mathbf{x} \in \mathbb{R}^n : \min_{t \in T} G(\mathbf{x},t) \leq 0\}) \end{aligned} \qquad [6.26]$$

This type of problem can be encountered when considering, for example, that the failed event is likely to occur at different points in a structure, as opposed to the often adopted simplifying assumption that only one point (or area) is critical to the structure.

In time-dependent reliability problems, the probability defined in equation [6.26] is referred to as the probability of cumulative failure over the interval T and is generally denoted by $P_{f,c}(T)$. The numerical evaluation of this probability requires in general discretizing the subset T into a finite number of instants (or points). Denoting by $T' = \{t_1,\ldots,t_j,\ldots,t_p\}$ the discretized set of T, the failure probability can be rewritten by equation [6.27]:

$$\begin{aligned} P_f &\approx \text{Prob}(\{\mathbf{x}\in \mathbb{R}^n : \exists t_j \in T', G(\mathbf{x},t_j) \le 0\}) \\ &\approx \text{Prob}(\{\mathbf{x}\in \mathbb{R}^n : \min_{t_j\in T'} G(\mathbf{x},t_j) \le 0\}) \end{aligned} \qquad [6.27]$$

By setting $G_j(\mathbf{x}) = G(\mathbf{x},t_j)$, equation [6.27] is equivalent to equations [6.18] and [6.20]. Therefore, it ultimately amounts to solving a serial system reliability problem where each component G_j of the system corresponds to the G function evaluated at time t_j of the discretized subset T'.

If we consider the context that each calculation $G_j(\mathbf{x}) = G(\mathbf{x},t_j)$, $\mathbf{x}\in \mathbb{R}^n$ and $t_j \in T'$ requires significant computing resources, the algorithm of the AK-SYS method presented in section 6.6.2 can then be applied. The number of metamodels to be calibrated corresponds to the number of elements of T'. During the training phase, the selection of the point $\mathbf{x}_a^{(*)}$ to be enriched and the metamodel $\hat{G}_s$ to be recalibrated is achieved using steps 4–6 of the AK-SYS algorithm. However, all metamodels can be updated simultaneously if the calculations of $G(\mathbf{x}_a^{(*)},t_j)$, $t_j \neq t_s$ do not generate any additional cost. This is the case, for example, when evaluating a performance function formulated from structural stresses obtained by finite element calculations. The stresses of the set of spatial coordinates $t_j \in T'$ are simultaneously calculated. Conversely, the evaluation of $G(\mathbf{x}_a^{(*)},t_s)$ may require the calculation of $G(\mathbf{x}_a^{(*)},t_j)$ such that $t_j < t_s$, and therefore generate additional costs if the call to the performance function t_j is not free of charge. This is particularly the case for problems where the cumulative aspect of the phenomenon is modeled (material degradation for example).

The declination of the AK-SYS method to time-dependent reliability problems has been proposed by Ahmadivala *et al.* (2019) and the method renamed AK-SYS-t. In the literature, the resolution of time-dependent reliability problems by Kriging metamodels has also recently been investigated by other authors. Among the proposed methods is the single-loop Kriging (SILK) surrogate modeling for the time-dependent reliability analysis method (Hu and Mahadevan 2016), which is based on the fundamentals of the AK-MCS method. The major difference with the AK-SYS-t method is that a single metamodel is calibrated on the space $\mathbb{R}^n \times T$ of dimension $n + 1$; in other words, the space of random variables increased by the time interval (indexing variable here). During the learning process, the choice of the pair $(\mathbf{x}_a^{(*)}, t_s)$ (point of the Monte Carlo population and instant of the finite subset T') is achieved in two steps. First of all, for each trajectory of the $\mathbf{x}^{(i)}$ performance function derived from the Monte Carlo population, one identifies the instant t_s of the discretization T' to be potentially enriched, i.e. which minimizes the function U. The corresponding value of U is denoted by $U_{min}(\mathbf{x}^{(i)})$. In the case where the trajectory $\mathbf{x}^{(i)}$ is identified as failing with a high probability, that is, $\exists t_j : \hat{G}(\mathbf{x}^{(i)}, t_j) \leq 0$ such that $U(\mathbf{x}^{(i)}, t_j) \geq 2$, then it is imposed that $U_{min}(\mathbf{x}^{(i)}) = 2$, which means that it is not necessary to enrich this trajectory since it is sufficiently well classified. The trajectory $\mathbf{x}^{(i)}$ of the Monte Carlo population that minimizes $U_{min}(\mathbf{x}^{(i)})$ is then selected and denoted here by $\mathbf{x}_a^{(*)}$. The associated value t_j, previously determined, is then denoted t_s. The performance function G is evaluated at the couple $(\mathbf{x}_a^{(*)}, t_s)$ and the metamodel $\hat{G}$ is recalibrated from the updated training set. A specific stopping criterion based on the error of the failure probability estimation is proposed by the authors.

6.7. The AK-HDMR1 method for high-dimensional problems

Metamodeling by Kriging is now widely integrated into reliability methods in order to reduce, often through iterative strategies, the number of calls to performance functions that can be complex to evaluate and require significant computation time. The family of AK methods presented earlier has been proposed with this context in mind and addresses many problems in moderate dimensions ($n \leq 12$).

Solving reliability problems that depend on a high number of random variables with an AK method can nevertheless become problematic if this leads to the estimation of a large number of Kriging hyperparameters (≥ 12) (see section 6.2.2). This is, for example, the case when an *anisotropic* kernel is used. The metamodel(s) to be calibrated then depends on a number of hyperparameters at least equal to dimension n of the random vector $\mathbf{X}(\omega)$.

The AK method for high-dimensional reliability problems (AK-HDMR1) was proposed by Lelièvre (2018) as an alternative to the AK-MCS method when the dimension is high. The idea is to express the performance function G as a sum of functions, of smaller dimensions compared to the initial dimension of the problem, then to substitute each function of the decomposition using a Kriging metamodel. The AK-MCS algorithm is then modified to incorporate this functional decomposition. The high-dimensional model representations (HDMR) decomposition is chosen here and truncated at order 1 to reduce the number of Kriging metamodels to be calibrated and thus make it possible to speed up the computations.

6.7.1. *HDMR functional decomposition*

The HDMR approach (Rabitz and Aliş 1999; Rabitz *et al.* 1999; Li *et al.* 2001) allows the exact decomposition of a function $G(\mathbf{x})$ of dimension n into a sum of functions of increasing dimension:

$$G(\mathbf{x}) = g_0 + \sum_{i=1}^{n} g_i(x_i) + \sum_{1 \le i < j \le n} g_{ij}(x_i, x_j) + \cdots + g_{12\ldots n}(x_1, \ldots, x_n) \qquad [6.28]$$

where g_0 is a constant. The functions $g_i(x_i), i = 1, \ldots, n,$ of order 1 model the contributions to the response of the variables x_i alone, that is, independently of each other. The functions $g_{ij}(x_i, x_j), i, j = 1, \ldots, n,$ represent the second-order components of $G(\mathbf{x})$ and reflect the cross influences of the variables x_i and x_j on the response.

More generally, the functions $g_{1\ldots k}(x_1, \ldots, x_k)$ $k = 1, \ldots, n,$ allow the k-order interactions between the variables to be expressed. There are two existing approaches, called ANOVA-HDMR and Cut-HDMR, to identify these functions. The Cut-HDMR approach is more adapted to reliability problems than the ANOVA-HDMR approach (Lelièvre 2018).

The principle of the Cut-HDMR approach consists of projecting the vector $\mathbf{x}$, of dimension n onto hyperplanes of smaller dimensions. The projection hyperplanes contain one (single) common point, denoted here $\bar{x}$ and called anchor point (or support point).

The functions of the Cut-HDMR decomposition are defined by:

$$\begin{array}{rcl} g_0 & = & G(\bar{\mathbf{x}}) \\ g_i(x_i) & = & G(x_i,\bar{\mathbf{x}}) - g_0 \\ g_{ij}(x_i,x_j) & = & G(x_i,x_j,\bar{\mathbf{x}}) - g_i(x_i) - g_j(x_j) - g_0 \\ & & \ldots \end{array} \quad [6.29]$$

where $G(x_i,\bar{\mathbf{x}}) = G(\bar{x}_1,\bar{x}_2,\ldots,\bar{x}_{i-1},x_i,\bar{x}_{i+1},\ldots,\bar{x}_n)$. By truncating the decomposition at the order 1, the number of functions to be identified is equal to $1+n$. It is equal to $1+n+\frac{n(n-1)}{2}$ at order 2.

6.7.2. *Presentation of the AK-HDMR1 method*

The AK-HDMR1 method is based on a first-order HDMR decomposition of the performance function. The AK-MCS algorithm is modified accordingly.

After generating a Monte Carlo population, an anchor point is selected for the HDMR decomposition. This is a key step in the approach that partly conditions its success. The influence of the anchor point is discussed in Lelièvre (2018), which also proposes an empirical strategy for its selection. A first experimental design for G is then defined and projected onto each dimension 1 hyperplane (for a first-order truncation).

This leads us to consider n designs of experiments, one for each function of the decomposition. n metamodels $\hat{G}_i(\cdot,\bar{\mathbf{x}}), i=1,\ldots,n,$ are then calibrated to substitute the functions $G_i(\cdot,\bar{\mathbf{x}}) = G(x_i,\bar{\mathbf{x}})$ and $\hat{G}$ is obtained as:

$$G(\mathbf{x}) \approx \hat{G}(\mathbf{x}) = g_0 + \sum_{i=1}^{n} \hat{g}_i(x_i) \quad [6.30]$$

with:

$$g_0 = G(\bar{\mathbf{x}}) \quad [6.31]$$

$$\hat{g}_i(x_i) = \hat{G}_i(x_i,\bar{\mathbf{x}}) - g_0 \quad [6.32]$$

Since every metamodel $\hat{G}_i(\cdot, \overline{\mathbf{x}})$ is independent by construction, prediction $\mu_{\hat{G}}$ and variance $\sigma^2_{\hat{G}}$ are defined at every point $\mathbf{x}^{(k)}$ in the population by:

$$\mu_{\hat{G}}(\mathbf{x}^{(k)}) = (1-n)g_0 + \sum_{i=1}^{n} \mu_{\hat{G}_i}(x_i^{(k)}, \overline{\mathbf{x}}) \qquad [6.33]$$

$$\sigma^2_{\hat{G}}(\mathbf{x}^{(k)}) = \sum_{i=1}^{n} \sigma^2_{\hat{G}_i}(x_i^{(k)}, \overline{\mathbf{x}}) \qquad [6.34]$$

The U learning function is defined here using equations [6.33] and [6.34]. The point $\mathbf{x}_a^{(*)}$ of the Monte Carlo population minimizing this function is selected. In order not to generate unnecessary numerical efforts, only the metamodel $\hat{G}_{i^*}(\cdot, \overline{\mathbf{x}})$ having the maximal variance is recalibrated from its enriched experimental design. The training of the set of metamodels stops when the probability of misclassifying points in the Monte Carlo population is higher than a predefined threshold P_{target} (namely $\min U \geq 2$). The probability of failure is finally evaluated in the same way as in the AK-MCS algorithm.

It is important to note that it is theoretically possible to consider the higher order functions of the HDMR decomposition to improve the representativeness of the performance function, but at the expense of the efficiency of the method as discussed in Lelièvre (2018).

The AK-HDMR1 method is illustrated using the modified *Oblate Spheroid* quadratic performance function (Waarts 2000), which is given as:

$$G(\mathbf{x}) = R - \sum_{i=1}^{n} \frac{x_i^2}{1+i/10} - \frac{p}{n} \sum_{i=1}^{n-1} \sum_{j=i+1}^{n} \frac{x_i x_j}{1+ij}$$

where R is a constant that is used to adjust the probability P_f and p is a parameter to control interactions.

All the random variables of the problem are independent and identically distributed; they follow a standard normal distribution. A reference calculation with the MCS method is performed with 10^5 samples. The results in Figure 6.7, obtained for coefficients of variation of the same order of magnitude, show that the AK-HDMR1 method is naturally very efficient in the absence of interactions ($p = 0$), but that the deviations become non-negligible when the interactions are preponderant in the assessment of the probability of failure.

However, AK-HDMR1 remains relevant if the problem exhibits weak interactions with an acceptable number of calls to the performance function. A judicious choice of the anchor point can reduce the impact of the interactions between the variables on the estimation of the probability of failure.

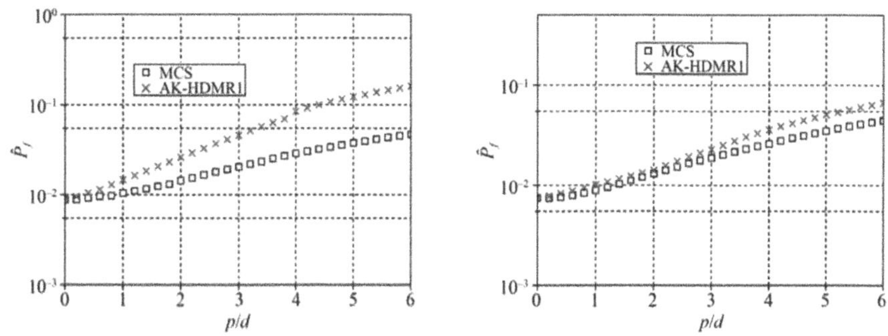

$d = 100$, $R = 35$ around 600 calls to the performance function

$d = 200$, $R = 42$ around 1200 calls to the performance function

Figure 6.7. *Modified oblate spheroid performance function. Influence of interactions on the efficacy of AK-HDMR1. For a color version of this figure, see www.iste.co.uk/gogu/uncertainties.zip*

6.8. Conclusion

This chapter presents adaptive methods based on Kriging for the propagation of uncertainties, and more specifically for the calculation of failure probabilities. AK methods are based on a classification principle, not regression, which makes them particularly well suited and therefore effective for calculating probabilities of failure. These methods, like many others, still suffer from some difficulties in achieving low probabilities and/or considering large problems. It can also be noted that, unlike regression methods, any change to a distribution parameter or a deterministic variable requires one to restart the computations of the performance function in order to, at the very least, complete the convergence of the method.

Finally, these methods are interesting in other contexts such as geometric conformity verification (Dumas *et al.* 2013) or stress management in optimization. The only restriction on the latter subject is the prohibition, or the possibility, taken with great care, of deriving a "classifier" of the Kriging-type resulting from the use of an AK-type method.

6.9. References

Ahmadivala, M., Mattrand, C., Gayton, N., Yalamas, T., Orcési, A., Causse, G. (2019). Time-variant reliability analysis based on AK-SYS. *13th International Conference on Applications of Statistics and Probability in Civil Engineering, ICASP13*, May 26–30, Seoul.

Au, S. and Beck, J. (2001). Estimation of small failure probabilities in high dimensions by subset simulation. *Probabilistic Engineering Mechanics*, 16(4), 263–277.

Bect, J., Ginsbourger, D., Li, L., Picheny, V., Vazquez, E. (2012). Sequential design of computer experiments for the estimation of a probability of failure. *Statistics and Computing*, 22(3), 773–793.

Bichon, B., Eldred, M., Swiler, L., Mahadevan, S., McFarland, J.M. (2008). Efficient global reliability analysis for nonlinear implicit performance functions. *AIAA Journal*, 46(10), 2459–2468.

Bichon, B., McFarland, J.M., Mahadevan, S. (2011). Efficient surrogate models for reliability analysis of systems with multiple failure modes. *Reliability Engineering and System Safety*, 96(10), 1386–1395.

Blatman, G. and Sudret, B. (2008). Sparse polynomial chaos expansions and adaptive stochastic finite elements using a regression approach. *Comptes rendus mécanique*, 336, 518–523.

Bucher, C. and Bourgund, U. (1990). A fast and efficient response surface approach for structural reliability problems. *Structural Safety*, 7, 57–66.

De Rocquigny, E. (2009). Structural reliability under monotony: A review of properties of form and associated simulation methods and a new class of monotonous reliability methods (MRM). *Structural Safety*, 31(5), 363–374.

Dubourg, V. (2011). Adaptive surrogate models for reliability analysis and reliability-based design optimization. PhD Thesis, Blaise Pascal University, Clermont-Ferrand.

Dumas, A., Echard, B., Gayton, N., Rochat, O., Dantan, J. (2013). AK-ILS: An active learning method based on Kriging for the inspection of large surfaces. *Precision Engineering*, 37(1), 1–9.

Echard, B., Gayton, N., Lemaire, M. (2011). AK-MCS: An active learning reliability method combining Kriging and Monte Carlo simulation. *Structural Safety*, 33(2), 145–154.

Echard, B., Gayton, N., Lemaire, M., Relun, N. (2013). A combined importance sampling and Kriging reliability method for small failure probabilities with time-demanding numerical models. *Reliability Engineering and System Safety*, 111, 232–240.

Fauriat, W. and Gayton, N. (2014). AK-SYS: An adaptation of the AK-MCS method for system reliability. *Reliability Engineering and System Safety*, 123, 137–144.

Forrester, A.I. and Keane, A.J. (2009). Recent advances in surrogate-based optimization. *Progress in Aerospace Sciences*, 45(1), 50–79.

Hansen, N., Müller, S., Koumoutsakos, P. (2003). Reducing the time complexity of the derandomized evolution strategy with covariance matrix adaptation (CMA-ES). *Evolutionary Computation*, 11(1), 1–18.

Hu, Z. and Mahadevan, S. (2016). A single-loop Kriging surrogate modeling for time-dependent reliability analysis. *Journal of Mechanical Design*, 138.

Hu, Z., Nannapaneni, S., Mahadevan, S. (2017). Efficient Kriging surrogate modeling approach for system reliability analysis. *Artificial Intelligence for Engineering Design, Analysis and Manufacturing*, 31, 143–160.

Huang, X., Chen, J., Zhu, H. (2016). Assessing small failure probabilities by AK-SS: An active learning method combining Kriging and Subset Simulation. *Structural Safety*, 59, 86–95.

Jones, D.R., Schonlau, M., Welch, W.J. (1998). Efficient global optimization of expensive black-box functions. *Journal of Global Optimization*, 13(4), 455–492.

Kaymaz, I. (2005). Application of Kriging method to structural reliability problems. *Structural Safety*, 27(2), 133–151.

Krige, D. (1951). A statistical approach to some basic mine valuation problems on the Witwatersrand. *Journal of the Chemical, Metallurgical and Mining Society of South Africa*, 52(6), 119–139.

Lelièvre, N. (2018). Développement des méthodes AK pour l'analyse de fiabilité. Focus sur les évènements rares et la grane dimension. PhD Thesis, Clermont-Auvergne University, Clermont-Ferrand.

Lelièvre, N., Beaurepaire, P., Mattrand, C., Gayton, N. (2018). AK-MCSi: A Kriging-based method to deal with small failure probabilities and time-consuming models. *Structural Safety*, 73, 1–11.

Lemaire, M. (2009). *Structural Reliability*. ISTE Ltd, London, and John Wiley & Sons, New York.

Li, G., Rosenthal, C., Rabitz, H. (2001). High dimensional model representations. *The Journal of Physical Chemistry*, 105(33), 7765–7777.

Lophaven, S., Nielsen, H., Søndergaard, J. (2002). Aspects of the Matlab toolbox DACE. Technical report, Technical University of Denmark, Kongens Lyngby.

Lv, A., Lu, Z., Wang, P. (2015). A new learning function for Kriging and its applications to solve reliability problems in engineering. *Computers and Mathematics with Applications*, 70(5), 1182–1197.

Ma, J., Ren, Z., Zhao, G., Zhang, Y., Koh, C.S. (2018). A new reliability analysis method combining adaptive Kriging with weight index Monte Carlo simulation. *IEEE Transactions on Magnetics*, 54(3), 1–4.

Matheron, G. (1973). The intrinsic random functions and their applications. *Advances in Applied Probability*, 5(3), 439–468.

Picheny, V., Ginsbourger, D., Roustant, O., Haftka, R., Kim, N.-H. (2010). Adaptive design of experiments for accurate approximation of a target region. *Journal of Mechanical Design*, 132.

Rabitz, H. and Aliş, Ö.F. (1999). General foundations of high-dimensional model representations. *Journal of Mathematical Chemistry*, 25, 197–233.

Rabitz, H., Aliş, Ö.F., Shorter, J., Shim, K. (1999). Efficient input-output model. *Computer Physics Communications*, 117, 11–20.

Romero, V.J., Swiler, L.P., Giunta, A. (2004). Construction of response surfaces based on progressive-lattice-sampling experimental designs with application to uncertainty propagation. *Structural Safety*, 26(2), 201–219.

Sudret, B. (2008). Global sensitivity analysis using polynomial chaos expansion. *Reliability Engineering and System Safety*, 93(7), 964–979.

Waarts, P.H. (2000). Structural reliability using finite element methods. Technical report, Delft University of Technology, Delft.

Wen, Z., Pei, H., Liu, H., Yue, Z. (2016). A sequential Kriging reliability analysis method with characteristics of adaptive sampling regions and parallelizability. *Reliability Engineering and System Safety*, 153, 170–179.

Xiong, B. and Tan, H. (2018). A robust and efficient structural reliability method combining radial-based importance sampling and Kriging. *Science China Technological Sciences*, 61(5), 724–734.

Yang, X., Liu, Y., Mi, C., Tang, C. (2018). System reliability analysis through active learning Kriging model with truncated candidate region. *Reliability Engineering and System Safety*, 169, 235–241.

7

Global Reliability-oriented Sensitivity Analysis under Distribution Parameter Uncertainty

Vincent CHABRIDON[1]**, Mathieu BALESDENT**[2]**,**
Guillaume PERRIN[3]**, Jérôme MORIO**[4]**, Jean-Marc BOURINET**[5]**,**
Nicolas GAYTON[5]

[1] *EDF R&D, Chatou, France*
[2] *ONERA/DTIS, Paris-Saclay University, Palaiseau, France*
[3] *CEA/DAM/DIF, Arpajon, France*
[4] *ONERA/DTIS, University of Toulouse, France*
[5] *SIGMA Clermont, Clermont-Ferrand, France*

7.1. Introduction

The design and analysis of *complex engineering systems* that one can find in major technical disciplines and industries (e.g. aerospace systems, nuclear power plants, bridges) rely intensely on both *modeling* and *simulation*. Modeling often implies starting from mathematical and physical equations in order to build relevant computer models. As for simulation, it aims at providing relevant and efficient numerical tools so as to simulate the systems' behavior, especially when physical phenomena cannot be directly observed or measured experimentally. The computer models commonly used are said to be *deterministic*, meaning that for two identical runs with the same input values, the same model output will be obtained twice (i.e. there is no intrinsic stochasticity within the computer model). Such models are typically encountered in industrial applications when dealing with ordinary or partial differential equations solvers (e.g. finite element simulations in structural analysis, finite volume schemes for computational fluid dynamics). As a result, these codes are often complex to run, potentially costly to evaluate and, sometimes, can be inserted in computational workflows. For these reasons, they can be considered as *black-box* models, i.e. models that are just considered by the user as an input–output function. Assessing the reliability in such a context requires taking the various sources of uncertainties that might affect the behavior of the system into account. Uncertainties may occur in the initial conditions, in the physical properties of the system (i.e. variability of the mechanical properties of materials due the manufacturing process), in some environmental variables (i.e. uncertain loads due to stochastic wind profiles) or, finally, in the modeling itself (e.g. imperfect mathematical models of the real underlying physical phenomena, limited numerical accuracy). In this chapter, one will consider a probabilistic framework for the input uncertainty modeling assuming sufficient information is available to construct a relevant input probabilistic model (i.e. through assigning a random variable structure to uncertain input variables). Once the sources of uncertainties have been clearly identified, it seems relevant to characterize them following whether the analyst can, in near-term and regarding a reasonable budget, reduce them (or not) so as to improve the accuracy of the predictions. From an engineering point of view, one can distinguish between two types of uncertainties:

– *aleatory uncertainty* refers to natural randomness or any *irreducible* uncertainty in a given context, regarding a given maximum allowable budget;

– *epistemic uncertainty* refers to the lack of knowledge of the analyst and is potentially *reducible* by acquiring more information (i.e. data, measurements, expert judgments).

Following Der Kiureghian and Ditlevsen (2009), such a distinction between aleatory and epistemic uncertainty should not be considered as a classification of phenomena, but only a pragmatic way of distinguishing between the uncertainties on which the analyst can allocate some of the budget for gaining knowledge and those for which it is impossible. Based on this classification, the analyst can take better information-based choices. For the interested reader, more information about this topic can be found in Chapter 1.

IMPORTANT.– In this chapter, one considers that aleatory uncertainty is taken into account through the use of an input probabilistic model, i.e. by assuming some probability distributions for the uncertain inputs (the type/shape of the law is supposed to be known). As for epistemic uncertainty, it is assumed to occur in the modeling of the input distributions (e.g. due to a lack of knowledge about the distribution parameters). This *distribution parameter uncertainty* potentially arises from statistical uncertainty (due to a limited amount of data) or could appear due to the absence of consensus in the expert elicitation process. As a result, one has to consider a *bi-level input uncertainty* composed of:

– the *aleatory uncertainty* modeled by random variables;

– the *epistemic distribution parameter uncertainty*.

Nonetheless, all these uncertainties have to be incorporated as soon as possible in the design process to ensure an optimized, reliable and safe design regarding standards and customers' requirements. This management of uncertainties can be achieved following a general *uncertainty quantification methodology* such as presented in Figure 7.1 (see, e.g. De Rocquigny *et al.* 2008).

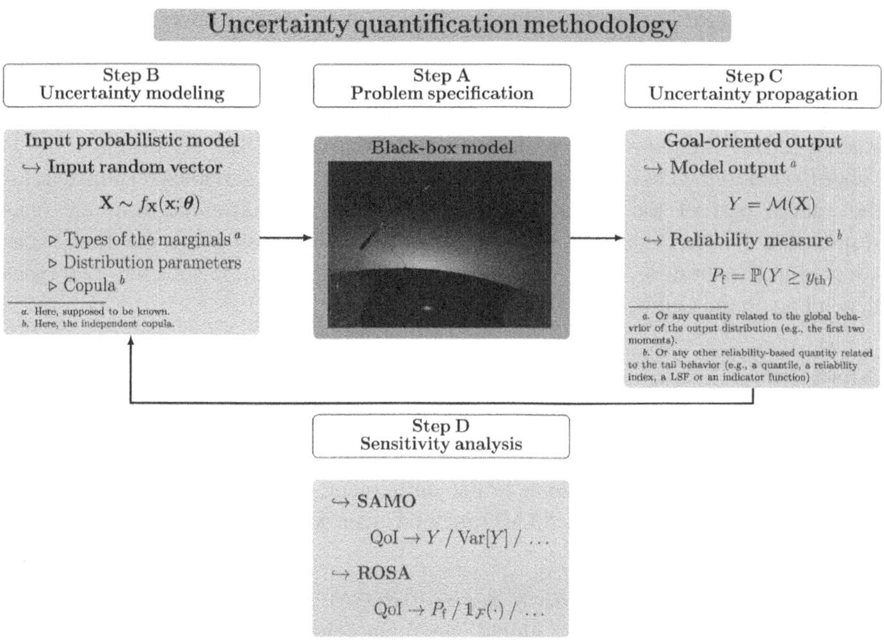

Figure 7.1. *Illustration of the uncertainty quantification methodology. For a color version of this figure, see www.iste.co.uk/gogu/uncertainties.zip*

Thus, starting from the black-box computer model and its specification (step A), one has to choose a mathematical framework to model the input uncertainty (step B). The input uncertainties can than be propagated through the computer model (step C) so as to get the desired *quantity of interest* (QoI). This step is mainly driven by the choice of the QoI to estimate: for instance, one might be interested in the global characterization of the model output distribution (e.g. by computing its first two moments), or in the estimation of a high-order quantile (e.g. in the context of risk analysis), or, finally, in estimating an exceedance probability with respect to (w.r.t.) a safety threshold for reliability assessment. Such a probability is also denoted "*failure probability*" in the reliability community. Usually, in many industrial contexts, the failure probability sought is associated with a rare event, and is thus difficult to estimate. Once the QoI has been statistically estimated, one might be interested in performing a *sensitivity analysis* (SA) so as to investigate the robustness of the estimated QoI regarding the input uncertainty (step D). As stressed by several authors, SA should be performed according to a range of general conceptual objectives, called "*SA settings*", properly defined in various references (see, e.g. Saltelli *et al.* 2008). As an example, one can mention the *factor prioritization* (FP-setting), which stipulates that the aim of SA could be to identify the key inputs driving the model behavior. Thus, a possible reduction of the uncertainty affecting these inputs might lead to the largest reduction of the output uncertainty. Another objective could be the *factor fixing* (FF-setting), which aims at identifying the non-influential inputs that could be fixed at some given values without any loss of information about the output. As already pointed out by several authors (see, e.g. Rachdi 2011; Lemaître 2014; Raguet and Marrel 2018), performing SA is subordinated to the choice of the QoI. Indeed, the key input variables responsible for the central part of the output distribution can be very different from the ones responsible for the tails (i.e. extreme quantiles). Thus, following the framework adopted in (Chabridon 2018)[1], one can consider that SA can be split into two types depending on the QoI being considered:

IMPORTANT.–

– *Sensitivity analysis of model output (SAMO)*: this involves gathering the set of methods for local, screening or global SA while considering the model output (e.g. focusing on the first moments or the entire range of the distribution) as the QoI. Comprehensive reviews about SAMO methods can be found in Iooss and Lemaître (2015) and Wei *et al.* (2015).

– *Reliability-oriented sensitivity analysis* (*ROSA*): this involves gathering the set of methods (from local ones to global ones) that constrain the SA to specific QoIs, such as the typical reliability measures (e.g. failure probability, reliability index) or, in a broader sense, any restrictive QoI (e.g. binary indicator function of a failure event, values of the limit-state function, high- or low-order quantile, union of failure

1 Note that, in the paper from Fort *et al.* (2016), the authors aim at developing a general approach for SA, named "*Goal Oriented Sensitivity Analysis*" (GOSA). The term "*Reliability-Oriented Sensitivity Analysis*" (ROSA) used in this work (not dedicated to a single method but to a wider class of methods) has been mainly borrowed and adapted from this reference. Note that, in the recent work of Perrin and Defaux (2019), a similar semantics is adopted.

events). Such a definition is close to the notion of "target sensitivity analysis" recently proposed by Raguet and Marrel (2018). More comprehensive reviews about ROSA methods can be found in Lemaître (2014) and Chabridon (2018).

To be more specific, the core differences between SAMO and ROSA can be briefly summarized as follows:

– the nature of the QoI: to make it simple, the idea is not to focus on the variability of the model output over its whole support (as done in SAMO), but on a more restrictive domain such as the left/right tail of the distribution (Raguet and Marrel 2018);

– the problem of rare event probability estimation: the second difference is related to the fact that performing ROSA implies first getting "access" to the reliability measure. In the context of rare event probability estimation, this specific task can be difficult and/or computationally expensive to achieve and may require dedicated techniques. Furthermore, we need to ensure the QoI has been accurately estimated so as not to introduce any strong bias in the ROSA. This part often implies using properly tuned rare event probability estimation algorithms (see, e.g. Chapter 6 for more information about this topic);

– a reinterpretation of the traditional SA settings: as highlighted in Zio and Pedroni (2012) and Lemaître (2014), the specific context of ROSA implies reinterpreting the traditional SA settings listed previously. As an example, the FF-setting should be carefully examined as a realization of an undesired failure event might be due to a combination of several input effects, rather than a strong effect from a sole input.

For a broader and deeper review of existing ROSA methods, the interested reader is invited to refer to Lemaître (2014), Raguet and Marrel (2018) and Chabridon (2018).

In this chapter, one will mainly focus on a specific class of ROSA methods and its extension to reliability problems involving two uncertainty levels in input. This chapter is therefore organized as follows. Section 7.2 introduces the theoretical framework by introducing the core notations used in the following sections. Section 7.3 aims at presenting the Sobol' indices on the indicator function. Section 7.4 proposes an extension of the previous indices to the case of a bi-level input uncertainty. Section 7.5 investigates an efficient strategy based on both simulation and kernel density estimation (KDE) so as to get efficient estimators for the indices. Section 7.6 illustrates the benefits of such a methodology on different test cases (two academic test cases and one black-box computer code issued from aerospace research). Finally, a conclusion gathering the most important outcomes and some perspectives of this chapter are given in section 7.7.

7.2. Theoretical framework and notations

The black-box computer model under study can be modeled as follows:

$$\mathcal{M} : \left| \begin{array}{l} \mathcal{D}_{\mathbf{X}} \subseteq \mathbb{R}^d \longrightarrow \mathcal{D}_Y \subseteq \mathbb{R} \\ \quad \mathbf{X}(\omega) \longmapsto Y = \mathcal{M}(\mathbf{X}(\omega)) \end{array} \right. \qquad [7.1]$$

where $\mathbf{X} = (X_1, \ldots, X_d)^\top$ is a d-dimensional second-order random vector gathering the input random variables. In the present study, these random variables are supposed to model the inputs affected by *aleatory uncertainty* (basically, the natural variability which is considered as irreducible). This continuous random vector is defined on a probability space $(\Omega, \mathcal{A}, \mathbb{P})$ with values in $\mathbb{R}^d$. In the rest of this chapter, the dependence over the sample space in $\mathbf{X}(\omega)$ will be omitted and the notation x will be used to denote a realization of the input random vector. One assumes that the joint probability density function (PDF) of $\mathbf{X}$ exists and is given by $f_{\mathbf{X}} : \mathcal{D}_{\mathbf{X}} \subseteq \mathbb{R}^d \to \mathbb{R}_+$. This joint PDF is basically made of the marginal distributions (denoted by $f_{X_i}, i \in \{1, \ldots, d\}$) and a stochastic dependence structure (i.e. a copula) between all these marginals. In this chapter, one assumes oneself to be in the independent copula case (Nelsen 2006; Lebrun 2013), meaning that the input variables are independent and, consequently, that the joint PDF is given by the product of the marginals. Moreover, it is assumed that the joint PDF belongs to a parametric family $\mathcal{P}_{\boldsymbol{\theta}}$ such that:

$$\mathcal{P}_{\boldsymbol{\theta}} = \{f_{\mathbf{X}}(\cdot; \boldsymbol{\theta}) \mid \boldsymbol{\theta} \in \mathcal{D}_{\boldsymbol{\theta}} \subseteq \mathbb{R}^{n_{\boldsymbol{\theta}}}\} \qquad [7.2]$$

where $\boldsymbol{\theta}$ stands for the vector of distribution parameters. In the following, the parametric dependence on $\boldsymbol{\theta}$ will be explicitly mentioned when necessary. Otherwise, it will be omitted for the sake of conciseness.

Due to the uncertainties in input and by propagating them through the model $\mathcal{M}(\cdot)$ (which is assumed to be deterministic), one obtains the model output, denoted by Y. This quantity is a random variable (assumed to be scalar in this chapter) and is characterized by its PDF $f_Y : \mathcal{D}_Y \subseteq \mathbb{R} \to \mathbb{R}_+$. The performance of the system of interest, modeled by $\mathcal{M}(\cdot)$, can be measured by a deterministic scalar function $g : \mathbb{R}^d \to \mathbb{R}$ called the *limit-state function* (LSF). A standard formulation for the LSF in the context of reliability assessment is given by:

$$g(\mathbf{X}) = y_{\text{th}} - \mathcal{M}(\mathbf{X}) = y_{\text{th}} - Y \qquad [7.3]$$

where $y_{\text{th}} \in \mathbb{R}$ is a characteristic threshold output value beyond the one the system is considered as in a failure state. For any realization x of the input random vector, one can distinguish between two canonical domains: the *failure domain* given by

$\mathcal{F} = \{\mathbf{x} \mid g(\mathbf{x}) \leq 0\}$ and the *safe domain* given by $\mathcal{S} = \{\mathbf{x} \mid g(\mathbf{x}) > 0\}$. The zero values of the LSF represent a hypersurface in $\mathbb{R}^d$ called the *limit-state surface* (LSS), which is defined by $\mathcal{F}^0 = \{\mathbf{x} \mid g(\mathbf{x}) = 0\}$. Thus, by definition, the LSS belongs to the failure domain.

Finally, a widely used QoI in reliability assessment is the *failure probability*, denoted by P_f, and given by the following d-fold integral:

$$\begin{aligned} P_\mathrm{f} &= \mathbb{P}(Y \geq y_\mathrm{th}) = \mathbb{P}\left(g(\mathbf{X}) \leq 0\right) \\ &= \int_{\mathcal{F}} f_\mathbf{X}(\mathbf{x})\mathrm{d}\mathbf{x} = \int_{\mathcal{D}_\mathbf{X}} \mathbb{1}_{\mathcal{F}}(\mathbf{x}) f_\mathbf{X}(\mathbf{x})\mathrm{d}\mathbf{x} = \mathbb{E}_{f_\mathbf{X}}\left[\mathbb{1}_{\mathcal{F}}(\mathbf{X})\right] \end{aligned} \quad [7.4]$$

where $\mathrm{d}\mathbf{x} = \mathrm{d}x_1 \ldots \mathrm{d}x_d$ and $\mathbb{1}_{\mathcal{F}}(\cdot)$ is the *indicator function* of the failure domain defined such that $\mathbb{1}_{\mathcal{F}}(\mathbf{x}) = 1$ if $\mathbf{x} \in \mathcal{F}$ and $\mathbb{1}_{\mathcal{F}}(\mathbf{x}) = 0$ otherwise. In the rest of this chapter, it will be assumed that the failure event is rare, i.e. the failure probability P_f is small (e.g. $P_\mathrm{f} \ll \frac{1}{N}$ with N the available simulation budget). In daily engineering practice, rare event failure probabilities often vary from 10^{-2} to 10^{-9} (De Rocquigny *et al.* 2008). The latter order of magnitude may often characterize highly-safe complex engineering systems for which the safety requirements are drastic (e.g. aerospace systems, nuclear power plants or civil engineering structures).

From a practical point of view, a failure probability as defined in equation [7.4] can be estimated using various techniques: on the one hand, using approximation-based techniques such as the first-/second-order reliability methods (FORM/SORM) or any metamodel-based technique such as Gaussian processes or kriging, support vector machines and polynomial chaos expansions (see Chapter 6 for a review of a few kriging-based adaptive strategies); on the other hand, sampling-based techniques such as crude Monte Carlo (CMC) sampling, importance sampling (IS), subset sampling (SS) (see Chapter 5 and Morio and Balesdent (2015) for more comprehensive reviews). In this chapter, the two algorithms that will be used are CMC sampling and SS[2].

For both CMC and SS, we need to consider a sample $\{\mathbf{X}^{(j)}\}_{j=1}^N$ of N *independent and identically distributed* (i.i.d.) copies of the input random vector $\mathbf{X}$ drawn according to $f_\mathbf{X}$ (in the following, this sentence will be denoted by $\{\mathbf{X}^{(j)}\}_{j=1}^N \overset{\text{i.i.d.}}{\sim} f_\mathbf{X}$). Therefore, one will consider the following *elite set* $\mathcal{E} = \{\mathbf{X}^{(j)}, 1 \leq j \leq N_\mathrm{fail} \mid \mathbb{1}_{\mathcal{F}}(\mathbf{X}^{(j)}) = 1\}$, i.e. the N_fail samples that lead to failure regarding the indicator function (or, in other words, the samples which belong to the failure domain $\mathcal{F}$). These elite points are distributed according to the theoretical

2 Note that the formulation of subset simulation used in this chapter is consistent with the one proposed by Au and Beck (2001). This technique is detailed in Chapter 5.

conditional PDF $f_{\mathbf{X}|\mathcal{F}}$, which is unknown in practice. As a remark, in the specific context of rare event probability estimation, this elite set only contains a few points as the failure domain might be difficult to attain.

Once all the core necessary ingredients have been presented to achieve reliability assessment (i.e. failure probability estimation), section 7.3 aims at presenting a class of global variance-based sensitivity indices adapted to ROSA.

7.3. Global variance-based reliability-oriented sensitivity indices

Among several global ROSA indices, variance-based ROSA indices play a major role as they have strong both theoretical and numerical properties and, depending on the context, an easier interpretability feature compared to their indices (Lemaître 2014; Chabridon 2018). Known limits of variance-based indices are often related to their estimation cost and their lack of ability to catch high-order relationships, which are not based on the variance (Da Veiga 2015). However, in the ROSA context, their use has not been widely investigated until very recent works as shown in the following sections.

7.3.1. *Introducing the Sobol' indices on the indicator function*

Following (Li *et al.* 2012), one can note the following link between failure probabilities and mathematical expectations:

$$P_{\mathrm{f}} - P_{\mathrm{f}|X_i} = \mathbb{E}[\mathbb{1}_{\mathcal{F}}(\mathbf{X})] - \mathbb{E}[\mathbb{1}_{\mathcal{F}}(\mathbf{X})|X_i] \quad [7.5]$$

where $P_{\mathrm{f}|X_i}$ is the conditional failure probability when X_i is fixed at its mean value. Then, by squaring and taking the mean w.r.t. X_i:

$$\mathbb{E}_{f_{X_i}}[(P_{\mathrm{f}} - P_{\mathrm{f}|X_i})^2] = \mathbb{E}_{f_{X_i}}\left[(\mathbb{E}[\mathbb{1}_{\mathcal{F}}(\mathbf{X})] - \mathbb{E}[\mathbb{1}_{\mathcal{F}}(\mathbf{X})|X_i])^2\right] \quad [7.6]$$

$$= \mathrm{Var}\left[\mathbb{E}[\mathbb{1}_{\mathcal{F}}(\mathbf{X})|X_i]\right]. \quad [7.7]$$

Finally, by dividing the last equation by the total variance $\mathrm{Var}\left[\mathbb{1}_{\mathcal{F}}(\mathbf{X})\right]$, we get the following set of Sobol' indices on the indicator function (proposed independently by Li *et al.* (2012) and Lemaître (2014)):

$$S_i^{\mathbb{1}_{\mathcal{F}}} = \frac{\mathrm{Var}\left[\mathbb{E}[\mathbb{1}_{\mathcal{F}}(\mathbf{X})|X_i]\right]}{\mathrm{Var}\left[\mathbb{1}_{\mathcal{F}}(\mathbf{X})\right]} \quad [7.8]$$

$$S_{T_i}^{\mathbb{1}_{\mathcal{F}}} = 1 - \frac{\mathrm{Var}\left[\mathbb{E}[\mathbb{1}_{\mathcal{F}}(\mathbf{X})|\mathbf{X}^{-i}]\right]}{\mathrm{Var}\left[\mathbb{1}_{\mathcal{F}}(\mathbf{X})\right]} \quad [7.9]$$

where $S_i^{\mathbb{1}_{\mathcal{F}}}$ is the first-order and and $S_{T_i}^{\mathbb{1}_{\mathcal{F}}}$ the total index associated with the variable X_i. Therefore, $\text{Var}\left[\mathbb{1}_{\mathcal{F}}(\mathbf{X})\right] = P_{\text{f}}(1 - P_{\text{f}})$ and $\mathbf{X}^{-i}$ stands for $\mathbf{X}$ without the ith component X_i. Moreover, note that, similarly to traditional Sobol' indices in the SAMO context, the total index can be rewritten as (Saltelli *et al.* 2008):

$$S_{T_i}^{\mathbb{1}_{\mathcal{F}}} = \frac{\mathbb{E}\left[\text{Var}\left[\mathbb{1}_{\mathcal{F}}(\mathbf{X})|\mathbf{X}^{-i}\right]\right]}{\text{Var}\left[\mathbb{1}_{\mathcal{F}}(\mathbf{X})\right]}. \qquad [7.10]$$

Advanced sampling-based estimation schemes for these Sobol' indices have been investigated in Wei *et al.* (2012). However, these numerical schemes still require a lot of simulations so as to achieve convergence. Very recently, another efficient estimation scheme using subset sampling has been proposed by Perrin and Defaux (2019). This work is detailed hereafter. In the following, one will denote the set of two indices given in equation [7.8] and equation [7.9] by the generic name "$S^{\mathbb{1}_{\mathcal{F}}}$-indices".

7.3.2. *Rewriting Sobol' indices on the indicator function using Bayes' theorem*

Following Perrin and Defaux (2019), it is proposed to rewrite the $S^{\mathbb{1}_{\mathcal{F}}}$-indices presented above as in proposition 1. Note that slightly similar derivations have been proposed in Wang *et al.* (2018) in the context of moment-independent reliability-oriented sensitivity indices.

PROPOSITION 1 (Perrin and Defaux, 2019).– The first and total order $S^{\mathbb{1}_{\mathcal{F}}}$-indices associated with each input X_i, $\forall i \in \{1\ldots, d\}$, can be rewritten as follows:

$$S_i^{\mathbb{1}_{\mathcal{F}}} = \frac{P_{\text{f}}}{1 - P_{\text{f}}}\text{Var}\left[\frac{f_{X_i|\mathcal{F}}(X_i)}{f_{X_i}(X_i)}\right] \qquad [7.11]$$

$$S_{T_i}^{\mathbb{1}_{\mathcal{F}}} = 1 - \frac{P_{\text{f}}}{1 - P_{\text{f}}}\text{Var}\left[\frac{f_{\mathbf{X}^{-i}|\mathcal{F}}(\mathbf{X}^{-i})}{f_{\mathbf{X}^{-i}}(\mathbf{X}^{-i})}\right] \qquad [7.12]$$

where:

$$f_{\mathbf{X}|\mathcal{F}}(\mathbf{x}) = \frac{\mathbb{1}_{\mathcal{F}}(\mathbf{x})f_{\mathbf{X}}(\mathbf{x})}{P_{\text{f}}} \qquad [7.13a]$$

$$f_{X_i|\mathcal{F}}(x_i) = \int_{\mathcal{D}_{\mathbf{X}^{-i}}} f_{\mathbf{X}|\mathcal{F}}(\mathbf{x}) \prod_{\substack{j=1\\j\neq i}}^{d} \mathrm{d}x_j, \quad \mathcal{D}_{\mathbf{X}^{-i}} = \bigtimes_{\substack{j=1\\j\neq i}}^{d} \mathcal{D}_{X_j} \qquad [7.13b]$$

$$f_{\mathbf{X}^{-i}|\mathcal{F}}(\mathbf{x}^{-i}) = \int_{\mathcal{D}_{X_i}} f_{\mathbf{X}|\mathcal{F}}(\mathbf{x})\,\mathrm{d}x_i. \qquad [7.13c]$$

PROOF.– A sketch of proof can be found in Perrin and Defaux (2019). However, more details are given below for the sake of clarity. Recalling that $\text{Var}[\mathbb{1}_{\mathcal{F}}(\mathbf{X})] = P_{\text{f}}(1 - P_{\text{f}})$ and applying Bayes' theorem to equation [7.7], we get:

$$\mathbb{E}_{f_{X_i}}[(P_{\text{f}} - P_{\text{f}|X_i})^2] = \mathbb{E}_{f_{X_i}}\left[\left(P_{\text{f}} - P_{\text{f}}\frac{f_{X_i|\mathcal{F}}(x_i)}{f_{X_i}(x_i)}\right)^2\right] \qquad [7.14a]$$

$$= P_{\text{f}}^2\,\mathbb{E}_{f_{X_i}}\left[\left(1 - \frac{f_{X_i|\mathcal{F}}(x_i)}{f_{X_i}(x_i)}\right)^2\right]. \qquad [7.14b]$$

Hence, for the first order index, by applying the total variance theorem, we get:

$$S_i^{\mathbb{1}_{\mathcal{F}}} = \frac{\mathbb{E}_{f_{X_i}}[(P_{\text{f}} - P_{\text{f}|X_i})^2]}{P_{\text{f}}(1 - P_{\text{f}})} \qquad [7.15a]$$

$$= \frac{P_{\text{f}}}{1 - P_{\text{f}}}\left(\text{Var}\left[1 - \frac{f_{X_i|\mathcal{F}}(x_i)}{f_{X_i}(x_i)}\right] + \mathbb{E}_{f_{X_i}}\left[1 - \frac{f_{X_i|\mathcal{F}}(x_i)}{f_{X_i}(x_i)}\right]^2\right) \qquad [7.15b]$$

$$= \frac{P_{\text{f}}}{1 - P_{\text{f}}}\left(\text{Var}\left[\frac{f_{X_i|\mathcal{F}}(x_i)}{f_{X_i}(x_i)}\right] + 1 - \mathbb{E}_{f_{X_i}}\left[\frac{f_{X_i|\mathcal{F}}(x_i)}{f_{X_i}(x_i)}\right]^2\right). \qquad [7.15c]$$

Moreover, one can note that:

$$\mathbb{E}_{f_{X_i}}\left[\frac{f_{X_i|\mathcal{F}}(x_i)}{f_{X_i}(x_i)}\right] = \int_{\mathcal{D}_{X_i}} \frac{f_{X_i|\mathcal{F}}(x_i)}{f_{X_i}(x_i)} f_{X_i}(x_i)\mathrm{d}x_i \qquad [7.16a]$$

$$= \int_{\mathcal{D}_{X_i}}\int_{\mathcal{D}_{\mathbf{X}^{-i}}} f_{\mathbf{X}|\mathcal{F}}(\mathbf{x}) \prod_{\substack{j=1\\ j\neq i}}^{d} \mathrm{d}x_j\,\mathrm{d}x_i \qquad [7.16b]$$

$$= \int_{\mathcal{D}_{\mathbf{X}}} \frac{1}{P_{\text{f}}}\mathbb{1}_{\mathcal{F}}(\mathbf{x}) f_{\mathbf{X}}(\mathbf{x})\mathrm{d}\mathbf{x} \qquad [7.16c]$$

$$= 1 \qquad [7.16d]$$

which finally allows us to get the expected result. Similar derivations for the total index can be achieved starting from:

$$\mathbb{E}_{f_{\mathbf{X}^{-i}}}[(P_{\text{f}} - P_{\text{f}|\mathbf{X}^{-i}})^2] = P_{\text{f}}^2\,\mathbb{E}_{f_{\mathbf{X}^{-i}}}\left[\left(1 - \frac{f_{\mathbf{X}^{-i}|\mathcal{F}}(\mathbf{x}^{-i})}{f_{\mathbf{X}^{-i}}(\mathbf{x}^{-i})}\right)^2\right] \qquad [7.17a]$$

and noticing that $\mathbb{E}_{f_{\mathbf{X}^{-i}}}\left[\frac{f_{\mathbf{X}^{-i}|\mathcal{F}}(\mathbf{x}^{-i})}{f_{\mathbf{X}^{-i}}(\mathbf{x}^{-i})}\right] = 1$.

As a result, the two estimators associated with the first-order and total indices are given by:

$$\widehat{S}_i^{\mathbb{1}_\mathcal{F}} = \frac{\widehat{P}_\mathrm{f}}{1-\widehat{P}_\mathrm{f}}\mathrm{Var}\left[\frac{\widehat{f}_{X_i|\mathcal{F}}(X_i)}{f_{X_i}(X_i)}\right] \qquad [7.18]$$

$$\widehat{S}_{T_i}^{\mathbb{1}_\mathcal{F}} = 1 - \frac{\widehat{P}_\mathrm{f}}{1-\widehat{P}_\mathrm{f}}\mathrm{Var}\left[\frac{\widehat{f}_{\mathbf{X}^{-i}|\mathcal{F}}(\mathbf{X}^{-i})}{f_{\mathbf{X}^{-i}}(\mathbf{X}^{-i})}\right] \qquad [7.19]$$

where $\widehat{f}_{X_i|\mathcal{F}}$ is the estimator of the marginal PDF of X_i at failure, while $\widehat{f}_{\mathbf{X}^{-i}|\mathcal{F}}$ is the estimator of the joint PDF of $\mathbf{X}^{-i}$ at failure, i.e. a $(d-1)$-dimensional PDF.

This set of estimators presents the main advantage of requiring only a postprocessing phase of a single elite set $\mathcal{E}$ obtained from a reliability analysis using any sampling-based technique mentioned in section 7.2. However, the difficulty remains the accurate estimation of the densities, especially for the total indices, which involve the estimation of a multivariate density at failure.

In the following, we propose to adapt the previous set of estimators and the efficient estimation strategy to the bi-level input uncertainty problem. This part consists the core technical contribution of this chapter.

7.4. Sobol' indices on the indicator function adapted to the bi-level input uncertainty

7.4.1. *Reliability analysis under distribution parameter uncertainty*

In this chapter, two levels of uncertainty are considered: the first one represents the variability in the basic input variables and thus affects the input random vector $\mathbf{X}$, while the second one represents the lack of knowledge affecting the distribution parameters. To do so, the following *Bayesian hierarchical model* (Der Kiureghian 2008) is considered:

$$\mathbf{X} \sim f_{\mathbf{X}|\boldsymbol{\Theta}}(\mathbf{x}|\boldsymbol{\theta}) : \mathcal{D}_\mathbf{X} \subseteq \mathbb{R}^d \to \mathbb{R}_+ \quad \text{(uncertainty level \#1)} \qquad [7.20a]$$

$$\boldsymbol{\Theta} \sim f_{\boldsymbol{\Theta}|\boldsymbol{\xi}}(\boldsymbol{\theta}|\boldsymbol{\xi}) : \mathcal{D}_{\boldsymbol{\Theta}} \subseteq \mathbb{R}^{n_\theta} \to \mathbb{R}_+ \quad \text{(uncertainty level \#2)} \qquad [7.20b]$$

$$\boldsymbol{\xi} = (\xi_1, \xi_2, \ldots, \xi_{n_\xi})^\top \in \mathbb{R}^{n_\xi} \quad \text{(deterministic level)} \qquad [7.20c]$$

In this hierarchical representation, we can distinguish three layers of inputs:

– the first layer is constituted by the random vector $\mathbf{X}$ gathering the basic stochastic variables. Based on prior knowledge, a probability distribution can be assumed through the choice of a parametric model;

– the second layer is constituted by uncertain and deterministic (i.e. supposed to be known accurately enough) distribution parameters. Adding such a layer is consistent with the Bayesian point of view of modeling either "uncertain" (in the sense of stochastic) or "unknown but fixed" parameters. In the present chapter, the prior distribution is mostly assumed to be derived from expert judgment or from a limited set of data. Thus, despite the fact that Bayes' theorem is not used as an updating procedure, we can consider that this prior distribution characterizes the epistemic uncertainty affecting the distribution parameters;

– the third layer is composed of fixed hyperparameters gathered in $\boldsymbol{\xi}$. These hyperparameters, which can be either moments or bounds, characterize the prior distributions of uncertain parameters $\boldsymbol{\Theta}$.

The treatment of distribution parameter uncertainty in reliability assessment has been originally described in several seminal works of Der Kiureghian (1989) and Hong (1996). Extensions of the previous works have been proposed in Pendola (2000), Der Kiureghian (2008), Der Kiureghian and Ditlevsen (2009). Therefore, in such a context of bi-level input uncertainty, an interesting reliability measure can be the *predictive failure probability*, defined such that:

$$\widetilde{P}_{\mathrm{f}}(\boldsymbol{\xi}) \stackrel{\text{def}}{=} \mathbb{E}_{f_{\boldsymbol{\Theta}|\boldsymbol{\xi}}}\left[P_{\mathrm{f}}(\boldsymbol{\Theta})\right] = \mathbb{E}_{f_{\boldsymbol{\Theta}|\boldsymbol{\xi}}}\left[\mathbb{E}_{f_{\mathbf{X}|\boldsymbol{\Theta}}}\left[\mathbb{1}_{\mathcal{F}}(\mathbf{X})|\boldsymbol{\Theta}\right]|\boldsymbol{\xi}\right] \qquad [7.21a]$$

$$= \int_{\mathcal{D}_{\boldsymbol{\Theta}}} P_{\mathrm{f}}(\boldsymbol{\theta}) f_{\boldsymbol{\Theta}|\boldsymbol{\xi}}(\boldsymbol{\theta}|\boldsymbol{\xi})\mathrm{d}\boldsymbol{\theta} \qquad [7.21b]$$

where $P_{\mathrm{f}}(\boldsymbol{\theta})$ is the *conditional failure probability*[3] given by:

$$P_{\mathrm{f}}(\boldsymbol{\theta}) = \mathbb{P}(g(\mathbf{X}) \leq 0 \mid \boldsymbol{\Theta} = \boldsymbol{\theta}) = \int_{\mathcal{D}_{\mathbf{X}}} \mathbb{1}_{\mathcal{F}}(\mathbf{x}) f_{\mathbf{X}|\boldsymbol{\Theta}}(\mathbf{x}|\boldsymbol{\theta})\mathrm{d}\mathbf{x} \qquad [7.22a]$$

$$= \mathbb{E}_{f_{\mathbf{X}|\boldsymbol{\Theta}}}\left[\mathbb{1}_{\mathcal{F}}(\mathbf{X}) \mid \boldsymbol{\Theta} = \boldsymbol{\theta}\right]. \qquad [7.22b]$$

As highlighted in Chabridon *et al.* (2017), an efficient way to estimate such a QoI can be achieved by considering a so-called "augmented" random vector $\mathbf{Z} \stackrel{\text{def}}{=} (\mathbf{X}, \boldsymbol{\Theta})^{\top}$ defined on $\mathcal{D}_{\mathbf{Z}} = \mathcal{D}_{\mathbf{X}} \times \mathcal{D}_{\boldsymbol{\Theta}}$ (where $\times$ is the Cartesian product) with joint PDF $f_{\mathbf{Z}|\boldsymbol{\xi}}(\mathbf{z}|\boldsymbol{\xi}) \stackrel{\text{def}}{=} f_{(\mathbf{X},\boldsymbol{\Theta})|\boldsymbol{\xi}}((\mathbf{x},\boldsymbol{\theta})|\boldsymbol{\xi}) = f_{\mathbf{X}|\boldsymbol{\Theta}}(\mathbf{x}|\boldsymbol{\theta}) f_{\boldsymbol{\Theta}|\boldsymbol{\xi}}(\boldsymbol{\theta}|\boldsymbol{\xi})$ such that the expression

3 Please note that the conditional failure probability defined here differs from the one defined in equation [7.5].

in equation [7.21b] can be rewritten as follows:

$$\widetilde{P}_{\mathrm{f}}(\boldsymbol{\xi}) = \int_{\mathcal{D}_{\boldsymbol{\Theta}}} \int_{\mathcal{D}_{\mathbf{X}}} \mathbb{1}_{\mathcal{F}}(\mathbf{x}) f_{\mathbf{X}|\boldsymbol{\Theta}}(\mathbf{x}|\boldsymbol{\theta}) f_{\boldsymbol{\Theta}|\boldsymbol{\xi}}(\boldsymbol{\theta}|\boldsymbol{\xi}) \mathrm{d}\mathbf{x} \mathrm{d}\boldsymbol{\theta} \quad [7.23a]$$

$$= \int_{\mathcal{D}_{\mathbf{Z}}} \mathbb{1}_{\mathcal{F}_{\mathbf{z}}}(\mathbf{z}) f_{\mathbf{Z}|\boldsymbol{\xi}}(\mathbf{z}|\boldsymbol{\xi}) \mathrm{d}\mathbf{z} \quad [7.23b]$$

$$= \mathbb{E}_{f_{\mathbf{Z}|\boldsymbol{\xi}}} \left[\mathbb{1}_{\mathcal{F}_{\mathbf{z}}}(\mathbf{Z}) \mid \boldsymbol{\xi} \right] \quad [7.23c]$$

where $\mathcal{F}_{\mathbf{z}} = \{\mathbf{z} \in \mathcal{D}_{\mathbf{Z}} \mid g(\mathbf{z}) \leq 0\}$. This formulation, called the "*augmented reliability approach*" (ARA), numerically implies estimating the expected value in equation [7.23c]. The coupling between the ARA framework to handle distribution parameter uncertainty and rare event probability estimation (e.g. ARA/CMC and ARA/SS) has been studied in Chabridon *et al.* (2017).

In the following, we will only focus on the first two layers in input, i.e. $\mathbf{X}$ and $\boldsymbol{\Theta}$. Thus, for the sake of conciseness, the third layer (i.e. the hyperparameters $\boldsymbol{\xi}$) will be omitted. The dependence on $\boldsymbol{\xi}$ is interesting in the particular case of failure probability robustness analysis using local ROSA, such as proposed in Chabridon *et al.* (2018).

7.4.2. *Bi-level input uncertainty: aggregated versus disaggregated types of uncertainty*

When dealing with the bi-level input uncertainty as contained in the hierarchical model described previously, two cases can be considered:

– on the one hand, one might be interested in the combined effects of both aleatory and epistemic uncertainties at failure, i.e. one might want to "*aggregate*" both sources of uncertainty;

– on the other hand, one might be instead interested in distinguishing between the contributions of aleatory and epistemic uncertainties, i.e. one might want to "*disaggregate*" (or "*separate*") both sources of uncertainty.

Concerning the first problem, adapting the previous $S^{\mathbb{1}_{\mathcal{F}}}$-indices can be achieved by considering the equivalent indices evaluated on the augmented vector $\mathbf{Z} = (\mathbf{X}, \boldsymbol{\Theta})^{\top}$. Such indices would reflect the contribution of both the variables $\mathbf{X}|\boldsymbol{\Theta}$ and distribution parameters $\boldsymbol{\Theta}$.

In the second problem, the idea is to study the impact of aleatory and epistemic uncertainties separately. Indeed, as stressed by several authors, such as Helton (1997) and Sankararaman and Mahadevan (2013), separating both contributions of aleatory and epistemic uncertainties may be of utmost importance if one wants to get a deeper insight about which type of uncertainty plays a major role on the system at failure (see

Chapter 1). Moreover, it allows us to build a well-informed decision process, in terms of budget allocation, to possibly reduce epistemic uncertainty if this plays a significant role on the variability of the reliability measure.

IMPORTANT.– In this chapter, one proposes to consider a *disaggregated* version of the augmented input vector to *separate both types of uncertainty*. Moreover, the effects of both types of uncertainty on the indicator function (in other words, "at failure") are studied and taken into account in the ROSA context, as presented in the following sections.

7.4.3. *Disaggregated random variables*

One possible manner to study the effects of distribution parameter uncertainty and input variability separately is to transform the input variables as proposed in Schöbi and Sudret (2017) (in the context of p-boxes). As a preliminary example, one can consider the Gaussian case as an illustrative example. Thus, under a single-level type of uncertainty, a Gaussian random variable X can be split as follows:

$$X = \mu_X + \sigma_X \, U_X \qquad [7.24]$$

where $\boldsymbol{\theta} = (\mu_X, \sigma_X)^\top$ are the distribution parameters and $U_X \sim \mathcal{N}(0,1)$ is a standard Gaussian variable, which characterizes the inherent variability of X. Note that such a transformation is not limited to the Gaussian case. One can build similar transformations for random variables following, for instance, lognormal, Gumbel or Weibull distributions. The interested reader should refer to Schöbi and Sudret (2017) for the presentation of these cases.

Under a bi-level input uncertainty, the idea is to consider a prior distribution $f_{\boldsymbol{\Theta}}$ over the uncertain distribution parameters $\boldsymbol{\Theta} = (M_X, S_X)^\top$. As a result, the previous decomposition can be rewritten as follows:

$$X = M_X + S_X \, U_X \qquad [7.25]$$

where $\mathbf{V}_{\text{dis}} = (M_X, S_X, U_X)^\top$ is a vector of independent variables denoted as the "*disaggregated augmented vector*". Note that, for some cases, only the mean value or the standard deviation could be considered as uncertain. In the following, the study will be limited to the Gaussian case as working in the $\mathbf{u}$-space will facilitate the calculations, without loss of generality. Finally, the augmented input vector under a bi-level uncertainty will be composed of two sorts of inputs: the inputs that are not affected by the bi-level uncertainty (gathered in the vector $\mathbf{X}_{\text{single}}$) and the disaggregated ones, such that $\mathbf{Z} = (\mathbf{V}_{\text{dis}}, \mathbf{X}_{\text{single}})^\top$ with $d_{\mathbf{Z}} = d_{\text{dis}} + d_{\text{single}}$ denoting the dimension of this vector. As a final remark, one should note that, in a multivariate setting, we have $f_{\mathbf{Z}} = f_{\mathbf{\Theta}} \times f_{\mathbf{U}} \times f_{\mathbf{X}_{\text{single}}}$.

As a remark and to avoid any confusion for the reader, one should note that, despite the disaggregated inputs, only the basic variables $\mathbf{X}$ play a role in the physical model $\mathcal{M}(\cdot)$. Thus, these disaggregated inputs are "re-aggregated" within the code such that only $\mathbf{X}$ variables play a role in the physical behavior. In the following, one will use the notation $\mathbf{Z}$ to deal with input variables, but the reader should be aware that only the $\mathbf{X}$ variables have a real physical sense from the failure point of view.

7.4.4. *Extension to the bi-level input uncertainty and pick-freeze estimators*

In this chapter, one proposes to extend the previous $S^{\mathbb{1}_{\mathcal{F}}}$-indices to the bi-level input uncertainty. To do so, the following pair of indices is proposed:

$$S_i^{\mathbb{1}_{\mathcal{F}}} = \frac{\text{Var}\left[\mathbb{E}[\mathbb{1}_{\mathcal{F}_{\mathbf{z}}}(\mathbf{Z})|Z_i]\right]}{\text{Var}\left[\mathbb{1}_{\mathcal{F}_{\mathbf{z}}}(\mathbf{Z})\right]} \quad [7.26]$$

$$S_{T_i}^{\mathbb{1}_{\mathcal{F}}} = 1 - \frac{\text{Var}\left[\mathbb{E}[\mathbb{1}_{\mathcal{F}_{\mathbf{z}}}(\mathbf{Z})|\mathbf{Z}^{-i}]\right]}{\text{Var}\left[\mathbb{1}_{\mathcal{F}_{\mathbf{z}}}(\mathbf{Z})\right]} = \frac{\mathbb{E}\left[\text{Var}\left[\mathbb{1}_{\mathcal{F}_{\mathbf{z}}}(\mathbf{Z})|\mathbf{Z}^{-i}\right]\right]}{\text{Var}\left[\mathbb{1}_{\mathcal{F}_{\mathbf{z}}}(\mathbf{Z})\right]}. \quad [7.27]$$

Pick-freeze estimators can be used to estimate these indices by MC simulations. In the following, these estimators will be used to get the reference results.

Computing $S_i^{\mathbb{1}_{\mathcal{F}}}$ and $S_{T_i}^{\mathbb{1}_{\mathcal{F}}}$ under a bi-level uncertainty using a single-loop CMC sampling can be achieved in four steps. The procedure presented below is adapted from the one proposed by Wei *et al.* (2012).

7.4.4.1. *Step #1*

Generate $2N$ copies of the augmented vector $\{\mathbf{Z}^{(j)}\}_{j=1}^{2N} \overset{\text{i.i.d.}}{\sim} f_{\mathbf{Z}}$. These samples are stored in the following two matrices:

$$\mathbf{A} = \begin{array}{c} \\ 1 \\ 2 \\ \vdots \\ N \end{array} \begin{array}{c} \begin{array}{cccccc} Z_1 & Z_2 & \dots & Z_i & \dots & Z_{d_{\mathbf{Z}}} \end{array} \\ \begin{pmatrix} z_1^{(1)} & z_2^{(1)} & \dots & z_i^{(1)} & \dots & z_{d_{\mathbf{Z}}}^{(1)} \\ z_1^{(2)} & z_2^{(2)} & \dots & z_i^{(2)} & \dots & z_{d_{\mathbf{Z}}}^{(2)} \\ \vdots & \vdots & \ddots & \vdots & \ddots & \vdots \\ z_1^{(N)} & z_2^{(N)} & \dots & z_i^{(N)} & \dots & z_{d_{\mathbf{Z}}}^{(N)} \end{pmatrix} \end{array} \quad [7.28]$$

$$\mathbf{B} = \begin{array}{c} N+1 \\ N+2 \\ \vdots \\ 2N \end{array} \overset{\begin{array}{cccccc} Z_1 & Z_2 & \dots & Z_i & \dots & Z_{d_\mathbf{Z}} \end{array}}{\begin{pmatrix} z_1^{(N+1)} & z_2^{(N+1)} & \dots & z_i^{(N+1)} & \dots & z_{d_\mathbf{Z}}^{(N+1)} \\ z_1^{(N+2)} & z_2^{(N+2)} & \dots & z_i^{(N+2)} & \dots & z_{d_\mathbf{Z}}^{(N+2)} \\ \vdots & \vdots & \ddots & \vdots & \ddots & \vdots \\ z_1^{(2N)} & z_2^{(2N)} & \dots & z_i^{(2N)} & \dots & z_{d_\mathbf{Z}}^{(2N)} \end{pmatrix}} \quad [7.29]$$

7.4.4.2. *Step #2*

Generate a set of matrices $\mathbf{C}^{(i)}$, with $i \in \{1, \dots, d_\mathbf{Z}\}$ where the ith column of $\mathbf{C}^{(i)}$ comes for $\mathbf{A}$ and all other $d_\mathbf{Z} - 1$ columns come from $\mathbf{B}$:

$$\mathbf{C}^{(i)} = \overset{\begin{array}{cccccc} Z_1 & Z_2 & \dots & Z_i & \dots & Z_{d_\mathbf{Z}} \end{array}}{\begin{pmatrix} z_1^{(N+1)} & z_2^{(N+1)} & \dots & z_i^{(1)} & \dots & z_{d_\mathbf{Z}}^{(N+1)} \\ z_1^{(N+2)} & z_2^{(N+2)} & \dots & z_i^{(2)} & \dots & z_{d_\mathbf{Z}}^{(N+2)} \\ \vdots & \vdots & \ddots & \vdots & \ddots & \vdots \\ z_1^{(2N)} & z_2^{(2N)} & \dots & z_i^{(N)} & \dots & z_{d_\mathbf{Z}}^{(2N)} \end{pmatrix}} \quad [7.30]$$

7.4.4.3. *Step #3*

Compute the indicator function values for each sample in the matrices $\mathbf{A}$, $\mathbf{B}$ and $\mathbf{C}^(i)$. Finally, we get the following set of N-dimensional column vectors:

$$\mathbb{1}_\mathbf{A} = \mathbb{1}_{\mathcal{F}_\mathbf{z}}(\mathbf{A}), \quad \mathbb{1}_\mathbf{B} = \mathbb{1}_{\mathcal{F}_\mathbf{z}}(\mathbf{B}), \quad \mathbb{1}_{\mathbf{C}^{(i)}} = \mathbb{1}_{\mathcal{F}_\mathbf{z}}(\mathbf{C}^{(i)}). \quad [7.31]$$

7.4.4.4. *Step #4*

The $S^{\mathbb{1}_\mathcal{F}}$-indices, for $i \in \{1, \dots, d_\mathbf{Z}\}$, are computed using the following CMC estimators:

$$\widehat{S}_i^{\mathbb{1}_\mathcal{F}} = \frac{\widehat{D}_i - \widehat{\widehat{P}}_\mathrm{f}^{\,2}}{\widehat{D}} \quad [7.32]$$

$$\widehat{S}_{T_i}^{\mathbb{1}_\mathcal{F}} = 1 - \frac{\widehat{D}_{-i} - \widehat{\widehat{P}}_\mathrm{f}^{\,2}}{\widehat{D}} \quad [7.33]$$

where:

$$\widehat{\widehat{P}}_{\mathrm{f}} = \frac{1}{2N}\sum_{j=1}^{N}(\mathbb{1}_{\mathbf{A}}^{(j)} + \mathbb{1}_{\mathbf{B}}^{(j)}) \qquad [7.34]$$

$$\widehat{\widehat{P}}_{\mathrm{f}}^{2} = \frac{1}{N}\sum_{j=1}^{N}\mathbb{1}_{\mathbf{A}}^{(j)}\mathbb{1}_{\mathbf{B}}^{(j)} \qquad [7.35]$$

$$\widehat{D} = \widehat{\widehat{P}}_{\mathrm{f}} - \widehat{\widehat{P}}_{\mathrm{f}}^{2} \qquad [7.36]$$

$$\widehat{D}_i = \frac{1}{N}\sum_{j=1}^{N}\mathbb{1}_{\mathbf{A}}^{(j)}\mathbb{1}_{\mathbf{C}^{(i)}}^{(j)} \qquad [7.37]$$

$$\widehat{D}_{-i} = \frac{1}{N}\sum_{j=1}^{N}\mathbb{1}_{\mathbf{B}}^{(j)}\mathbb{1}_{\mathbf{C}^{(i)}}^{(j)} \qquad [7.38]$$

where $\mathbb{1}_{\mathbf{A}}^{(j)}$, $\mathbb{1}_{\mathbf{B}}^{(j)}$ and $\mathbb{1}_{\mathbf{C}^{(i)}}^{(j)}$ are, respectively, the jth component of the column vector $\mathbb{1}_{\mathbf{A}}$, $\mathbb{1}_{\mathbf{B}}$ and $\mathbb{1}_{\mathbf{C}^{(i)}}$.

In the following, reference values for $S^{\mathbb{1}_{\mathcal{F}}}$-indices will be estimated using this four-step procedure. Note that, following Wei *et al.* (2012), an extra step could be added to compute analytical formulas to get the estimation error. However, in the following, this error is controlled by use of repetitions of the whole procedure. At this point, one should note that this single-loop CMC sampling procedure may become inefficient in the context of rare event probability estimation as most of the indicator functions in equation [7.31] could be equal to zero. Moreover, the simulation cost associated with this single-loop procedure is $N_{\mathrm{cost}} = N(d_{\mathbf{Z}} + 2)$ calls to the LSF. To avoid such a computational burden, Wei *et al.* (2012) proposes to use other importance sampling algorithms (in the context of single-level ROSA). In the next section, we investigate another way to efficiently obtain the indices under a bi-level uncertainty, using an efficient estimation scheme coupling the estimators presented in section 7.3.2, a subset sampling technique and KDE.

7.5. Efficient estimation using subset sampling and KDE

7.5.1. *The problem of estimating the optimal distribution at failure*

By adapting the expressions given in equation [7.18] and equation [7.19] to the bi-level uncertainty, we get the following estimators:

$$\widehat{S}_i^{\mathbb{1}_{\mathcal{F}}} = \frac{\widehat{\widehat{P}}_{\mathrm{f}}}{1 - \widehat{\widehat{P}}_{\mathrm{f}}}\mathrm{Var}\left[\frac{\widehat{f}_{Z_i|\mathcal{F}}(Z_i)}{f_{Z_i}(Z_i)}\right] \qquad [7.39]$$

$$\widehat{S}_{T_i}^{\mathbb{1}_\mathcal{F}} = 1 - \frac{\widehat{\widetilde{P}}_\mathrm{f}}{1 - \widehat{\widetilde{P}}_\mathrm{f}} \mathrm{Var}\left[\frac{\widehat{f}_{\mathbf{Z}^{-i}|\mathcal{F}}(\mathbf{Z}^{-i})}{f_{\mathbf{Z}^{-i}}(\mathbf{Z}^{-i})}\right] \qquad [7.40]$$

where $\widehat{\widetilde{P}}_\mathrm{f}$ is the estimated value of the predictive failure probability as presented in section 7.4.1, obtained here by ARA/SS. As suggested by these estimators adapted to the bi-level uncertainty, the first and total sensitivity indices require the estimation of $f_{Z_i|\mathcal{F}}$ and $f_{\mathbf{Z}^{-i}|\mathcal{F}}$. These densities correspond, respectively, to the marginal distribution w.r.t. Z_i and to the "joint minus one" (abbreviated as "joint-1", the one corresponding to the ith marginal) distribution of the failure points. The estimation of these two quantities exhibits a number of challenges that are detailed below in the dedicated paragraphs.

REMARK.– As a remark, one should note that these densities $f_{Z_i|\mathcal{F}}$ and $f_{\mathbf{Z}^{-i}|\mathcal{F}}$ are closely related to the theoretical optimal densities required to achieve optimal variance reduction in the context of importance sampling and subset sampling (see, e.g. Dubourg 2011). Indeed, such densities are known to be challenging to estimate due to many reasons. Therefore, KDE techniques are very popular for tackling this kind of issue (see, e.g. (Silverman 1986)). A brief review of multivariate KDE is given in Box 1.

In the multivariate setting, one can assume that a set $\mathcal{X} = \{\mathbf{x}^{(1)}, \dots, \mathbf{x}^{(N)}\}$ of N i.i.d. observations of a d-dimensional random vector, drawn from an unknown parent density denoted by $f_\mathbf{X}$, is available. One would then like to estimate this unknown parent density. Its kernel density estimator is given by:

$$\widehat{f}_\mathbf{X}(\mathbf{x}) = \frac{\det(\mathbf{H})^{-1/2}}{N} \sum_{j=1}^{N} K_d\left(\mathbf{H}^{-1/2}(\mathbf{x} - \mathbf{x}^{(j)})\right) \qquad [7.41]$$

where $\det(\cdot)$ is the determinant operator, $K_d(\cdot)$ is a multivariate (d-dimensional) kernel and $\mathbf{H}$ is a $(d \times d)$-dimensional positive definite symmetric matrix, often named the "*bandwidth matrix*". Among several widely used kernels, one can mention the Gaussian one, defined such that:

$$K_d(\mathbf{x}) = \frac{1}{(2\pi)^{d/2}} \exp\left[-\frac{1}{2}\mathbf{x}^\top \mathbf{x}\right]. \qquad [7.42]$$

Using a Gaussian kernel corresponds to the traditional "*Gaussian kernel density estimation*" (G-KDE). Although the choice of the kernel $K_d(\cdot)$ may influence the performance of the estimation in some specific cases, the most influential parameter remains the bandwidth matrix $\mathbf{H}$. This matrix has to be tuned so as to find an admissible trade-off between a global smoothing and capturing the peaks of the density. It can be optimized following some optimality criteria, such that the *mean square integrated error* (MISE) given by:

$$\mathrm{MISE}(\mathbf{H}) = \mathbb{E}\left[\int_{\mathbb{R}^d} \left(\widehat{f}_{\mathbf{X}}(\mathbf{x}) - f_{\mathbf{X}}(\mathbf{x})\right)^2 \mathrm{d}\mathbf{x}\right]. \qquad [7.43]$$

In common engineering practice (and when d increases), the Gaussian assumption on $f_{\mathbf{X}}$ is often considered. Thus, a widely used approximation for $\mathbf{H}$ is derived from "*Silverman's rule of thumb*" (Silverman 1986) which assumes that $\mathbf{H}$ takes the following form:

$$\mathbf{H} = \eta_{\mathrm{Silv}}^2 \begin{bmatrix} \widehat{\sigma}_1^2 & 0 & \dots & 0 \\ 0 & \widehat{\sigma}_2^2 & \ddots & \vdots \\ \vdots & \ddots & \ddots & 0 \\ 0 & \dots & 0 & \widehat{\sigma}_d^2 \end{bmatrix} \qquad [7.44]$$

where $\widehat{\sigma}_i^2$ for $i = 1, \dots, d$ is the empirical estimation of the variance of X_i and η_{Silv} is a scalar factor given by:

$$\eta_{\mathrm{Silv}} = \left(\frac{1}{N}\frac{4}{(d+2)}\right)^{\frac{1}{d+4}}. \qquad [7.45]$$

This factor only depends on the input dimension and the number of available samples.

Box 7.1. *Basics of multivariate kernel density estimation*

In the above-defined context, i.e. global ROSA with $S^{\mathbb{1}_{\mathcal{F}}}$-indices under a bi-level input uncertainty, we need to recall the major challenges arising from the estimation of the optimal distribution at failure using KDE, and the new challenges induced by the use of the estimators in equation [7.39] and equation [7.40].

7.5.1.1. *Challenge #1: input dimension*

It is known that the performance of KDE deteriorates with the input dimension. However, in the estimators presented above, this issue predominantly affects the total index $\widehat{S}_{T_i}^{\mathbb{1}_{\mathcal{F}}}$ as it requires the estimation of the joint-1 density $f_{\mathbf{Z}^{-i}|\mathcal{F}}$ at failure. However, this issue is intrinsically affected by the fact that, under a bi-level uncertainty, one considers the augmented vector $\mathbf{Z} = (\mathbf{V}_{\mathrm{dis}}, \mathbf{X}_{\mathrm{single}})^\top$ of size $d_{\mathbf{Z}} > d$ due to the disaggregated variables contained in $\mathbf{V}_{\mathrm{dis}}$. This increased input dimension is the counterpart of taking the bi-level input uncertainty into account.

7.5.1.2. *Challenge #2: complex shape of the optimal density at failure*

Another key issue arises from the fact that the shape of the optimal density at failure may be complex. By complex, we mean that it is a truncated (i.e. truncated by the LSS) distribution, which can be multimodal and may thus arise from a complex combination of the inputs. Note that the multimodality can be either smooth (close modes) or very sharp, especially in the context of disconnected failure regions of almost equal importance.

7.5.1.3. *Challenge #3: limited number of failure points in the elite set*

In the context of rare event probability estimation, the number of failure points, i.e. samples that lie in the true failure domain, is often limited (e.g. with the CMC technique). Even if variance reduction techniques such as IS or SS can be used and may significantly increase the number of failure points in the last iteration, the elite set $\mathcal{E}$ is often quite small compared to the total number of simulations used to get an estimate of the failure probability. Finally, one can also add more samples by using *resampling* within this failure domain by using, for instance, Markov chain Monte Carlo (MCMC) sampling techniques such as the Metropolis–Hastings algorithm (or a modified version, as proposed in Au and Beck (2001)). A more detailed presentation of these algorithms is given in Chapter 6. In the context of costly-to-evaluate computer codes, resampling may even become untractable. As a result, the learning procedure of $f_{Z_i|\mathcal{F}}$ and $f_{\mathbf{Z}^{-i}|\mathcal{F}}$ by KDE has to be performed over a limited elite set composed of, at least, a few hundred to a few thousand failure points.

7.5.1.4. *Challenge #4: accuracy of the KDE*

A last challenge concerns the fact that, in equation [7.39] and equation [7.40], not only are the estimations of $f_{Z_i|\mathcal{F}}$ and $f_{\mathbf{Z}^{-i}|\mathcal{F}}$ required but also their evaluations on samples Z_i and $\mathbf{Z}^{-i}$. Moreover, these indices are based on the variance of ratios of these density evaluations. Thus, we need to ensure that the KDE provides sufficiently accurate estimations for the densities to avoid large estimation errors. Consequently, if a coarse KDE can be sufficient to get a sampling density able to generate samples within a specific region such as a failure region (see, e.g. the non-parametric importance sampling techniques as proposed by Zhang (1996)), getting a very accurate estimation of a multivariate density over a limited set of samples is far more challenging.

If the traditional KDE formulation such as that presented in Box 7.1 may be sufficient for basic "*estimation-sampling*" tasks used in reliability assessment (e.g. as used in non-parametric importance sampling algorithms), it is definitely not efficient enough regarding the four challenges listed above. To do this, one needs to focus on dedicated tools developed w.r.t. these challenges. Recently, a modified formulation of a traditional G-KDE has been proposed in Perrin *et al.* (2018). This formulation is briefly reviewed below.

7.5.2. *Data-driven tensorized KDE*

Introduced by Perrin *et al.* (2018), a new *data-driven tensorized* G-KDE has been proposed so as to overcome a few difficulties of traditional G-KDE. For the sake of clarity and conciseness, only a few prerequisites and the core modifications are presented below. The interested reader is invited to refer to Perrin *et al.* (2018) for further information.

7.5.2.1. *Context and prerequisites*

This data-driven tensorized G-KDE relies on the assumption that the maximum available information consists of a set of independent realizations $\{\mathbf{Z}\}_{j=1}^{N}$ of a random vector $\mathbf{Z}$. In the reliability context, such a dataset can typically be the elite set denoted by $\mathcal{E}_{\mathbf{z}}$, i.e. a N-dimensional vector of realizations of the input vector whose coordinates fall in the failure domain.

Another prerequisite consists of assuming that the unknown underlying distribution $f_{\mathbf{Z}}$ of these data samples might be concentrated on an unknown subset of $\mathbb{R}^d$. This assumption may be particularly true when the distribution of $\mathbf{Z}$ may exhibit a strong stochastic dependence structure. In the reliability context, this case typically corresponds to the case of the optimal density at failure for complex LSFs (even if the input vector of basic variables $\mathbf{X}$ is made of independent variables). This theoretical problem of probability concentration is strongly linked with mathematical concepts related to dimension reduction via the *diffusion maps theory* (see, e.g. Coifman *et al.* 2015). The basic idea of diffusion maps is to identify the underlying manifold upon which the data are embedded. Nonetheless, this theoretical topic is beyond the scope of this chapter and the interested reader is invited to refer to the references mentioned above for further information.

7.5.2.2. *Main features of the data-driven tensorized G-KDE*

Regarding the potential constraints and challenges mentioned above, Perrin *et al.* (2018) propose two main modifications to the traditional G-KDE. These two features are detailed as follows:

– The *data-driven* feature: while considering a dataset $\mathcal{E}_{\mathbf{z}} = \{\mathbf{Z}^{(1)}, \ldots, \mathbf{Z}^{(N_{\text{fail}})}\}$, the idea is to modify the traditional G-KDE estimator so that the mean and covariance matrix are equal to the empirical moments $\widehat{\mathbf{m}}_{\mathcal{E}_{\mathbf{z}}}$ and $\widehat{\boldsymbol{\Sigma}}_{\mathcal{E}_{\mathbf{z}}}$ estimated from the dataset $\mathcal{E}_{\mathbf{z}}$. Moreover, it is proposed that we replace "Silverman's rule of thumb" (see Box 7.1) characterized by the bandwidth matrix given in equation [7.44], parameterized by a single scalar bandwidth η_{Silv} given in equation [7.45]. Instead, it is proposed that we use the maximum likelihood estimate $\mathbf{H}^{\text{ML}}$ based on the available data.

– The *tensorization* feature: to take into account the possible complex dependence structure of the underlying distribution of interest, it is proposed that we consider a *block-by-block decomposition* to separate the components of $\mathbf{Z}$ that can be reasonably assumed to be independent from those which can be assumed to be dependent. Thus, two dependent inputs should belong to the same block. This leads us to consider a tensorized version of the bandwidth matrix $\mathbf{H}^{\mathrm{ML}}$ parameterized by a set of bandwiths η_l with $l \in \{1, \ldots, n_{\text{block}}\}$.

7.5.3. *Methodology based on subset sampling and data-driven tensorized G-KDE*

The proposed methodology can be summarized as proposed in Figure 7.2. Thus, it can be decomposed into two major phases:

– **Phase #1**: the first phase (see the blue blocks in Figure 7.2) corresponds to a *reliability analysis phase*. The idea is to perform a rare event probability estimation using ARA/SS. To do so, we first need to set the augmented problem by constructing the augmented vector using the disaggregated strategy. Then, one performs an augmented SS (i.e. ARA/SS), which involves setting and tuning several inner parameters algorithms for ARA/SS (see Chapter 6) so as to get an estimate of the predictive failure probability and an elite set. A possible final step consists, if necessary, of resampling within the failure domain (using a dedicated MCMC algorithm, as described in Chapter 6).

– **Phase #2**: the second phase (see the orange blocks in Figure 7.2) corresponds to a *learning and* SA *phase*. To do this, we first need to estimate the empirical moments based on the available elite set. Then one can run the procedure of *data-driven tensorized G-KDE* as set in Perrin *et al.* (2018). Note that, in this chapter, the algorithm used to find the block-dependent structure is a greedy algorithm. Another possibility could be to use a genetic algorithm, as achieved in Perrin *et al.* (2018). Finally, one can evaluate the $S^{\mathbb{1}_{\mathcal{F}}}$-indices using estimators given in equation [7.39] and equation [7.40].

In the following section, this methodology is applied and tested on two academic test cases and one realistic aerospace test case.

7.6. Application examples

The numerical applications presented in this section are based on the following tools:

– the rare event probability estimation has been implemented in Matlab® and performed using an in-house rare event simulation toolbox developed at ONERA;

– a Python implementation of the data-driven tensorized G-KDE procedure proposed by Perrin *et al.* (2018).

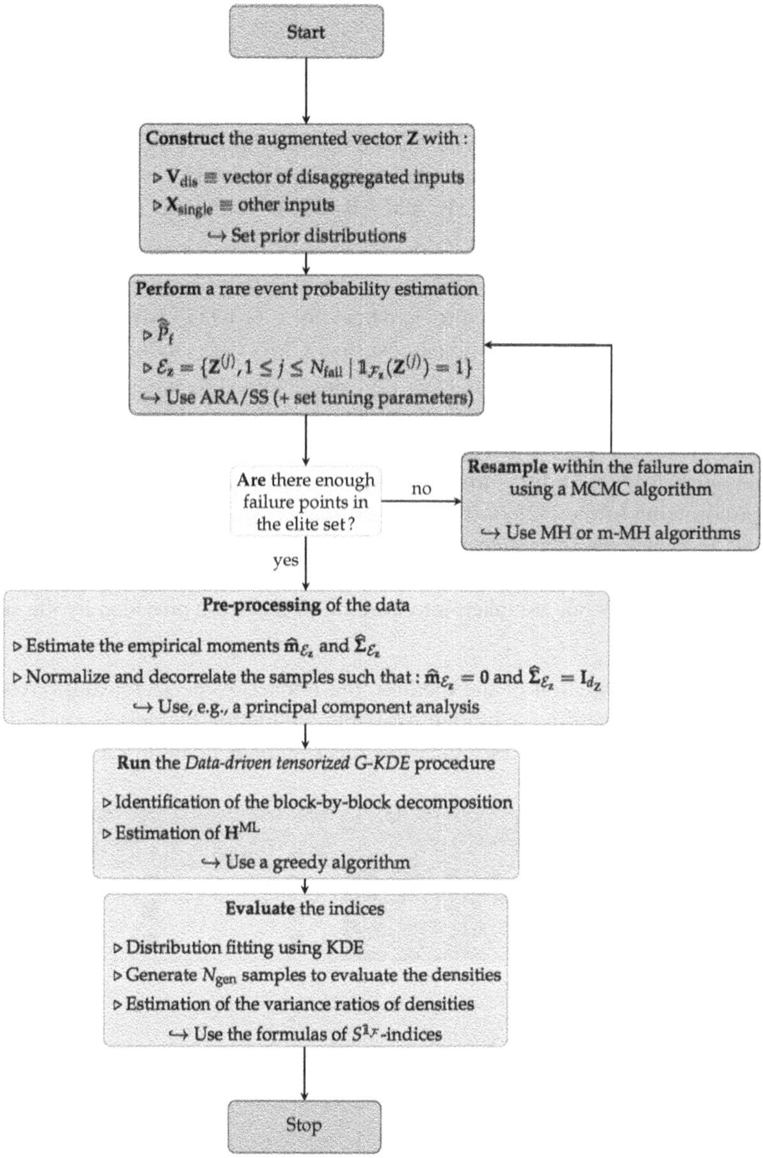

Figure 7.2. *Flowchart of the proposed methodology to compute $S^{\mathbb{1}_{\mathcal{F}}}$-indices under a bi-level uncertainty. For a color version of this figure, see www.iste.co.uk/gogu/uncertainties.zip*

The numerical testing of the methodology proposed in this chapter, as summed up in Table 7.1, relies on the following settings:

– First, the idea is to highlight the fact that dealing with bi-level uncertainty, especially with disaggregated random variables, may provide more information about the way epistemic uncertainty plays a role on the variability of the indicator function. To do so, one proposes a comparison of $S^{\mathbb{1}_{\mathcal{F}}}$-indices under single- and bi-level uncertainty, obtained using both CMC and ARA/CMC with large sample sizes.

– Second, one proposes to efficiently estimate the bi-level $S^{\mathbb{1}_{\mathcal{F}}}$-indices using the methodology described in section 7.5.3 (i.e. the coupling between ARA/SS and the data-driven tensorized G-KDE, abbreviated as "SS + G-KDE" below for the sake of conciseness), at a lower computational cost than CMC (see section 7.4.4). To do so, one considers that several elite sets $\{\mathcal{E}_{\mathbf{Z}}^{(j)}\}_{j=1}^{n_{\text{set}}}$ are obtained by repeating n_{set} times the ARA/SS algorithm. Then, based on these elite sets, one can then repeat the overall procedure proposed in section 7.5.3 so as to estimate the indices. However, due to possible numerical instabilities in the estimation of the $S^{\mathbb{1}_{\mathcal{F}}}$-indices (mainly due to the approximation of the PDFs), the mean estimate is given together with a "*success rate*" that indicates the percentage of estimated values that have been kept to get the mean estimate. The rejected ones correspond to negative values of the indices that are considered as outliers but are taken into account in the score provided by the success rate.

– Third, after validation of the method, the impact of the increasing rareness of the failure event (regarding a limited available simulation budget) is studied as an extension of the first test-case (see the three black stars ★ in Table 7.1).

Test-case		Single-level	Bi-level	
		CMC	Ref. ARA/CMC	ARA/SS
Example #1: Polynomial function[a]	(see 7.6.1)	■★	■★	■★
Example #2: Truss structure[b]	(see 7.6.2)	■	■	■
Example #3: Launcher stage fallout[c]	(see 7.6.3)	■	■	■

[a] $d_{\mathbf{Z}} = 5$.
[b] $d_{\mathbf{Z}} = 7$.
[c] 6 basic variables, 2 uncertain parameters, $g(\cdot)$ nonlinear.

Table 7.1. *Overall strategy for the numerical tests of the proposed methodology*

As shown in Table 7.1, the proposed methodology is first applied on two academic test cases: a polynomial toy case and the failure of a roof structure modeled by a truss (the black squares stand for the calculations performed). Finally, a numerical application on a realistic aerospace test case is then proposed so as to illustrate both the performance and the limits of the proposed method.

7.6.1. *Example #1: a polynomial function toy-case*

7.6.1.1. *Description*

In a first example issued from Perrin and Defaux (2019), one considers an analytical toy case made up of a polynomial function whose failure is given by the following LSF:

$$g(\mathbf{X}) = y_{\text{th}} - \mathcal{M}(\mathbf{X}) = y_{\text{th}} - (1 + X_1)(5 + X_2)(10 + X_3) \qquad [7.46]$$

where the X_i, for $i = 1, \ldots, d$, are three independent standard Gaussian variables. The failure is supposed to occur as soon as the output value exceeds the threshold y_{th}. In the following, two cases are treated:

– $y_{\text{th}} = 250$: in this case, the reference failure probability estimated using CMC (with $N = 10^6$ samples and $N_{\text{rep}} = 100$ replications) is $P_{\text{f,ref}} = 8.55 \times 10^{-4}$;

– $y_{\text{th}} = 350$: in this case, the rareness of the failure event is increased so as to reach a reference failure probability under a single-level uncertainty of $P_{\text{f,ref}} = 1.17 \times 10^{-5}$ (estimated using CMC with $N = 10^7$ samples and $N_{\text{rep}} = 100$ replications). For this case, the values of the standard deviation of the prior distributions are increased so as to get stronger sensitivities (see the values marked with the symbol ★ in Table 7.2).

Under a bi-level uncertainty, we consider that the probabilistic model of the first input variable X_1 is not perfectly known (see Table 7.2). Thus, one considers the following decomposition:

$$X_1 = M_{X_1} + S_{X_1} U_{X_1} \qquad [7.47]$$

where M_{X_1} and S_{X_1} follow, respectively, some prior distributions described in Table 7.2. As for U_{X_1}, it represents the natural variability of the input X_1.

Variable X_i	Distribution	Mean μ_{X_i}	S.d. σ_{X_i}
X_1	Normal	μ_{X_1} ***uncertain***[a]	σ_{X_1} ***uncertain***
X_2	Normal	0	1
X_3	Normal	0	1
M_{X_1}	Normal	0	0.1/0.7(★)
S_{X_1}	Normal	1	0.1/0.7(★)
U_{X_1}	Normal	0	1

[a]For fixed values $\mu_{X_1} = 0$ and $\sigma_{X_1} = 1$, $P_{\text{f,ref}} = 8.55 \times 10^{-4}$.

Table 7.2. *Input probabilistic model for Example #1*

7.6.1.2. *Results*

Figure 7.3(a) and Figure 7.3(b) provide the reference results for the estimation of both first-order and total $S^{\mathbb{1}_\mathcal{F}}$-indices under a single-level uncertainty. Based on these plots, one can notice, first, that X_1 is the most influential variable on the indicator function, and second, that all three variables show high values for the total indices, which indicate that these variables present strong interactions at failure (which is a particular feature of the $S^{\mathbb{1}_\mathcal{F}}$-indices already pointed out by (Lemaître 2014, Chapter 1)).

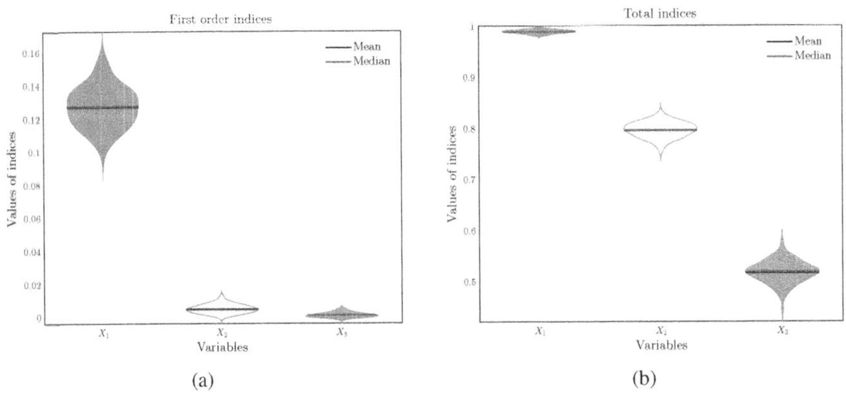

Figure 7.3. *Reference $S^{\mathbb{1}_\mathcal{F}}$-indices estimated for Example #1 under a single-level uncertainty (CMC of $N_{sim} = 10^6$ samples and $N_{rep} = 100$ repetitions, with $p_{\mathrm{f,ref}} = 8.55 \times 10^{-4}$). For a color version of this figure, see www.iste.co.uk/gogu/uncertainties.zip*

Now, one can compare these results with the reference ones obtained under a bi-level uncertainty using the ARA/CMC pick-freeze estimators presented in section 7.4.4. The most influential input variable X_1 has been disaggregated. Thus, Figure 7.4(a) and Figure 7.4(b) show that it is U_{X_1} which is the most influential regarding first-order indices. Then, looking at total indices, one can note that, even if the ranking is still preserved ($U_{X_1} > X_2 > X_3$), S_{X_1} is more influential than M_{X_1} and is of the same order of magnitude as X_3. Consequently, considering an extra level of uncertainty plays a role, not only on the reliability assessment (here, the failure probability slightly increases from 8.55×10^{-4} to 1.0×10^{-3}), but also on the relative influence of the random variables at failure. Moreover, these results show that epistemic uncertainty affecting distribution parameters might play a non-negligible role compared to the aleatory uncertainty of the basic variables.

Numerical results obtained by the proposed methodology are given in Table 7.3. These results have been obtained for the following settings: $n_{\mathrm{set}} = 10$ elite sets of $N_{\mathrm{fail}} = 2.5 \times 10^3$ failure samples have been obtained by repetition of the ARA/SS algorithm. In addition to that, we generate $N_{\text{gen}} = 5 \times 10^4$ samples on which one evaluates the densities and computes the variance of the ratio. Thus, by comparing the second and third columns, one can note that the first-order indices are rather correctly estimated, at a relative moderate cost and with a high success rate. One can still note that the index associated with U_{X_1} is a little underestimated. As for the total indices, even if the orders of magnitude are almost correct, most of them are a little overestimated.

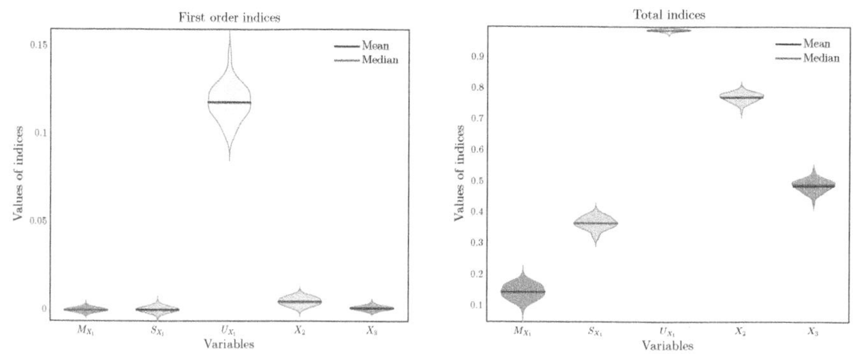

(a) (b)

Figure 7.4. *Reference $S^{1_{\mathcal{F}}}$-indices estimated for Example #1 under a bi-level uncertainty (CMC of $N_{sim} = 10^6$ samples and $N_{rep} = 100$ repetitions, with $\widetilde{P}_{\text{f,ref}} = 1.0 \times 10^{-3}$). For a color version of this figure, see www.iste.co.uk/gogu/uncertainties.zip*

Variable	First order indices			Total indices		
	Ref. CMC	SS + G-KDE	Success rate	Ref. CMC	SS + G-KDE	Success rate
M_{X_1}	1.4×10^{-4}	8.4×10^{-5}	(100 %)	0.145	0.224	(90 %)
S_{X_1}	4×10^{-4}	5.8×10^{-4}	(100 %)	0.366	0.476	(100 %)
U_{X_1}	0.118	0.069	(100 %)	0.987	0.981	(100 %)
X_2	5.1×10^{-3}	4.1×10^{-3}	(100 %)	0.772	0.806	(100 %)
X_3	1.4×10^{-3}	1.5×10^{-3}	(100 %)	0.487	0.604	(100 %)

Table 7.3. *Results for Example #1*

Considering the influence of the failure event rareness (i.e. for $y_{\text{th}} = 350$), numerical results are given in Table 7.4. For this case, the reference predictive failure probability is such that $\widetilde{P}_{\text{f,ref}} = 2.05 \times 10^{-5}$. Reference values for $S^{1_{\mathcal{F}}}$-indices have been obtained by ARA/CMC, with $N = 10^8$ samples and $N_{\text{rep}} = 100$ replications. Numerical results for the proposed methodology have been obtained for the following settings: $n_{\text{set}} = 10$ elite sets of $N_{\text{fail}} = 500$ failure samples have been obtained by repetition of the SS algorithm (with 5×10^3 samples/step, an

α_{SS}-quantile set to 0.75 and a final MCMC-based resampling step of 5×10^3 samples whose first 10% of samples are kept only). In addition, we generate $N_{\text{gen}} = 5 \times 10^5$ samples, on which we evaluate the densities and compute the variance of the ratio. Thus, one can note that the first-order indices are rather correctly estimated, at a relative moderate cost, compared to CMC, and high success rates. As for the total indices, almost all of them have been correctly estimated. However, the smallest one, associated with M_{X_1}, is overestimated, at a low success rate of 50%.

Variable	First order indices			Total indices		
	Ref. CMC	SS + G-KDE	Success rate	Ref. CMC	SS + G-KDE	Success rate
M_{X_1}	7.8×10^{-8}	2.6×10^{-6}	(100 %)	0.176	0.394	(50 %)
S_{X_1}	2.0×10^{-5}	3.5×10^{-5}	(100 %)	0.539	0.517	(100 %)
U_{X_1}	0.026	0.016	(100 %)	0.998	0.990	(100 %)
X_2	4.6×10^{-4}	1.3×10^{-3}	(100 %)	0.902	0.906	(100 %)
X_3	7.4×10^{-5}	8.7×10^{-5}	(100 %)	0.659	0.700	(100 %)

Table 7.4. *Results for Example #1 (★)*

7.6.2. *Example #2: a truss structure*

7.6.2.1. *Description*

The second example (issued from Wei *et al.* (2012)) is a roof structure whose behavior is modeled by a truss, as sketched in Figure 7.5. For the bars, two different materials are assumed: the top boom and the compression bars are reinforced by concrete (denoted by the subscript "c" in the mechanical characteristics) while the bottom boom and the tension bars are made up of steel (denoted by the subscript "s" in the mechanical characteristics). A uniformly distributed load q is applied to the roof. As a result of the modeling, one can reduce it to nodal loads $P = ql/4$, applied, respectively, on nodes C, D and F. The failure of such a structure can be attained if the deflection of node C (denoted by Δ_C) reaches a given threshold Δ_{th}. Such a scenario is given by the following LSF:

$$g(\mathbf{X}) = \Delta_{\text{th}} - \mathcal{M}(\mathbf{X}) = 0.025 - \Delta_C. \qquad [7.48]$$

The analytical formulation of Δ_C can be derived from the basic principles of structural mechanics applied to trusses. As a result, one has the following deflection formula (Wei *et al.* 2012):

$$\Delta_C = \frac{ql^2}{2}\left(\frac{3.81}{A_c E_c} + \frac{1.13}{A_s E_s}\right) \qquad [7.49]$$

where l denotes the total length of the basis of the truss, A_c and A_s the sectional areas of, respectively, the concrete and steel bars, and finally E_c and E_s the Young's modulus of concrete and steel. Thus, one has to consider a 6-dimensional input random vector, such that $\mathbf{X} = (q, l, A_s, A_c, E_s, E_c)^\top$, composed of independent Gaussian variables, as described in Table 7.5.

Considering a single-level of uncertainty in input, the reference failure probability estimated using CMC (with $N = 10^6$ samples and $N_{\text{rep}} = 100$ replications) is $P_{\text{f,ref}} = 1.27 \times 10^{-2}$. Under a bi-level uncertainty, one assumes that the probabilistic model of the first input variable X_1 is not perfectly known (see Table 7.2). Thus, we consider the following decomposition:

$$X_1 = M_{X_1} + \sigma_{X_1} U_{X_1} \qquad [7.50]$$

where M_{X_1} follow, respectively, a prior distribution, described in Table 7.5, while σ_{X_1} is supposed to be well known. As for U_{X_1}, it represents the natural variability of the input X_1 and is thus modeled by a standard Gaussian random variable.

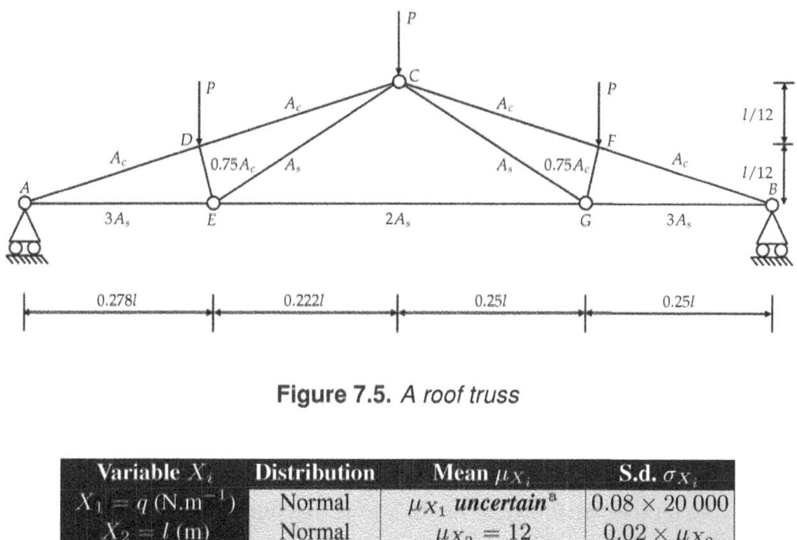

Figure 7.5. *A roof truss*

Variable X_i	Distribution	Mean μ_{X_i}	S.d. σ_{X_i}
$X_1 = q$ (N.m^{-1})	Normal	μ_{X_1} ***uncertain***[a]	$0.08 \times 20\,000$
$X_2 = l$ (m)	Normal	$\mu_{X_2} = 12$	$0.02 \times \mu_{X_2}$
$X_3 = A_s$ (m^2)	Normal	$\mu_{X_3} = 9.82 \times 10^{-4}$	$0.06 \times \mu_{X_3}$
$X_4 = A_c$ (m^2)	Normal	$\mu_{X_4} = 0.04$	$0.2 \times \mu_{X_4}$
$X_5 = E_s$ (N.m^{-2})	Normal	$\mu_{X_5} = 1.2 \times 10^{11}$	$0.07 \times \mu_{X_5}$
$X_6 = E_c$ (N.m^{-2})	Normal	$\mu_{X_6} = 3 \times 10^{10}$	$0.08 \times \mu_{X_6}$
M_{X_1}	Normal	20 000	$0.05 \times 20\,000$
U_{X_1}	Normal	0	1

[a]For the fixed value $\mu_{X_1} = 20\,000$, $P_{\text{f,ref}} = 1.27 \times 10^{-2}$.

Table 7.5. *Input probabilistic model for Example #2*

7.6.2.2. *Results*

Figure 7.6(a) and Figure 7.6(b) provide the reference results for the estimation of both first-order and total $S^{\mathbb{1}_{\mathcal{F}}}$-indices under a single-level uncertainty. Based on these plots, one can note, first, that X_4 presents the largest first order index, and second, that the total indices indicate a partition into three groups of inputs: (X_4, X_1) show the strongest indices, then (X_2, X_3, X_5) have quasi-similar values for their indices, and finally X_6 (these results are coherent with those obtained by Wei *et al.* (2012)).

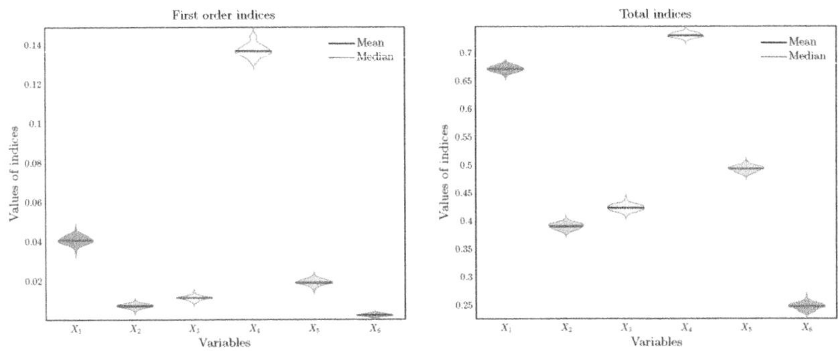

(a) (b)

Figure 7.6. *Reference $S^{\mathbb{1}_{\mathcal{F}}}$-indices estimated for Example #2 under a single-level uncertainty (CMC of $N_{sim} = 10^6$ samples and $N_{rep} = 100$ repetitions, with $P_{\mathrm{f,ref}} = 1.26 \times 10^{-2}$) For a color version of this figure, see www.iste.co.uk/gogu/uncertainties.zip*

Under a bi-level uncertainty, it is assumed that the mean of X_1 is affected by epistemic uncertainty. Thus, X_1 has been disaggregated into M_{X_1} and U_{X_1}. As a result, Figure 7.7(a) and Figure 7.7(b) show that it is U_{X_1} that is the second most influential input regarding first-order indices. As for total indices, one can see that the group (X_4, U_{X_1}) is the most influential, followed by (M_{X_1}, X_2, X_3, X_5) and, finally, X_6. Thus, again, the epistemic uncertainty affecting the mean of X_1 plays a non-negligible role on the variability of the indicator function, mainly due to interactions with other inputs.

The numerical results obtained by the proposed methodology are given in Table 7.6. These results have been obtained for the following settings: $n_{\mathrm{set}} = 50$ elite sets of $N_{\mathrm{fail}} = 2 \times 10^3$ failure samples have been obtained by repetition of the ARA/SS algorithm. In addition, we generate $N_{\mathrm{gen}} = 1 \times 10^4$ samples on which one evaluates the densities and computes the variance of the ratio. Thus, by comparing the second and third columns, one can note that the first-order indices are almost perfectly estimated (only the strongest index associated with X_4 is moderately overestimated), at a lower cost, compared to ARA/CMC, and with the highest success rates for all the indices. As for the total indices, some of them are perfectly estimated while others show a relative bias. Moreover, the ranking is slightly modified between M_{X_1} and X_3. However, the reference values are also very close. Thus, the proposed methodology is able to grasp, at least, accurate orders of magnitude. Finally, one can see that the success rates of total indices do not reach values under 82%.

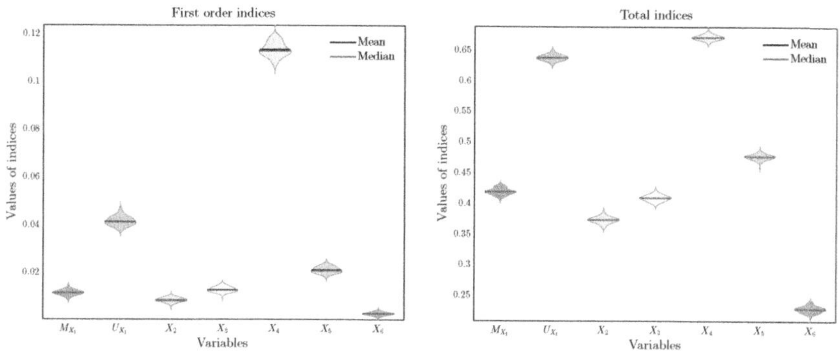

(a) (b)

Figure 7.7. *Reference $S^{1_{\mathcal{F}}}$-indices estimated for Example #2 under a bi-level uncertainty (CMC of $N_{sim} = 10^6$ samples and $N_{rep} = 100$ repetitions, with $\widetilde{P}_{f,ref} = 1.65 \times 10^{-2}$) For a color version of this figure, see www.iste.co.uk/gogu/uncertainties.zip*

Variable	First order indices			Total indices		
	Ref. CMC	SS + G-KDE	Success rate	Ref. CMC	SS + G-KDE	Success rate
M_{X_1}	0.012	0.012	(100 %)	0.420	0.346	(98 %)
U_{X_1}	0.042	0.046	(100 %)	0.639	0.567	(100 %)
X_2	8.8×10^{-3}	9.1×10^{-3}	(100 %)	0.374	0.297	(94 %)
X_3	0.013	0.015	(100 %)	0.410	0.365	(94 %)
X_4	0.114	0.145	(100 %)	0.672	0.643	(100 %)
X_5	0.021	0.023	(100 %)	0.478	0.444	(100 %)
X_6	3.2×10^{-3}	3.4×10^{-3}	(100 %)	0.229	0.207	(82 %)

Table 7.6. *Results for Example #2*

7.6.3. *Example #3: application to a launch vehicle stage fallback zone estimation*

7.6.3.1. *Description*

The role of a *launch vehicle* (a.k.a. *space launcher*) is to carry a payload (e.g. a satellite) from the Earth's surface to a given orbit. A traditional *expendable* space launcher is composed of multiple stages, equipped with their propulsion systems. During the flight, uncertainties can affect several variables in multiple disciplines (e.g. on the perturbation during trajectory or propellant combustion). For instance, focusing on the optimal trajectory assessment leads us to consider the *separation point* as a key point in terms of uncertainty analysis (see, e.g. Derennes *et al.* (2019) for an illustration). In practice, separation altitudes may be lower than the *Kármán line* (which denotes the boundary between the Earth's atmosphere and outer space, at around 100 km high). Therefore, wind perturbations might affect the system dynamics. As a result, the handling of uncertainties (e.g. arising from dynamic perturbations, error measurements, due to sensors or varying unburned propellant left mass) plays a crucial role in the comprehension and prediction of the global system behavior. This is the reason why it is of prime importance to take it into account during the reliability analysis and the prediction of the fallout zone. A misestimation can have dramatic consequences in terms of launcher safety, human safety and environmental impact.

The goal of this chapter is to present a simplified, but representative, fallout trajectory simulation model. Indeed, this model is representative of the phenomena encountered, but with a reduced simulation cost (e.g. use of mass point model), while remaining challenging enough regarding the use of advanced rare event estimation techniques and SA methods. For more realistic problems, one can refer to Ronse and Mooij (2014) and Hoogendoorn *et al.* (2018). In these studies, other parameters are considered and investigated concerning the dynamics of the vehicle (e.g. perturbation of atmospheric density, winds). In this chapter, they are not taken into account for the sake of simplicity and interpretation of the given results, but they remain parameters that should be incorporated in the model to get a full high-fidelity simulation model.

The launcher stage fallout simulation computer code may be represented as an input–output black-box model. The input variables are, among others, some characteristics of the launcher and some conditions (initial or arising during the flight) of the fallout phase. These inputs are affected by uncertainties and are gathered in a random vector with a given PDF. It is assumed that this PDF is described by a parametric model of density. The output corresponds to the position of fallout and is also a random variable because of the input randomness. A typical safety measure can be the probability that a stage (e.g. the first stage) falls at a distance greater than a given safety limit. Indeed, this estimation is strategic for the qualification of such vehicles.

The simulation model used in this chapter can be considered as a black-box model denoted by $\mathcal{M} : \mathbb{R}^{d=6} \to \mathbb{R}$. Here, it is a simplified trajectory simulation code of the dynamic fallout phase of a generic launcher first stage. The advantage of a black-box model is to broaden the applicability of the proposed statistical approaches illustrated in this chapter to any test cases in this range of models. As a matter of fact, the following methods proposed in this chapter are said to be "non-intrusive" w.r.t. the model under study. The d-dimensional (here $d = 6$) input vector of the simulation code, denoted by $\mathbf{X}$, contains the following basic variables (i.e. physical variables) modeling some initial conditions, environmental variables and launch vehicle characteristics:

– X_1: stage altitude perturbation at separation Δa (m);

– X_2: velocity perturbation at separation Δv (m.s^{-1});

– X_3: flight path angle perturbation at separation $\Delta\gamma$ (rad);

– X_4: azimuth angle perturbation at separation $\Delta\psi$ (rad);

– X_5: propellant mass residual perturbation at separation Δm (kg);

– X_6: drag force error perturbation ΔC_d (dimensionless).

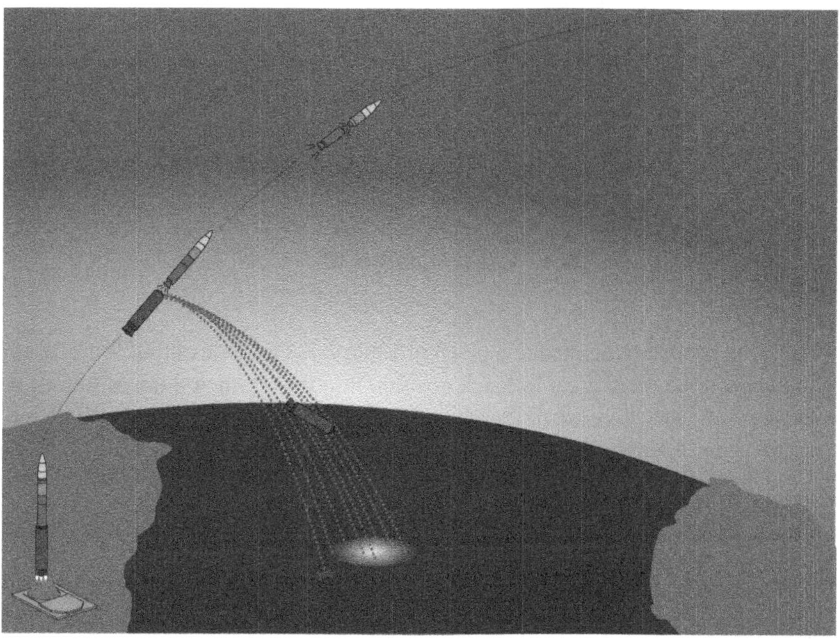

Figure 7.8. *Illustration scheme of a launch vehicle first stage fallout phase into the Atlantic Ocean. Multiple fallout trajectories are drawn (red dotted lines), leading to the impact zone (yellow circular surface). Due to uncertainties, one fallout trajectory may lead to a failure impact point (red star) For a color version of this figure, see www.iste.co.uk/gogu/uncertainties.zip*

These variables are assumed to be independent for the sake of simplicity. As an output, the code will give back the scalar distance $Y = \mathcal{M}(\mathbf{X})$, which represents the distance D_{code} between the theoretical fallout position into the ocean and the estimated one due to the uncertainty propagation.

In the context of the launch vehicle fallout case, the input variables are known to be affected by uncertainties (e.g. due to the natural variability of wind or due to lack of knowledge). Thus, applying UQ methodology leads to consider a probabilistic model for the input vector $\mathbf{X}$, i.e. by assuming the existence of a joint PDF $f_{\mathbf{X}} : \mathcal{D}_{\mathbf{X}} \subseteq \mathbb{R}^d \to \mathbb{R}_+$. Since the input variables are assumed to be independent, this joint PDF corresponds to the product of the marginal PDFs f_{X_i} of the input variables X_i, $i \in \{1, \dots, d\}$. The input probabilistic model for the launch vehicle fallout case is given in Table 7.7. Note that the numerical values used in this test case are hypothetical and should not be used for industrial applications.

Variable X_i [a]	Distribution	Mean μ_{X_i}	S.d. σ_{X_i}
$X_1 = \Delta a$ (m)	Normal	0	1650
$X_2 = \Delta v$ (m.s^{-1})	Normal	0	3.7
$X_3 = \Delta\gamma$ (rad)	Normal	0	0.001
$X_4 = \Delta\psi$ (rad)	Normal	0	0.0018
$X_5 = \Delta m$ (kg)	Normal	0	70
$X_6 = \Delta C_d$ (1)	Normal	0	0.1

[a] The input variables are independent.

Table 7.7. *Input probabilistic model*

The input probabilistic model is given in Table 7.7. The code output is the distance D_{code} between the theoretical fallback position into the ocean and the estimated one. For the sake of clarity, we recall that the LSF $g(\cdot)$ can be written as follows:

$$g(\mathbf{X}) = d_{\text{safe}} - \mathcal{M}(\mathbf{X}) = d_{\text{safe}} - D_{\text{code}} \qquad [7.51]$$

for which the rareness of the failure event depends on the safety threshold distance d_{safe}.

Due to the presence of a bi-level input uncertainty and following the methodology proposed in section 7.4.3, it is proposed that we consider the following disaggregated inputs:

$$X_2 = M_{X_2} + \sigma_{X_2} U_{X_2} \qquad [7.52]$$

$$X_3 = M_{X_3} + \sigma_{X_3} U_{X_3} \qquad [7.53]$$

where M_{X_2} and M_{X_3} follow the same distributions as presented in Table 7.8, U_{X_2} and U_{X_3} are two standard Gaussian variables, and σ_{X_2} and σ_{X_3} are the two standard deviations, defined in Table 7.8. Finally, one considers the following augmented input vector:

$$\mathbf{Z} = (\mathbf{V}_{\text{dis}}, \mathbf{X}_{\text{single}})^{\top} = (X_1, M_{X_2}, U_{X_2}, M_{X_3}, U_{X_3}, X_4, X_5, X_6)^{\top} \quad [7.54]$$

Variable[a]	Distribution	Mean	S.d.
$X_1 = \Delta h$ (m)	Normal	$\mu_{X_1} = 0$	$\sigma_{X_1} = 1650$
$X_2 = \Delta v$ (m.s^{-1})	Normal	μ_{X_2} ***uncertain***[b]	$\sigma_{X_2} = 3.7$
$X_3 = \Delta\gamma$ (rad)	Normal	μ_{X_3} ***uncertain***	$\sigma_{X_3} = 0.001$
$X_4 = \Delta\psi$ (rad)	Normal	$\mu_{X_4} = 0$	$\sigma_{X_4} = 0.0018$
$X_5 = \Delta m$ (kg)	Normal	$\mu_{X_5} = 0$	$\sigma_{X_5} = 70$
$X_6 = \Delta C_d$ (1)	Normal	$\mu_{X_6} = 0$	$\sigma_{X_6} = 0.1$
$\Theta_2 = \mu_{X_2}$ (m.s^{-1})	Normal	$\xi_1 = \mu_{\mu X_2} = 0$	$\xi_2 = \sigma_{\mu X_2} = 3.7$
$\Theta_3 = \mu_{X_3}$ (rad)	Normal	$\xi_3 = \mu_{\mu X_3} = 0$	$\xi_4 = \sigma_{\mu X_3} = 0.001$

[a] The basic variables are independent.
[b] For fixed values $\mu_{X_2} = 0$ and $\mu_{X_3} = 0$, one has:
▷ for $d_{\text{safe}} = 15$ km, $P_{\text{f,ref}} = 1.36 \times 10^{-4}$;
▷ for $d_{\text{safe}} = 20$ km, $P_{\text{f,ref}} = 2.31 \times 10^{-7}$.

Table 7.8. *Input probabilistic model under bi-level input uncertainty*

For the numerical experiments, simulation settings have been defined as follows. First, concerning the LSF, two cases are treated to investigate the influence of the rareness of the failure event:

– in the first place, the safety threshold distance is set to $d_{\text{safe}} = 11$ km (associated with a reference failure probability under single-level input uncertainty such that $P_{\text{f,ref}} = 6.10 \times 10^{-3}$). This case is used to obtain reference results (using CMC) in order to have a better insight of what is at stake in terms of sensitivity indices;

– then, the safety threshold distance is set to $d_{\text{safe}} = 15$ km. In such a case, only reference results in bi-level are provided and compared to those obtained from the proposed methodology.

Finally, concerning the proposed methodology, mean estimates of the sensitivity indices are provided together with a "success rate" which indicates, as a percentage of the total number of repetitions, the number of experiments that have provided positive values of indices. If the index value is negative, the numerical experiment is considered as "failed" and the estimated value is removed.

7.6.3.2. *Results*

Figure 7.9(a) and Figure 7.9(b) provide the reference results for the estimation of both first-order and total $S^{\mathbb{1}_{\mathcal{F}}}$-indices under single-level uncertainty. Based on these plots, one can note, first, that X_2 presents the largest first order index, and second, the total indices indicate a partition into two groups of inputs: (X_2, X_3, X_6) show the strongest indices (above 0.50) while (X_1, X_4, X_5) have rather moderate values (around 0.15). Moreover, one can see that first-order indices are, for the most part, very low, while total order indices are much stronger. This clearly indicates that variables interact a lot at failure but do not contribute that much on their own to the variability of the indicator function. Such a remark has been already pointed out in the work of Lemaître (2014).

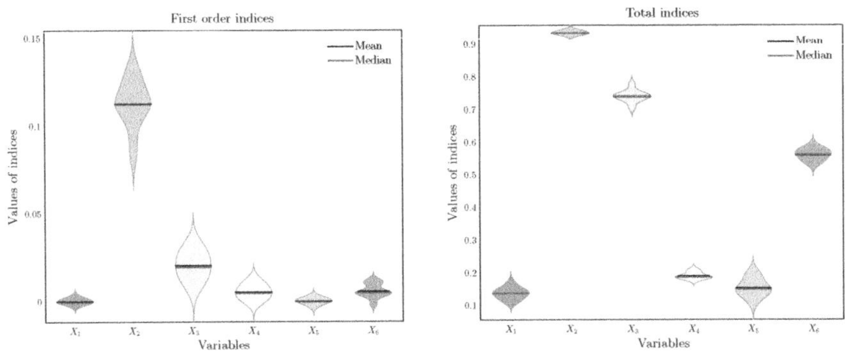

(a) (b)

Figure 7.9. *$S^{1_{\mathcal{F}}}$-indices estimated under single-level uncertainty (CMC of $N_{sim} = 10^5$ samples and $N_{rep} = 10$ repetitions, with $P_{\mathrm{f,ref}} = 6.10 \times 10^{-3}$). For a color version of this figure, see www.iste.co.uk/gogu/uncertainties.zip*

Under bi-level uncertainty, Figure 7.10(a) shows that, for X_2, U_{X_2} is more influential than M_{X_2}, while for X_3, it is M_{X_3} which is more influential than U_{X_3}. This clearly highlights that, due to the disaggregated version of the augmented vector, one can analyze both effects from aleatory and epistemic uncertainties. Again, the first-order indices are low, which indicates a poor contribution of each variable to the overall variability of the indicator. As for total indices displayed in Figure 7.10(b), one can observe the same ranking as for first-order indices. However, one can see the epistemic uncertainties affecting the mean value of X_2 and X_3 play a major role on the variability of the indicator function and thus have to be taken into account.

When considering the case where d_{safe} is set to 15 km, numerical results obtained by the proposed methodology are given in Table 7.9. These results have been obtained for the following settings:

– $n_{\mathrm{set}} = 35$ elite sets obtained after 35 repetitions of the SS algorithm;

– each elite set contains $N_{\mathrm{fail}} = 2 \times 10^3$ failure samples resampled at the end of each application of SS;

– in addition, we generate $N_{\text{gen}} = 1 \times 10^5$ samples on which we evaluate the densities and compute the variance of the ratio;

– the reference CMC is obtained by $N_{\text{rep}} = 10$ repetitions of $N_{\text{sim}} = 10^5$ samples;

– the "success rate" is obtained by counting the number of negative-valued indices removed from the statistics.

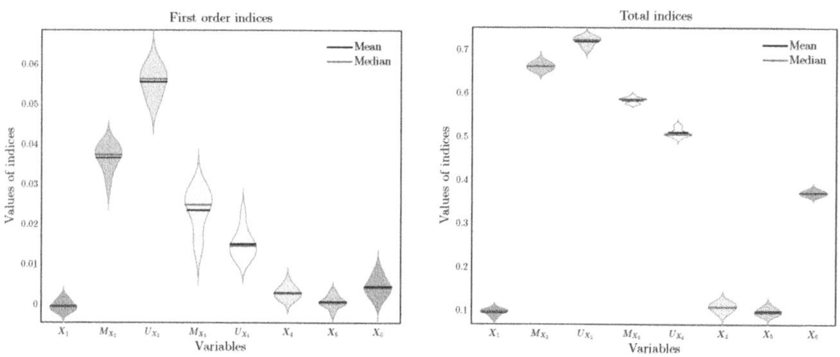

(a) (b)

Figure 7.10. *$S^{1_\mathcal{F}}$-indices estimated under bi-level uncertainty (ARA/CMC of $N_{sim} = 10^5$ samples and $N_{rep} = 10$ repetitions, with $\widetilde{P}_{\text{f,ref}} = 3.98 \times 10^{-2}$). For a color version of this figure, see www.iste.co.uk/gogu/uncertainties.zip*

Variable	First order indices			Total indices		
	Ref. CMC	SS + G-KDE	Success rate	Ref. CMC	SS + G-KDE	Success rate
X_1	2.3×10^{-3}	1.1×10^{-4}	(100 %)	0.155	0.135	(3 %)
M_{X_2}	0.014	0.022	(100 %)	0.776	0.598	(66 %)
U_{X_2}	0.028	0.038	(100 %)	0.792	0.511	(77 %)
M_{X_3}	9.3×10^{-3}	0.012	(100 %)	0.697	0.523	(60 %)
U_{X_3}	4.7×10^{-3}	5.6×10^{-3}	(100 %)	0.620	0.540	(60 %)
X_4	0	5.3×10^{-4}	(100 %)	0.137	0.608	(63 %)
X_5	0	1.2×10^{-4}	(100 %)	0.132	0.407	(49 %)
X_6	2.3×10^{-3}	2.0×10^{-3}	(100 %)	0.524	0.518	(51 %)

Table 7.9. *Results for step #3*

Concerning the first-order indices, one can see that most of them are correctly estimated (despite the fact that they have small values). The strongest first-order indices (i.e. those associated with M_{X_2} and U_{X_2}) are rather well estimated. As for the total order indices, the results show that the indice associated with X_1 can be correctly estimated but with a very low success rate. As for X_6, its total index is well estimated but with a moderate success rate. The group $(M_{X_2}, U_{X_2}, M_{X_3}, U_{X_3})$ are estimated in terms of order of magnitude but not accurately. Finally, the indices associated with (X_4, X_5) are completely overestimated.

7.6.4. *Summary about numerical results and discussion*

As a summay, and according to the numerical results presented previously, one can highlight a few characteristics and perspectives about the proposed methodology:

– the proposed methodology allows us to investigate the impact of the relative parts of epistemic uncertainty and aleatory uncertainty to the variability of the indicator function (which is directly linked to the variability of the failure probability as stated in equation [7.7]);

– the coupling between ARA/SS and the data-driven tensorized G-KDE seems to provide very promising results for global ROSA under a bi-level uncertainty, especially concerning the reduction in terms of computational cost compared to CMC, since this methodology only requires a posttreatment of an elite set obtained after a reliability analysis;

– the methodology seems able to handle moderate dimensions, at least, greater dimensions for which traditional KDE fails to correctly estimate multivariate densities;

– this methodology is driven by several tuning parameters that may drastically influence the results: the quality of the elite set (influenced by the tuning of the SS algorithm), the number N_{fail} of failure samples in the elite set and the number N_{gen} of generated samples to compute the variance. As for n_{set}, it is important to note that, in practice, for costly-to-evaluate computer models (and rare event probability estimation), it might not be possible to obtain either multiple or large sample size elite sets.

7.7. Conclusion

In this chapter, a new methodology is proposed to estimate, at a lower computational cost than CMC, a set of $S^{\mathbb{1}_{\mathcal{F}}}$-indices, in the context of bi-level input uncertainty. This methodology relies on the combination of three main components:

– first, a disaggregated structure of the inputs that are affected by epistemic uncertainty;

– second, the use of the SS algorithm to get an estimate of the predictive failure probability and, jointly, to use the final elite set of failure samples to estimate the $S^{\mathbb{1}_{\mathcal{F}}}$-indices;

– third, a data-driven tensorized G-KDE which allows us to improve the density estimation required to compute the indices.

The combination of these three tools allows us to propose a methodology whose efficiency has been observed on two numerical test cases. As for the complex aerospace test case, very promising results have been obtained, even if several difficulties remain in the estimation accuracy of small indices. Moreover, the influence of the failure event rareness has also been investigated. However, this work only provides promising preliminary results. Indeed, as mentioned in the results, some numerical experiments can lead to negative values of the $S^{\mathbb{1}_{\mathcal{F}}}$-indices as indicated by the "success rate" scores. Thus, the proposed methodology might not be suited to being a one-shot methodology. Its robustness should be improved, mainly by investigating the potential reasons for such a failure in terms of estimation (e.g. possible insufficiency in terms of budget allocation/inaccuracy of the KDE/inefficiency of the resampling phase using MCMC creating correlated samples).

7.8. Acknowledgments

The first author would like to thank ONERA – The French Aerospace Lab and SIGMA Clermont as this work has been mostly performed under a 3-year long PhD program co-funded by these two institutions. Moreover, the first author would like to thank several people: Émilien Flayac for several valuable discussions about technical details that really helped to get this work achieved, and Dr. Bertrand Iooss (EDF R&D) and Prof. Carsten Proppe (KIT, Karlsruhe) for their advice and consideration about this work.

7.9. References

Au, S.-K. and Beck, J.L. (2001). Estimation of small failure probabilities in high dimensions by subset simulation. *Probabilistic Engineering Mechanics*, 16(4), 263–277.

Chabridon, V. (2018). Reliability-oriented sensitivity analysis under probabilistic model uncertainty – Application to aerospace systems. PhD Thesis, University of Clermont Auvergne.

Chabridon, V., Balesdent, M., Bourinet, J.-M., Morio, J. and Gayton, N. (2017). Evaluation of failure probability under parameter epistemic uncertainty: Application to aerospace system reliability assessment. *Aerospace Science and Technology*, 69, 526–537.

Chabridon, V., Balesdent, M., Bourinet, J.-M., Morio, J. and Gayton, N. (2018). Reliability-based sensitivity estimators of rare event probability in the presence of distribution parameter uncertainty. *Reliability Engineering and System Safety*, 178, 164–178.

Coifman, R.R., Lafon, S., Lee, A.B., Maggioni, M., Nadler, B., Warner, F. and Zucker, S.W. (2015). Geometric diffusions as a tool for harmonic analysis and structure definition of data: Diffusion maps. *Proceedings of the National Academy of Sciences of the United States of America*, 102(21), 7426–7431.

Da Veiga, S. (2015). Global sensitivity analysis with dependence measures. *Journal of Statistical Computation and Simulation*, 85(7), 1283–1305.

De Rocquigny, E., Devictor, N. and Tarantola, S. (2008). *Uncertainty in Industrial Practice: A Guide to Quantitative Uncertainty Management*. Wiley, Chichester.

Der Kiureghian, A. (1989). Measures of structural safety under imperfect states of knowledge. *Journal of Structural Engineering ASCE*, 115(5), 1119–1140.

Der Kiureghian, A. (2008). Analysis of structural reliability under parameter uncertainties. *Probabilistic Engineering Mechanics*, 23(4), 351–358.

Der Kiureghian, A. and Ditlevsen, O. (2009). Aleatory or epistemic? Does it matter? *Structural Safety*, 31(2), 105–112.

Derennes, P., Chabridon, V., Morio, J., Balesdent, M., Simatos, F., Bourinet, J.-M. and Gayton, N. (2019). Nonparametric importance sampling techniques for sensitivity analysis and reliability assessment of a launcher stage fallout. In *Modeling and Optimization in Space Engineering: State of the Art and New Challenges*. Springer International Publishing, Basel, 59–86.

Dubourg, V. (2011). Adaptive surrogate models for reliability analysis and reliability-based design optimization. PhD Thesis, Blaise Pascal University – Clermont II.

Fort, J.-C., Klein, T. and Rachdi, N. (2016). New sensitivity analysis subordinated to a contrast. *Communications in Statistics – Theory and Methods*, 45(15), 4349–4364.

Helton, J.C. (1997). Uncertainty and sensitivity analysis in the presence of stochastic and subjective uncertainty. *Journal of Statistical Computation and Simulation*, 57, 3–76.

Hong, H.P. (1996). Evaluation of the probability of failure with uncertain distribution parameters. *Civil Engineering Systems*, 13, 157–168.

Hoogendoorn, R., Mooij, E. and Geul, J. (2018). Uncertainty propagation for statistical impact prediction of space debris. *Advances in Space Research*, 61, 167–181.

Iooss, B. and Lemaître, P. (2015). A review on global sensitivity analysis methods. In *Uncertainty Management in Simulation-Optimization of Complex Systems: Algorithms and Applications*, Dellino, G. and Meloni, C. (eds). Springer US, Boston.

Lebrun, R. (2013). Contributions to stochastic dependence modeling. PhD Thesis, Paris Diderot University.

Lemaître, P. (2014). Sensitivity analysis in structural reliability. PhD Thesis, University of Bordeaux.

Li, L., Lu, Z., Jun, F. and Bintuan, W. (2012). Moment-independent importance measure of basic variable and its state dependent parameter solution. *Structural Safety*, 38, 40–47.

Morio, J. and Balesdent, M. (2015). *Estimation of Rare Event Probabilities in Complex Aerospace and Other Systems: A Practical Approach*. Woodhead Publishing, Elsevier.

Nelsen, R.B. (2006). *An Introduction to Copulas*, 2nd edition. Springer-Verlag, New York.

Pendola, M. (2000). Fiabilité des structures en contexte d'incertitudes statistiques et d'écarts de modélisation. PhD Thesis, Blaise Pascal University – Clermont II.

Perrin, G. and Defaux, G. (2019). Efficient evaluation of reliability-oriented sensitivity indices. *Journal of Scientific Computing*, 79, 1433–1455.

Perrin, G., Soize, C. and Ouhbi, N. (2018). Data-driven kernel representations for sampling with an unknown block dependence structure under correlation constraints. *Journal of Computational Statistics and Data Analysis*, 119, 139–154.

Rachdi, N. (2011). Apprentissage statistique et computer experiments – Approche quantitative du risque et des incertitudes en modélisation. PhD Thesis, University of Toulouse III – Paul Sabatier.

Raguet, H. and Marrel, A. (2018). Target and conditional sensitivity analysis with emphasis on dependence measures. *ArXiv e-prints*, 1–48.

Ronse, A. and Mooij, E. (2014). Statistical impact prediction of decaying objects. *Journal of Spacecraft and Rockets*, 51(6), 1797–1810.

Saltelli, A., Ratto, M., Andres, T., Campolongo, F., Cariboni, J., Gatelli, D., Saisana, M. and Tarantola, S. (2008). *Global Sensitivity Analysis. The Primer.* Wiley, Chichester.

Sankararaman, S. and Mahadevan, S. (2013). Separating the contributions of variability and parameter uncertainty in probability distributions. *Reliability Engineering and System Safety*, 112, 187–199.

Schöbi, R. and Sudret, B. (2017). Global sensitivity analysis in the context of imprecise probabilities (p-boxes) using sparse polynomial chaos expansions. *Reliability Engineering and System Safety*, 148, 129–141.

Silverman, B.W. (1986). *Density Estimation for Statistics and Data Analysis.* Chapman and Hall, Boca Raton.

Wang, Y., Xiao, S. and Lu, Z. (2018). A new efficient simulation method based on Bayes' theorem and importance sampling Markov chain simulation to estimate the failure-probability-based global sensitivity measure. *Aerospace Science and Technology*, 79, 364–372.

Wei, P., Lu, Z., Hao, W., Feng, J. and Wang, B. (2012). Efficient sampling methods for global reliability sensitivity analysis. *Computer Physics Communications*, 183, 1728–1743.

Wei, P., Lu, Z. and Song, S. (2015). Variable importance analysis: A comprehensive review. *Reliability Engineering and System Safety*, 142, 399–432.

Zhang, P. (1996). Nonparametric importance sampling. *Journal of the American Statistical Association*, 91(425), 1245–1253.

Zio, E. and Pedroni, N. (2012). Monte Carlo simulation-based sensitivity analysis of the model of a thermal-hydraulic passive system. *Reliability Engineering and System Safety*, 107, 90–106.

8

Stochastic Multiobjective Optimization: A Descent Algorithm

Quentin MERCIER[1] **and Fabrice POIRION**[2]
[1] *Télécom Paris, Palaiseau, France*
[2] *ONERA, Paris-Saclay University, Palaiseau, France*

8.1. Introduction

In the industrial world, the following paradigm prevails: how can the most efficient systems be designed for the lowest (production and operating) cost? One tool that is capable of simultaneously providing answers to these antagonistic questions is *multiobjective optimization*. The field of optimization and operational research has undergone very important development with the appearance of computers in the 1950s. The works of Collette and Siarry (2011) and Miettinen (1999) offer an exhaustive overview of numerical methods dedicated to this field of mathematics. Regarding the multiobjective optimization subdomain, the numerical techniques proposed are most often derived from those existing in single-objective optimization, for which an adaptation to the multicriteria framework has been carried out. The first proposed approaches transform multiobjective problems into conventional single-criteria optimization problems, for which a large number of efficient solving algorithms exist. However, the downside of this rewriting is that it loses the multiobjective aspect of the problem, and tends to propose only a single solution among the set of possible solutions: among them, preference is given to the objectives, this is referred to as *a priori* optimization. More recently, the adaptation of heuristic algorithms has experienced an important development. These methods have the advantage of being relatively simple to implement; their adaptation only requires changing the way performing individuals are denoted or selected. Among the most widespread genetic algorithms in the literature can be found the NSGA-II (Deb *et al.* 2002), the SPEA (Zitzler and Thiele 1999) and the PAES (Knowles and Corne 1999) algorithms. Although widely used when the objectives are simple to compute, the number of calculations required for their convergence (when convergence theorems exist) is very high, making them difficult to apply to problems requiring high-performance or numerically expensive computations. The extension of gradient methods, which are very efficient in the single-objective case, only takes place later in multiobjective optimization, but compensates for the lack of speed of the previously mentioned algorithms. In differentiable convex multiobjective optimization, the INRIA has developed the multiple gradient descent algorithm (MGDA) (Désidéri 2009) which, when the gradients of the different objectives of the problem are known at a given point, provides a means of constructing a descent direction common to the set of the objectives, regardless of their number. When the gradients are linearly independent, the descent direction provided is such that the directional derivatives of the criteria in this direction are equal. With the exception of the second order, the objectives therefore decrease by the same amount with each iteration. When the provided descent direction is null, a stationary Pareto point of the multiobjective problem is reached. We refer the reader to a website[1] where an online software platform can be used to test this algorithm.

When addressing optimization problems involving complex structures from real life – aircraft, nuclear power plant, etc. – one is often confronted with working with an imperfectly known description of these structures. This may be due to manufacturing defects, lack of knowledge of the mechanical or physical properties of certain parts, or an uncertain environment. A conventional approach to address these uncertainties is to model them by means of random variables. The problem then becomes much more difficult to solve. A first difficulty is to give mathematical meaning to the optimization problem, which then depends on parameters of infinite dimension: the random variables introduced. There are many possible interpretations and formulations: robust formulation, in expectation, in probability, reliable, etc. The difficulty becomes critical when optimization variables are considered to be random; as a matter of fact, the domain for finding solutions then becomes a space of infinite dimension, because it contains the space of the random variables. In practice, this situation rarely comes to pass. A review of these formulations of the optimization problem with random parameters is given in Lelièvre *et al.* (2016). Optimization in the presence of uncertainties has progressed significantly since the middle of the 20th century with, in particular, the work of Bellman and

1 https://mgda.inria.fr.

Zadeh (1970), Dantzig (1955) and Klamroth *et al.* (2017), where various formulations have been proposed, in particular the robust formulation that includes a certain number of concepts such as the *worst case* or the formulation in expectation that has a variance part (Gabrel *et al.* 2014) often utilized in financial mathematics. A wide choice of numerical methods exists in the literature (Sahinidis 2004; Roy *et al.* 2008) and two families of approaches can be distinguished: the first is based on stochastic approximations, such as Robbins–Monro methods (Robbins and Monro 1951; Ermoliev and Wets 1988; Eskandari and Geiger 2009), and the second is based on scenario methods (Shapiro 2003; Nemirovski and Shapiro 2006).

As far as we are concerned, we shall focus on the formulation of the optimization problem *in expectation*, that is to say, we shall consider the problem stated with the mean objective functions. We shall then finally show how it is possible to reformulate a reliability-based optimization problem (RBDO) into a multiobjective stochastic optimization problem.

8.2. Mathematical refresher

Throughout the whole work, the notation $\langle \cdot,\cdot \rangle$ will be used to refer to the Euclidean dot product of $\mathbb{R}^n$. Similarly, the notation $||.||$ is used for the norm associated with the dot product on $\mathbb{R}^n$.

8.2.1. *Stochastic processes*

We shall study a recursive sequence of random variables. We recall here, without developing them further, the few notions of probability necessary to justify the construction of the proposed algorithm.

The details of the processes recalled in this section can be found in any probability book, for example in that of Durrett (2010), and the definitions concerning probability theory are given in Chapter 1 of this book.

DEFINITION 8.1.– *A discrete-time random process is a sequence of random variables* $(X_k)_{k\in\mathbb{N}}$ *defined on* $(\Omega,\mathcal{T},\mathbb{P})$ *with values in* $\mathbb{R}^n$.

DEFINITION 8.2.– *Let* $\{X_k\}_{k\in\mathbb{N}}$ *be a family of random variables on* $(\Omega,\mathcal{T},\mathbb{P})$ *and with values in the measurable spaces* $(F_k,\mathcal{F}_k)_{k\in\mathbb{N}}$. *One calls σ-algebra generated by* $\{X_k\}_{k\in\mathbb{N}}$ *the smallest σ-algebra containing the union of the reciprocal image σ-algebra* $X_k^{-1}(\mathcal{F}_k)$ *and it is denoted by* $\sigma(X_k, k\in\mathbb{N})$.

DEFINITION 8.3.– *A discrete-time random process* $(X_k)_{k\in\mathbb{N}}$ *is said to converge in norm* L^2 *toward a random variable* X_∞ *if at any time* k*, the random variable* X_k *is an integrable square and:*

$$\lim_{k\to+\infty}\mathbb{E}\left[\| X_k - X_\infty \|^2\right] = 0$$

DEFINITION 8.4.– *A random process* $(X_k)_{k\in\mathbb{N}}$ *is said to converge almost surely to a random variable* X_∞ *if:*

$$\mathbb{P}\left(\left\{\lim_{k\to+\infty} X_k = X_\infty\right\}\right) = 1$$

8.2.2. *Convex analysis*

Descent algorithms require the construction of a common descent direction for each objective of the optimization problem. The construction of such a direction, if it exists, is based on convex analysis results.

We shall now recall the few notions that are indispensable for the understanding of this construction. We refer the reader to Bonnans *et al.* (2006), Clarke (1990) and Miettinen (1999) from which the following definitions and results are drawn.

8.2.2.1. *Convex set and convex envelope*

DEFINITION 8.5.– *A set* $C\subseteq\mathbb{R}^n$ *is said to be convex if for any element* $\mathbf{x}$ *and* $\mathbf{y}$ *of C, the segment* $[\mathbf{x},\mathbf{y}]$ *is entirely contained in* C: $\forall t\in[0,1], t\mathbf{x}+(1-t)\mathbf{y}\in C$.

DEFINITION 8.6.– *Let A be a subset of* $E\subseteq\mathbb{R}^n$. *The convex envelope of A is the intersection of all the convex parts of E containing A. The convex envelope of A will be denoted by* $\mathrm{Conv}(A)$.

PROPOSITION 8.1.– Let A be a subset of $E \subseteq \mathbb{R}^n$. The convex envelope of A is the set of convex combinations of families of points of A:

$$\text{Conv}(A) = \left\{ \mathrm{x} \in E : \exists \{\mathbf{a}_i\}_{i \in [1,p]}, \mathbf{a}_i \in A ; \mathrm{x} = \sum_{i=1}^{p} \lambda_i \mathbf{a}_i, \lambda_i \geq 0, \sum_{i=1}^{p} \lambda_i = 1 \right\}$$

8.2.2.2. *Subdifferentials and directional derivatives*

When the different objectives of the optimization problem are regular, the common descent direction is built from the derivatives of the objectives. When this is no longer the case, a generalization of the derivative must be used: the subderivative. Nonetheless, the objective functions will have to be locally Lipschitzian.

DEFINITION 8.7.– *The function* $f : \mathbb{R}^n \to \mathbb{R}$ *is said to be Lipschitzian if there exists* K *strictly positive such that for all* $(\mathbf{x}, \mathbf{y})$ *in* $(\mathbb{R}^n)^2$, *we have the following inequality*:

$$| f(\mathbf{y}) - f(\mathbf{x}) | \leq K | \mathbf{y} - \mathbf{x} |$$

DEFINITION 8.8.– *The function* $f : \mathbb{R}^n \to \mathbb{R}$ *is said to be locally Lipschitzian at point* $\mathbf{x}$ *if there exists a ball of center* $\mathbf{x}$ *and a positive radius* ρ *such that for every* $(\mathbf{y}, \mathbf{z})$ *in* $\mathcal{B}(\mathbf{x}, \rho)$, *we have the following inequality:*

$$| f(\mathbf{z}) - f(\mathbf{y}) | \leq K | \mathbf{z} - \mathbf{y} |, K > 0$$

DEFINITION 8.9.– *The directional derivative at point* $\mathbf{x}$ *following the direction* $\mathbf{v}$ *in* $\mathbb{R}^n$ *of a function* $f : \mathbb{R}^n \to \mathbb{R}$ *is defined by the limit:*

$$f'(\mathbf{x}, \mathbf{v}) = \lim_{t \downarrow 0} \frac{f(\mathbf{x} + t\mathbf{v}) - f(\mathbf{x})}{t}$$

When f *is differentiable, we have:*

$$f'(\mathbf{x}, \mathbf{v}) = \langle \nabla f(\mathbf{x}), \mathbf{v} \rangle$$

where $\nabla f(\mathbf{x}) = (\frac{\partial}{\partial \mathbf{x}_1}(f(\mathbf{x})), \ldots, \frac{\partial}{\partial \mathbf{x}_n}(f(\mathbf{x})))$ *designates the gradient of the function* f *at point* $\mathbf{x}$.

DEFINITION 8.10.– *A function* $f:\mathbb{R}^n \to \mathbb{R}$ *is said to be convex if for all* $\mathbf{x}$ *and* $\mathbf{y}$ *in* $\mathbb{R}^n$ *and* λ *in* $[0,1]$ *the following inequality is verified:*

$$f(\lambda\mathbf{x}+(1-\lambda)\mathbf{y}) \leq \lambda f(\mathbf{x})+(1-\lambda)f(\mathbf{y})$$

DEFINITION 8.11.– *The subdifferential of a convex function* $f:\mathbb{R}^n \to \mathbb{R}$ *at point* $\mathbf{x}$ *is defined by the set:*

$$\partial f(\mathbf{x}) = \left\{\mathbf{s} \in \mathbb{R}^n : f'(\mathbf{x},\mathbf{d}) \geq \langle \mathbf{s},\mathbf{d}\rangle, \forall \mathbf{d} \in \mathbb{R}^n\right\}$$

This set is non-empty, convex and closed. Each vector of $\partial f(\mathbf{x})$ *is called subgradient of* f *at point* $\mathbf{x}$, *and the subdifferential is reduced to the gradient* $\nabla f(\mathbf{x})$ *when the function* f *is differentiable.*

Figure 8.1 illustrates the notion of subdifferential of a convex function. In this example, the function $f(x)$ is continuous and convex but is non-differentiable at the points x_1 and x_2. At these points, the subdifferential corresponds to the set of unit vectors of the straight lines, which pass through the point $(x_i)_{i\in[1,2]}$ while remaining under the curve representing the function $f(x)$.

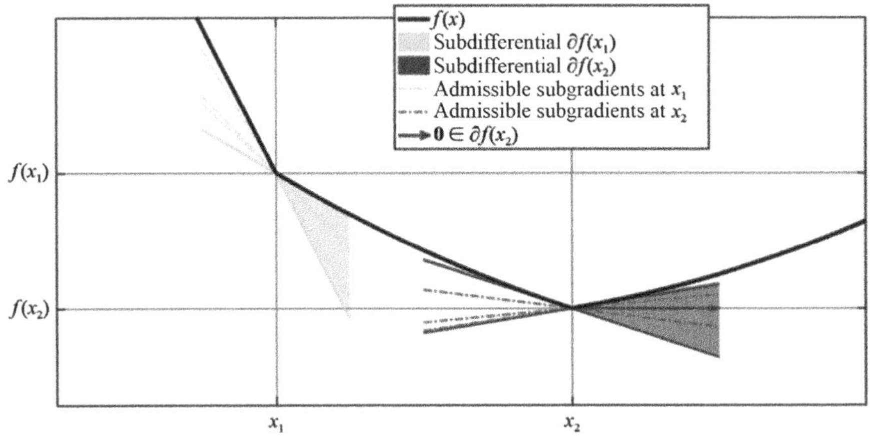

Figure 8.1. *Example: calculation of a subdifferential of a convex function. For a color version of this figure, see www.iste.co.uk/gogu/uncertainties.zip*

The notions of directional derivative and subdifferential defined above do not necessarily exist when the function is no longer convex but locally Lipschitzian. In order to take this situation into account for the construction of a common descent vector, it will be necessary to generalize these two notions. To this end, we shall follow the approach given in Clarke (1990), which we recall.

DEFINITION 8.12.– *Let* $f:\mathbb{R}^n \to \mathbb{R}$ *be a locally continuous Lipschitzian function. The generalized directional derivative of the function* f *at the point* $\mathbf{x}$ *in the direction* $\mathbf{v}$ *of* $\mathbb{R}^n$ *is defined by:*

$$f^{\circ}(\mathbf{x},\mathbf{v}) = \limsup_{\mathbf{y}\to\mathbf{x}\ t\downarrow 0} \frac{f(\mathbf{y}+t\mathbf{v})-f(\mathbf{y})}{t}$$

DEFINITION 8.13.– *A locally continuous Lipschitzian function* $f:\mathbb{R}^n \to \mathbb{R}$ *is said to be regular for subdifferentiability at a point* $\mathbf{x}\in\mathbb{R}^n$ *if for all* $\mathbf{d}\in\mathbb{R}^n$ *the directional derivative* $f'(\mathbf{x},\mathbf{d})$ *exists and if one has the equality:*

$$f'(\mathbf{x},\mathbf{d}) = f^{\circ}(\mathbf{x},\mathbf{d})$$

DEFINITION 8.14.– *A locally continuous Lipschitzian function* $f:\mathbb{R}^n \to \mathbb{R}$ *is* f°*-quasi-convex if:*

$$\forall(\mathbf{x},\mathbf{y})\in(\mathbb{R}^n)^2, f(\mathbf{y})\le f(\mathbf{x}) \Rightarrow f^{\circ}(\mathbf{x},\mathbf{y}-\mathbf{x})\le 0$$

DEFINITION 8.15.– *A locally continuous Lipschitzian function* $f:\mathbb{R}^n \to \mathbb{R}$ *is* f°*-pseudo-convex if:*

$$\forall(\mathbf{x},\mathbf{y})\in(\mathbb{R}^n)^2, f(\mathbf{y})< f(\mathbf{x}) \Rightarrow f^{\circ}(\mathbf{x},\mathbf{y}-\mathbf{x})< 0$$

THEOREM 8.1 (Mäkelä *et al.* 2010).– A f°-pseudo-convex function is f°-quasi-convex.

DEFINITION 8.16.– *Let* $f:\mathbb{R}^n \to \mathbb{R}$ *be a locally continuous Lipschitzian function. The Clarke subdifferential of* f *at point* $\mathbf{x}$ *is the set* $\partial^C f(\mathbf{x})$ *defined by:*

$$\partial^C f(\mathbf{x}) = \left\{\mathbf{s}\in\mathbb{R}^n : f^{\circ}(\mathbf{x},\mathbf{v}) \ge \langle \mathbf{s},\mathbf{v}\rangle\ ;\forall \mathbf{v}\in\mathbb{R}^n\right\}$$

PROPOSITION 8.2 (Mäkelä *et al.* 2010).– Convex functions are regular for subdifferentiability. Let $f:\mathbb{R}^n \to \mathbb{R}$ be a convex function; at a point $\mathbf{x} \in \mathbb{R}^n$, we have:

$$\partial f(\mathbf{x}) = \partial^C f(\mathbf{x})$$

THEOREM 8.2 (Mäkelä *et al.* 2010).– Let m locally Lipschitzian functions $f_i:\mathbb{R}^n \to \mathbb{R}$ be continuous at $\mathbf{x} \in \mathbb{R}^n$. And let $\{\lambda_i\}_{i\in[1,m]}$ be a family of real numbers. The function defined by:

$$h(\mathbf{x}) = \sum_{i=1}^{m} \lambda_i f_i(\mathbf{x})$$

is locally Lipschitzian and continuous in $\mathbf{x}$; in addition, its subdifferential has the following property:

$$\partial h(\mathbf{x}) \subseteq \sum_{i=1}^{m} \lambda_i \partial f_i(\mathbf{x})$$

Moreover, when the functions f_i are regular for the subdifferential and the family of the $\lambda_i, i = 1,\ldots,m$ is only composed of real positives, then h is also regular for the subdifferential, and we have:

$$\partial h(\mathbf{x}) = \sum_{i=1}^{m} \lambda_i \partial f_i(\mathbf{x})$$

Clarke gives an equivalent definition of the subdifferential, a definition that will be useful to construct an approximation: it is based on Rademacher Federer's theorem (1996) that says that any Lipschitzian function on a set $E \subseteq \mathbb{R}^n$ is differentiable almost everywhere on E.

DEFINITION 8.17.– *Let $f:\mathbb{R}^n \to \mathbb{R}$ be a locally continuous Lipschitzian function and Q a subset of $\mathbb{R}^n$ contained in a neighborhood of $\mathbf{x} \in \mathbb{R}^n$ where f is differentiable. Then:*

$$\partial^C f(\mathbf{x}) = \mathrm{Conv}\left(\left\{ \lim_{\mathbf{x}_r \to \mathbf{x}} \nabla f(\mathbf{x}_r) : \mathbf{x}_r \in Q \right\}\right)$$

8.2.2.3. *Numerical construction of the Clarke subdifferential*

Based on this definition of the Clarke subdifferential, Burke *et al.* (2002) proposed a numerical approach to construct an approximation of this subdifferential. First of all, the Clarke subdifference can be defined in the following equivalent fashion: let $\mathcal{B} = \{\mathbf{x} : \|\mathbf{x}\| \leq 1\}$ be the unit ball and $\varepsilon > 0$.

Consider the convex set:

$$G_{\varepsilon}(\mathbf{x}) = \text{Conv}\left(\{\nabla f(\mathbf{x} + \varepsilon\mathcal{B}) \cap Q\}\right)$$

then the Clarke subdifferential can still be written as:

$$\partial^{C} f(\mathrm{x}) = \bigcap_{\varepsilon>0} G_{\varepsilon}(\mathrm{x})$$

For a fixed $\delta > 0$, the proposed approach then consists of carrying out uniform and independent draws $x_i, i = 1,\ldots,k$ in the set $(\mathbf{x} + \delta\mathcal{B}) \cap Q$, then in building the sequence of gradients $G_i = \nabla f(\mathbf{x}_i)$, as well as the increasing sequence of closed convex sets:

$$C_k = \text{Conv}\left(\{G_1,\ldots,G_k\}\right)$$

THEOREM 8.3.– If f is a local Lipschitzian function in the neighborhood of $\mathbf{x}$, then for δ small enough and for every $\varepsilon > 0$, we have:

$$\lim_{k \to +\infty} \mathbb{P}\left(\{\partial^{C} f(\mathbf{x}) \subset C_k + \varepsilon\mathcal{B}\}\right) = 1$$

Figure 8.2 illustrates the construction of an approximation of the Clarke subdifferential for a function that is not differentiable at point x_0. A radius neighborhood ε is chosen around x_0, and the gradient of the function is calculated for some points x_i randomly selected in the neighborhood. The convex envelope of these samples constitutes the used approximation of the subdifferential. In this simple example, one can observe a difference between the true subdifferential plotted in green and its approximation in brown.

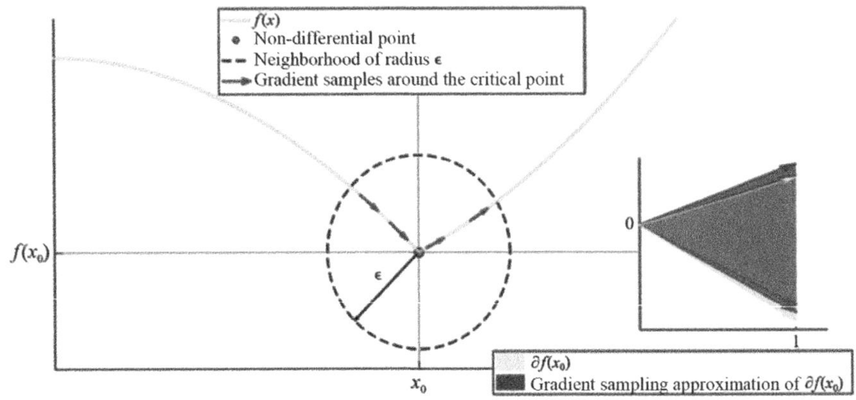

Figure 8.2. *Approximation of the subdifferential of a non-convex function. For a color version of this figure, see www.iste.co.uk/gogu/uncertainties.zip*

8.3. Multiobjective optimization and common descent vector

First, one recalls the mathematical formulation of an optimization problem in general, as well as the dominance relation according to Pareto, which is used in order to achieve multiobjective optimization without *a priori*. The results and notions on the Pareto dominance relation are taken from Ehrgott (2006).

8.3.1. *Binary relations*

The optimization corresponds to the search for the best element(s) of a set, the best term having to be specified.

In order to be able to identify which elements are considered to be the best performers, it is necessary to be able to compare each of the elements present in the searching set. This section recalls the mathematical tools that can be used to implement the comparison methods between the elements of a set.

It is first necessary to define the notion of relationship between elements of a set or of several sets.

DEFINITION 8.18.– *A binary relation $\mathcal{R}$ of a set $\mathcal{X}$ between $\mathcal{Y}$ is defined by a subset $\mathcal{G} \subseteq \mathcal{X} \times \mathcal{Y}$. It is said that an element $x \in \mathcal{X}$ is related to $y \in \mathcal{Y}$ if the couple $(x, y) \in \mathcal{G}$ and it is denoted $x\mathcal{R}y$. When $\mathcal{X} = \mathcal{Y}$, $\mathcal{R}$ is said to be a relation on $\mathcal{X}$.*

DEFINITION 8.19.– *Let $\mathcal{R}$ be a relation on a set $\mathcal{X}$, it is said that $\mathcal{R}$ is:*

1) *reflexive (resp. non-reflexive) if:* $\forall x \in \mathcal{X}, x\mathcal{R}x$ (resp. $x \not{\mathcal{R}} x$);

2) *symmetric if:* $\forall (x,y) \in \mathcal{X}^2, x\mathcal{R}y \Rightarrow y\mathcal{R}x$;

3) *weakly antisymmetric if:* $\forall (x,y) \in \mathcal{X}^2, x\mathcal{R}y \wedge y\mathcal{R}x \Rightarrow x = y$;

4) *strongly antisymmetric if:* $\forall (x,y) \in \mathcal{X}^2, x\mathcal{R}y \Rightarrow y \not{\mathcal{R}} x$;

5) *transitive if:* $\forall (x,y,z) \in \mathcal{X}^3, x\mathcal{R}y \wedge y\mathcal{R}z \Rightarrow x\mathcal{R}z$;

6) *total if:* $\forall (x,y) \in \mathcal{X}^2, x\mathcal{R}y \vee y\mathcal{R}x$;

where $\vee$ *(resp.* $\wedge$*) designates logical "or" (resp. the "and"). From binary relations, it is possible to construct comparative relations between different elements of a given set.*

DEFINITION 8.20.– *It is said that a binary relation $\mathcal{R}$ on $\mathcal{X}$ defines:*

1) *a preorder on $\mathcal{X}$ when the relation is $\mathcal{R}$ reflexive and transitive. The couple $(\mathcal{X},\mathcal{R})$ is said to constitute a preordered space. The notation $\preceq_{\mathcal{X}}$ is used to designate a preorder on $\mathcal{X}$;*

2) *an equivalence on $\mathcal{X}$ when the relation is $\mathcal{R}$ reflexive, transitive and symmetric and will be denoted by $\equiv_{\mathcal{X}}$*

3) *an order on $\mathcal{X}$ when the relation $\mathcal{R}$ is reflexive, transitive and antisymmetric. The couple $(\mathcal{X},\mathcal{R})$ is said to constitute an ordered space. The notation $\leq_{\mathcal{X}}$ will be used;*

4) *a strict order on $\mathcal{X}$ when the relation is $\mathcal{R}$ not reflexive, transitive and antisymmetric. The notation $<_{\mathcal{X}}$ will be used.*

Optimization naturally uses the notions of order and preorder to compare different elements of a set and to deduce which elements are the most efficient. It is clear that the elements of a set considered to be the best performing depends strongly on the binary relations used: different comparison relations will give different best performing elements.

In the following section, the principle of optimization and the notion of minimum of an ordered and preordered set are formally introduced. In particular, the Pareto preorder is presented which is one of the binary relations used in the framework of multiobjective optimization.

8.3.2. *Multiobjective optimization, Pareto preorder*

An optimization problem can always be mathematically represented as a minimization problem of a function called *cost function* f defined on a set called *variable space* $\aleph$ and with values in Υ which is called the *target space*:

$$\begin{aligned} f: \quad & \aleph \rightarrow \Upsilon \\ & \mathbf{x} \mapsto f(\mathbf{x}) \end{aligned}$$

The space $\mathcal{X} \subseteq \aleph$ on which minimization is carried out corresponds to the *admissible solutions* of the problem, and the solutions take into account the possible constraints. The image by the *cost function* f of the set $\mathcal{X}$ is denoted $\mathcal{Y} \subseteq \Upsilon$ and this space is called the *achievable set*. Minimizing f on $\mathcal{X}$ is in fact tantamount to finding the set of minima of $\mathcal{Y}$ endowed with an order $\leq_{\mathcal{Y}}$:

$$\min_{\mathbf{x} \in \mathcal{X}} f(\mathbf{x}) = \min(\mathcal{Y}, \leq_{\mathcal{Y}}) \qquad [8.1]$$

and to identify the elements of $\mathcal{X}$ whose images are minima of $(\mathcal{Y}, \leq_{\mathcal{Y}})$:

$$\operatorname*{argmin}_{\mathbf{x} \in \mathcal{X}} f(\mathbf{x}) = \left\{ \mathbf{x} \in \mathcal{X}; f(\mathbf{x}) \in \min(\mathcal{Y}, \leq_{\mathcal{Y}}) \right\}$$

DEFINITION 8.21.– *Let f be a function of $\mathcal{X} \subseteq \aleph$ with values in $\mathcal{Y} \subseteq \Upsilon$ endowed with a total order $\leq_{\mathcal{Y}}$. The set of the elements of $\mathcal{Y}$ which minimizes the ordered space $(\mathcal{Y}, \leq_{\mathcal{Y}})$ is given by:*

$$\min(\mathcal{Y}, \leq_{\mathcal{Y}}) = \left\{ \mathbf{y}^* \in \mathcal{Y}; \forall \mathbf{y} \in \mathcal{Y}, \mathbf{y}^* \leq_{\mathcal{Y}} \mathbf{y} \right\}$$

In case the *cost function* f has a value in $\mathbb{R}$, it is generally associated with the usual order $\leq$ of real numbers. This is then referred to as single-objective optimization.

When the *cost function* f has values in $\mathbb{R}^m$, each component f_i of f is associated with a function with values in $\mathbb{R}$ which is called an *objective function*:

$$\begin{aligned} \forall i \in [1, m], f_i: \quad & \mathbb{R}^n \rightarrow \mathbb{R} \\ & \mathbf{x} \mapsto f_i(\mathbf{x}) \end{aligned}$$

This is then referred to as multiobjective optimization and problem [8.1] can be interpreted as:

$$\min_{\mathbf{x}\in\mathcal{X}} f(\mathbf{x}) = \min_{\mathbf{x}\in\mathcal{X}}\{f_1(\mathbf{x}), f_2(\mathbf{x}), \dots, f_m(\mathbf{x})\} \qquad [8.2]$$

The direct extension of the above definition with a function $f = (f_1, \dots, f_m)$ having values in $\mathbb{R}^m$ with m an integer strictly greater than 1 is not possible since the resulting product order $\prod^m(\mathbb{R}, \leq)$ is not a total order. A first approach in multiobjective optimization is to consider a space $\mathcal{Z}$ by means of a function:

$$\begin{aligned} \theta:\ & \mathcal{Y} \to \mathcal{Z} \\ & f(\mathbf{x}) \mapsto \theta \circ f(\mathbf{x}) \end{aligned}$$

for which there exists a total order $\leq_{\mathcal{Z}}$. The most conventional way is to use a weighted sum of the objectives so as to be brought back to a real-valued *cost function*. This allows the order $\leq$ of the real numbers to be preserved.

Problem [8.2] can then be rewritten as:

$$\min_{\mathbf{x}\in\mathcal{X}}\left(\theta \circ f(\mathbf{x})\right) = \min_{\mathbf{x}\in\mathcal{X}}\left(\sum_{i=1}^{m} p_i f_i(\mathbf{x})\right) \qquad [8.3]$$

where $p = (p_1, \dots, p_m) \in [0,1]^m$ is the weight vector such that $\sum_{i=1}^{m} p_i = 1$. However, reducing the multiobjective problem to a single-objective problem endowed with the usual order $\leq$ implies that preferences between objectives have been added by the choice of weights p_i. When the multiobjective problem is modified by a function θ in order to fall back onto a new totally ordered space, this is referred to as *a priori* optimization.

When seeking to completely preserve the multiobjective nature of the initial problem [8.2], a new order must be constructed that allows *feasible solutions* to be compared. It can be considered that one solution is better than another if it is as good with all of the objectives and is strictly better for at least one objective of the problem: this is the Pareto dominance.

DEFINITION 8.22.– *For optimization problem [8.2], a point in the space of optimization variables* $\mathbf{x}_1$ *is said to dominate, following Pareto's definition, a point*

$\mathbf{x}_2$, *if the point* $\mathbf{x}_1$ *is at least as good as* $\mathbf{x}_2$ *with the set of objectives, and if* $\mathbf{x}_1$ *is better than* $\mathbf{x}_2$ *for at least one objective:*

$$\begin{cases} \forall i \in [1,m],\ f_i(\mathbf{x}_1) \leq f_i(\mathbf{x}_2) \\ \exists \ell \in [1,m],\ f_\ell(\mathbf{x}_1) < f_\ell(\mathbf{x}_2) \end{cases} \qquad [8.4]$$

Let $(\mathbf{x}_1, \mathbf{x}_2)$ be two *admissible solutions* and $(\mathbf{y}_1, \mathbf{y}_2)$ their respective images by the cost function f with values in $\mathbb{R}^m$. If $\mathbf{x}_1$ dominates the point $\mathbf{x}_2$ according to Pareto's definition, it can be written that:

$$\mathbf{x}_1 \overset{f}{\prec_{\mathcal{P}}} \mathbf{x}_2$$

and in the objective space, it is equivalently written using the symbol $\prec_{\mathcal{P}}$ that:

$$f(\mathbf{x}_1) \prec_{\mathcal{P}} f(\mathbf{x}_2)$$

It is then directly obtained that:

$$\mathbf{y}_1 \prec_{\mathcal{P}} \mathbf{y}_2$$

PROPOSITION 8.3.– The binary dominance relation $\prec_{\mathcal{P}}$ according to Pareto defines a strict order in the space $\mathbb{R}^m$ for the optimization problem [8.2].

Using this principle of dominance, it is no longer possible to order all solutions in relation to one another, and the point is then to determine the *admissible solutions* such that their images by the cost function f are not dominated in the *space of objectives*:

$$\mathcal{P}_O = \left\{ \mathbf{y}^\star \in \mathcal{Y}; \nexists \mathbf{y} \in \mathcal{Y}, \mathbf{y} \prec_{\mathcal{P}} \mathbf{y}^\star \right\} \qquad [8.5]$$

The set of non-dominated achievable points is called the *Pareto front*. The reciprocal image of the Pareto front by the cost function f is called the set of *Pareto solutions* and is denoted $\mathcal{P}_D$:

$$\mathcal{P}_D = \left\{ \mathbf{x} \in \mathcal{X}; f(\mathbf{x}) \in \mathcal{P}_O \right\}$$

The set of *Pareto solutions* thus corresponds to the set of *admissible solutions* such that there is no other *admissible solution* that is simultaneously better in the whole set of objectives that constitute the optimization problem: it is impossible to improve a *Pareto solution* on one of the objectives without degrading another one.

One way of interpreting the solutions is to say that the *Pareto front* gives access to the set of the best trade-offs between the different objectives of the optimization problem.

Once the Pareto front has been determined, it becomes possible to choose a particular solution by giving preference to a particular objective. This is called *a posteriori* optimization.

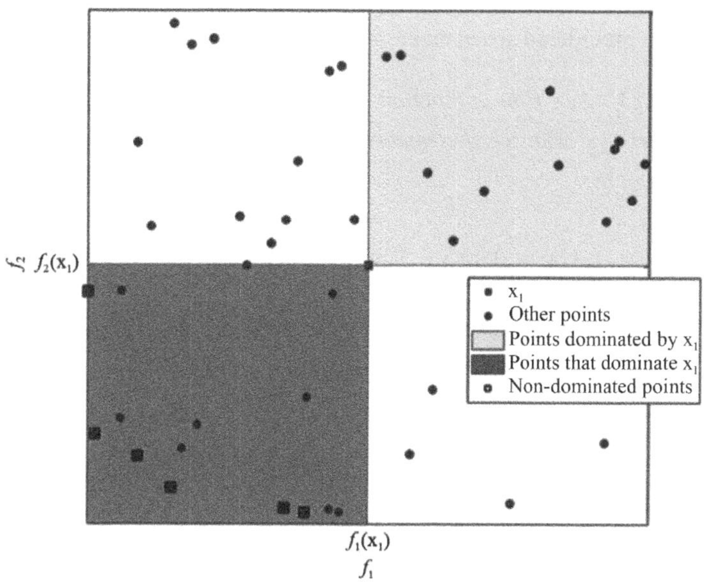

Figure 8.3. *Example of dominance according to Pareto for a two-objective problem $min_x\{f_1(x),f_2(x)\}$. For a color version of this figure, see www.iste.co.uk/gogu/uncertainties.zip*

As an illustration of Pareto's notion of dominance, some points can be observed, in Figure 8.3, belonging to a *feasible set* of a two-objective optimization problem:

$$\min_{\mathbf{x}}\{f_1, f_2\}$$

Taking the point $\mathbf{y}_1$ as a reference, and using the Pareto definition of dominance, it is possible to determine the regions for which:

– any point $\mathbf{y}$ is dominated by $\mathbf{y}_1$ (green region);

– any point $\mathbf{y}$ dominates $\mathbf{y}_1$ (brown region).

In this figure, it is also possible to find the non-dominated points (white region) for the sample of the *feasible set* being represented.

Based on Pareto's dominance relation, it is not possible to create an order compatible with the definition of a minimum. A second, more general definition of a minimum on a preordered space makes it possible to work around this difficulty.

DEFINITION 8.23.– *Let f be a function of $\mathcal{X} \subseteq \aleph$ with values in $\mathcal{Y} \subseteq \Upsilon$ endowed with a preorder $\preceq_{\mathcal{Y}}$. The set of elements of $\mathcal{Y}$ which minimize the preordered space $(\mathcal{Y}, \preceq_{\mathcal{Y}})$ is given by:*

$$\min(\mathcal{Y}, \preceq_{\mathcal{Y}}) = \left\{\mathbf{y}^\star \in \mathcal{Y} \,; \forall \mathbf{y} \in \mathcal{Y}, \mathbf{y} \preceq_{\mathcal{Y}} \mathbf{y}^\star \Rightarrow \mathbf{y} \equiv_{\mathcal{Y}} \mathbf{y}^\star\right\}$$

DEFINITION 8.24.– *Let f be a function of $\aleph$ in Υ, $\mathcal{X} \subseteq \aleph$ the set of admissible solutions of the associated optimization problem, $\mathcal{Y} \subseteq \Upsilon$ the image of $\mathcal{X}$ by f and $\preceq_{\mathcal{Y}}$ a preorder on $\mathcal{Y}$. It is said that $x^\star$ is a local minimum of f on $\mathcal{X}$ for the preorder $\preceq_{\mathcal{Y}}$ if and only if there exists an open ball $\mathcal{B}(\mathbf{x}^\star, \rho)$ of center $x^\star$ and radius $\rho > 0$ such that:*

$$\operatorname{argmin}(\mathcal{Y}, \preceq_{\mathcal{Y}}) = \left\{\mathbf{x}^\star \in \mathcal{X} \,; \forall \mathbf{x} \in \mathcal{X} \cap \mathcal{B}(\mathbf{x}^\star, \rho), f(\mathbf{x}) \preceq_{\mathcal{Y}} f(\mathbf{x}^\star) \Rightarrow f(\mathbf{x}) \equiv_{\mathcal{Y}} f(\mathbf{x}^\star)\right\}$$

In the following, the aim is to construct a preorder derived from Pareto's dominance, compatible with the definition of the set of elements that minimize a preordered space. A new binary relation is created and denoted as $\preceq_{\mathcal{P}}$ such that for any element $(\mathbf{y}_1, \mathbf{y}_2)$, the relation $\mathbf{y}_1 \preceq_{\mathcal{P}} \mathbf{y}_2$ implies that $\mathbf{y}_1$ dominates $\mathbf{y}_2$ according to Pareto or that $\mathrm{y}_1 = \mathrm{y}_2$:

$$\forall (\mathbf{y}_1, \mathbf{y}_2) \in \mathcal{Y}^2, \mathbf{y}_1 \preceq_{\mathcal{P}} \mathbf{y}_2 \begin{cases} \mathbf{y}_1 \prec_{\mathcal{P}} \mathbf{y}_2 \\ \quad \text{or} \\ \mathbf{y}_1 = \mathbf{y}_2 \end{cases}$$

PROPOSITION 8.4.– The binary relation $\preceq_{\mathcal{P}}$ is a preorder on $\mathbb{R}^m$.

IMPORTANT.– The preorder resulting from Pareto's dominance and the relation of equality corresponds in fact to the created order of $\mathbb{R}^m$.

PROPOSITION 8.5.– Let f be a cost function with values in $\mathbb{R}^m$, and let $\preceq_{\mathcal{P}}$ be the Pareto preorder on $\mathbb{R}^m$:

1) any element of the Pareto front $\mathcal{P}_O$ defined by relation [8.5] is a minimum element of $\left(\mathbb{R}^m, \preceq_{\mathcal{P}}\right)$;

2) any element that minimizes $\left(\mathbb{R}^m, \preceq_{\mathcal{P}}\right)$ belongs to the Pareto front $\mathcal{P}_O$.

When two points $(\mathbf{y}_1, \mathbf{y}_2)$ of the feasible set are incomparable, in other words when:

$$\mathbf{y}_1 \not\prec_{\mathcal{P}} \mathbf{y}_2 \text{ and } \mathbf{y}_2 \not\prec_{\mathcal{P}} \mathbf{y}_1 \quad [8.6]$$

it is said, misleadingly, that the two points $\mathbf{y}_1$ and $\mathbf{y}_2$ are equivalent following Pareto, although the binary relationship described in [8.6] is not transitive and is not a relation of equivalence. This relation is denoted by:

$$\mathbf{y}_1 \equiv_{\mathcal{P}} \mathbf{y}_2$$

PROPOSITION 8.6.– The elements of the Pareto front $\mathcal{P}_O$ are incomparable with each other.

Solving a minimization problem for a cost function with values in $\mathbb{R}^m$ endowed with Pareto's preorder thus corresponds to finding the set of incomparable solutions such that they cannot be dominated by any element of the *feasible set*.

The next section of the chapter aims to define the notion of the common descent vector. More specifically, several methods are proposed in order to construct such a vector according to the regularity of the objective functions. In the following, an optimization problem described by relation [8.2] is considered, where the cost function $f = (f_1, \ldots, f_m)$ is a function of $\mathcal{X} \subseteq \mathbb{R}^n$ with values in $\mathcal{Y} \subseteq \mathbb{R}^m$ endowed with the Pareto preorder $\preceq_{\mathcal{P}}$ on $\mathcal{Y}$.

8.3.3. *Common descent vector*

This section is dedicated to giving definitions and different properties of the common descent vectors that are essential ingredients of the algorithms proposed in the remainder of the chapter.

It is assumed that the objective functions f_i are at least f°-pseudo-convex and locally Lipschtizian.

DEFINITION 8.25.– *Let* **d** *be a vector of* $\mathbb{R}^n$*, it is said that* **d** *is a descent vector of the function* $f_i, i \in [1,m]$*, at point* $\mathbf{x} \in \mathcal{X}$ *if the function* f_i *decreases monotonically on a direction segment* **d**:

$$\exists \ell_0 \in \mathbb{R}, \forall \ell \in [0, \ell_0]; f_i(\mathbf{x} + \ell \mathbf{d}) \le f_i(\mathbf{x})$$

PROPOSITION 8.7.– Let **d** be a descent vector for f_i, $i \in [1,m]$, at the point $\mathbf{x} \in \mathcal{X}$. For a f_i°-pseudo-convex function, the generalized directional derivative following the direction **d** is negative:

$$f_i^{\circ}(\mathbf{x}, \mathbf{d}) \le 0$$

DEFINITION 8.26.– *Let* **d** *be a vector of* $\mathbb{R}^n$*, and* **d** *is said to be the descent vector common to the functions* $\{f_i\}$ *with* $i \in [1,m]$ *in* $\mathbf{x} \in \mathcal{X}$ *if the functions* f_i *all decrease monotonically over a direction segment* **d**:

$$\exists \ell_0 \in \mathbb{R}, \forall \ell \in [0, \ell_0]; f_i(\mathbf{x} + \ell \mathbf{d}) \le f_i(\mathbf{x}), \forall i \in [1,m]$$

PROPOSITION 8.8.– Let **d** be a descent vector common to the functions $\{f_i\}$ with $i \in [1,m]$ and at the point $\mathbf{x} \in \mathcal{X}$. The generalized directional derivatives of the $\{f_i\}$ following the direction **d** are negative:

$$\forall i \in [1,m], f_i^{\circ}(\mathbf{x}, \mathbf{d}) \le 0$$

In the following, the notation $\mathcal{C}(\mathbf{x})$ is used to designate the convex envelope of the union of the subdifferentials calculated at the point $\mathbf{x}$:

$$\mathcal{C}(\mathbf{x}) = \mathrm{Conv}\left(\bigcup_{i=1}^{m} \partial^C f_i(\mathbf{x}) \right) \qquad [8.7]$$

PROPOSITION 8.9.– Let $\mathbf{x}$ be an admissible solution and $\mathcal{C}(\mathbf{x})$ the convex envelope of the union of the Clarke subdifferentials. Any element of $\mathcal{C}(\mathbf{x})$ can be written as a convex combination of an element of each of the subdifferentials:

$$\forall \mathbf{c} \in \mathcal{C}(\mathbf{x}), \exists \{\Lambda_i\}_{i\in[1,m]}, \mathbf{c} = \sum_{i=1}^{m} \Lambda_i T^i, T^i \in \partial^C f_i(\mathbf{x})$$

where the sum of the Λ_i are positive or zero and of sum equal to 1, and $\Lambda_i \geq 0$ are positive.

DEFINITION 8.27.– *A point* $\mathbf{x}^{\star}$ *of the optimization variable space is said to be a stationary Pareto point if the null vector belongs to the convex envelope of Clarke subdifferentials:*

$$\mathbf{0} \in \mathcal{C}(\mathbf{x}^{\star})$$

IMPORTANT.– When the objective functions are convex, the Pareto-stationarity definition then becomes:

$$\mathbf{0} \in \mathrm{Conv}\left(\bigcup_{i=1}^{m} \partial f_i(\mathbf{x})\right)$$

In addition, when the functions are convex and differentiable, we obtain:

$$\mathbf{0} \in \mathrm{Conv}\left(\{\nabla f_i(\mathbf{x})\}_{i\in[1,m]}\right)$$

DEFINITION 8.28.– *A vector* $\mathbf{x}^{\star}$ *of* $\mathbb{R}^n$ *is said to be a weak Pareto solution for the optimization problem [8.2], if there exists no* $\mathbf{x}$ *of* $\mathbb{R}^n$ *such that:*

$$f_i(\mathbf{x}) < f_i(\mathbf{x}^{\star}), \forall i \in [1,m]$$

Obviously, it can already be said that any Pareto solution is also a weak Pareto solution.

THEOREM 8.4 (Mäkelä *et al.* 2014).– If a point is a local weak Pareto solution, then the null vector belongs to the convex envelope of the Clarke subdifferentials:

$$\mathbf{0} \in \mathcal{C}(\mathbf{x}^{\star}) \qquad [8.8]$$

When the objective functions are all at least f°-pseudo-convex, then condition [8.8] becomes a sufficient condition to have a weak global Pareto solution.

LEMMA.– Let $\mathbf{x}$ be an admissible solution. There exists a single element of minimal norm in $\mathcal{C}(\mathbf{x})$. In addition, the minimal norm element $\xi = \text{argmin}\{\|\mathbf{v}\|, \mathbf{v} \in \mathcal{C}(\mathbf{x})\}$ verifies the following property:

$$\forall \mathbf{c} \in \mathcal{C}(\mathbf{x}), \langle \mathbf{c}, \xi \rangle \geq \| \xi \|^2$$

THEOREM 8.5.– Let $\mathbf{x}$ be an admissible solution and $\mathcal{C}(\mathbf{x})$ the convex envelope of the union of the Clarkes subdifferentials. The opposite of the minimal norm element $-\xi$ in $\mathcal{C}(\mathbf{x})$ if non-zero, is a common descent vector:

$$\xi = \arg\min\{\|\mathbf{c}\|, \mathbf{c} \in \mathcal{C}(\mathbf{x})\} \qquad [8.9]$$

From the convex envelope of the union of the Clarke subdifferentials, it is possible to construct a descent vector common to all of the objectives. This construction of a descent direction allows classical descent algorithms to be generalized to problems where the image of the cost function is larger than 1. In the following, a descent algorithm is presented, compatible with the definition of a minimum on $\mathbb{R}^m$ endowed with the Pareto preorder.

8.4. Descent algorithm for multiobjective optimization and its extension to the stochastic framework

8.4.1. *Multiple gradient descent algorithm*

The MGDA was first introduced in the technical report of Désidéri (2009) to solve multiobjective optimization problems composed of differentiable functions:

$$\min_{\mathrm{x} \in \mathcal{X}} \{f_1(\mathrm{x}), \ldots, f_m(\mathrm{x})\}$$

The MGDA is a descent algorithm based on the existence and construction of a descent vector common to the set of iteration objectives. The algorithm makes use of the common descent vector defined by relation [8.9], and is included in the framework of objectives that are assumed to be convex and differentiable.

Consequently, the convex envelope of the union of the subdifferentials $\mathcal{C}$ at a point $\mathbf{x}$ is reduced to:

$$\mathcal{C}(\mathrm{x}) = \mathrm{Conv}\left(\{\nabla f_i(\mathrm{x})\}_{i\in[1,m]}\right)$$

The descent vector calculated is then used as the descent direction for a given step. Conventional descent step selection methods can be utilized to find an optimal step. It is possible to prove the convergence of such an algorithm to stationary Pareto solutions that, in the convex framework, are also Pareto solutions.

The steps of the MGDA algorithm are as follows:

1) initialization: one sets:

– a starting point $\mathbf{x}_0$;

– a threshold for the stopping criterion σ;

– the choice of a descent step ε;

– a counter $k = 0$;

2) iteration: while $\| \xi_{k-1} \| > \sigma$ $\{\nabla f_i(\mathbf{x}_{k-1})\}_{i\in[1,m]}$ is evaluated;

3) descent: the common descent vector is constructed $-\xi_k$ by solving:

$$\xi_k = argmin\{\| \mathbf{v} \|; \mathbf{v} \in \mathcal{C}(\mathbf{x}_{k-1})\}$$

4) update: the current point is modified: $\mathbf{x}_k = \mathbf{x}_{k-1} - \varepsilon_k \xi_k$ and $k \to k+1$;

5) construction of the optimal value of the objective functions: $\{f_i(\mathbf{x}_k)\}_{i\in[1,m]}$ is evaluated at the convergence of the algorithm.

The stopping conditions of the algorithm will be driven by the value of the norm of the descent vector being used. As a matter of fact, if the norm of this descent vector is zero, it is known that the current point is a stationary Pareto point. In practice, a threshold will be imposed on the norm of the descent vector, below which the point will be considered as a stationary Pareto point.

THEOREM 8.6 (Désidéri 2009).– When the objectives of the optimization problem [8.2] are convex, a subsequence of the iterative sequence of the $\{\mathbf{x}_k\}_{k\in\mathbb{N}}$ described in the MGDA algorithm converges to a point $\mathbf{x}^*$, which is the Pareto solution of the optimization problem [8.2].

The MGDA is therefore a descent algorithm compatible with the minimization in $\mathbb{R}^m$ endowed with the Pareto preorder. The algorithm is based on the existence of a descent vector common to all of the objectives and offers convergence results toward the solutions of problem [8.2] when the objectives of the problem are convex and differentiable at any point of the set of *admissible solutions*.

In the remainder of the chapter, we shall focus on the case where objective functions depend on uncertain parameters that will be modeled by random variables. In the context of this work, we consider that the randomness does not concern the optimization variables. We shall then show how to adapt the MGDA algorithm. First, we shall formulate the random multicriteria problem before presenting the algorithm.

8.4.2. *Stochastic multiple gradient descent algorithm*

In this section, we propose to modify the optimization problem [8.2] initially discussed. In the following, it is now considered that the cost function f of the problem is a random function with values in $\mathbb{R}^m$.

8.4.2.1. *Multiobjective stochastic optimization problem*

Let $(\Omega, \mathcal{A}, \mathbb{P})$ be a probabilized space, and $W : \Omega \to \mathbb{R}^d, \omega \mapsto W(\omega)$ a random vector assumed to be given. The distribution of the random vector W is denoted by μ and its image $W(\Omega) \subseteq \mathbb{R}^d$ by $\mathcal{W}$. Let $W_1, \ldots, W_p, \ldots$ be independent copies of the random variable W that will be used to generate independent random draws with the distribution μ.

Consider m functions $f_i : \mathbb{R}^n \times \mathcal{W} \to \mathbb{R},\ i = 1, \ldots, m$. The problem here is to solve the average multiobjective problem, which can be written as follows:

$$\min_{\mathbf{x} \in \mathbb{R}^n} \{\mathbb{E}[f_1(\mathbf{x}, W), \mathbb{E}[f_2(\mathbf{x}, W)], \ldots, \mathbb{E}[f_m(\mathbf{x}, W)]\} \qquad [8.10]$$

More precisely, one aims to build a set of Pareto solutions points. As it is presented, problem [8.10] is a deterministic problem. However, the expectations of the objectives are generally not known. A conventional approach consists of replacing each expectation by an approximation, constructed by means of an expectation estimator, using independent random draws w_k of the random variable W (Fliege and Xu 2011; Bonnel and Collonge 2014). This approach is numerically costly since it brings out two loops nested in the resolution algorithms. As for the stochastic gradient algorithm, the algorithm that is proposed in this part of the chapter does not require the computation of expectations; it is solely based on the computation of the Clarke subdifferentials for random variable draws.

8.4.2.2. *Algorithm and theoretical results*

Consider the following problem:

$$\min_{\mathbf{x}\in\mathbb{R}^n} \{\mathbb{E}[f_1(\mathbf{x},W), \mathbb{E}[f_2(\mathbf{x},W)],...,\mathbb{E}[f_m(\mathbf{x},W)]\}$$

Three classes of locally Lipschitz objective functions will be considered:

1) the class of convex and differentiable functions;

2) the class of non-differentiable convex functions;

3) the class of f°-pseudo-convex functions.

The minimal norm element of the convex envelope $C(\mathbf{x},W(\omega))$ will be denoted $\xi(\mathbf{x},W(\omega))$:

$$\exists\{\alpha_i(\mathbf{x},W(\omega))\}_{i\in[1,m]}, \xi(\mathbf{x},W(\omega)) = \sum_{i=1}^{m}\alpha_i(\mathbf{x},W(\omega))\mathbf{s}_i(\mathbf{x},W(\omega))$$

where $\mathbf{s}_i(\mathbf{x},W(\omega))\in\partial^C f_i(\mathbf{x},W(\omega))$, $\alpha_i(\mathbf{x},W(\omega))$ are positive and $\sum_i \alpha_i(\mathbf{x},W(\omega)) = 1$ almost surely.

The algorithm proposed in the remainder of the chapter consists of achieving a random draw w_k of a copy W_k of the random variable W and in calculating the minimal norm element in the envelope $C(\mathbf{x},w_k)$ thus generated. Its opposite is a descent vector common to the objectives of the problem for the particular draw w_k. As a descent step, one uses the elements of a σ-sequence defined as follows:

DEFINITION 8.29.– *A sequence* $(\varepsilon_i)_{i\in\mathbb{N}}$ *is said to be a* σ*-sequence if its first-order series diverges, and its second-order series converges:*

$$\sum_{i=1}^{+\infty}\varepsilon_i = \infty \text{ and } \sum_{i=1}^{+\infty}\varepsilon_i^2 < \infty$$

EXAMPLE.– For (A, B, C, D) real positives such that $C \in]\frac{1}{2}, 1]$, sequences such as:

$$\varepsilon_k = \frac{A}{B \cdot k^C + D}, \forall k \in \mathbb{N}$$

are σ - sequences.

8.4.2.2.1. SMGDA algorithm

1) Initialization: given:

– a starting point $\mathbf{X}_0$;

– a number of iterations N;

– a σ -sequence $(\varepsilon_i)_{i \in \mathbb{N}}$;

– a counter $k = 0$.

2) Iteration:

– generation of a random draw w_{k+1} of the random variable W_{k+1};

– construction of the convex subgradient envelopes: $C_k(\mathbf{X}_k, w_{k+1})$;

– construction of the common descent vector $-\xi_k(\mathbf{X}_k, w_{k+1})$ by solving:

$\xi_k = \text{argmin}\{\| \mathbf{v} \|; \mathbf{v} \in C(\mathbf{X}_k, w_{k+1})\}$

– update: $\mathbf{X}_{k+1} = \mathbf{X}_k - \varepsilon_k \xi_k$.

3) Construction of the optimal value of the objective functions: $\{\mathbb{E}[f_i(\mathbf{X}_N, W)]\}_{i \in [1,m]}$ are evaluated at the end of the iterations.

The SMGDA algorithm never requires the estimation of objective expectations during the optimization process. Expectations are evaluated only after the last iteration of the algorithm, for the last value of the point $\mathbf{X}_N$, in order to build the Pareto front in the objective space.

The recursive sequence described by the equation:

$$\mathbf{X}_{k+1}(\omega) = \mathbf{X}_k(\omega) - \varepsilon_k \xi(\mathbf{X}_k(\omega), W_{k+1}(\omega))$$

defines a discrete-time stochastic process.

From a single starting point $\mathbf{x}_0$, the SMGDA algorithm provides a single solution belonging to the Pareto front at the end of the optimization process. Therefore, in order to have a complete description of the Pareto front, it is necessary to run the algorithm from a set of starting points. This can be done by parallelization by performing an optimization for a different starting point per available core.

In this algorithm, no stopping criteria are proposed. The existence of a convergence indicator will be discussed later in the chapter after giving some results.

We introduce the notation $\mathbf{x}^\star$ that designates one of the Pareto solution points among those closest to a current point $\mathbf{x}$:

$$\forall k \in \mathbb{N}, \mathbf{x}^\star \in \operatorname{argmin}_{\mathbf{v}\in\mathcal{P}_D} \|\mathbf{x}-\mathbf{v}\|$$

The convergence results of the algorithm are based on the set of assumptions presented below.

8.4.2.2.2. $\mathcal{H}$ assumptions

1) The set of Pareto solutions is non-empty $\mathcal{P}_D \neq \varnothing$

2) The functions $\mathbf{x} \mapsto f_i(\mathbf{x}, W)$ belong to one of the three classes (1,2,3) of functions for every $i \in [1, m]$.

3) Random variables $\omega \mapsto f_i(\mathbf{x}, W(\omega))$ are integrable for any $i \in [1, m]$.

4) The norm of any vector $\mathbf{s}_i \in \partial^C f_i(\mathbf{x}, W(\omega))$ is almost certainly increased by a real number M_i:

$$\|\mathbf{s}_i\| \leq M_i, \; p.s. \, \forall \mathbf{x} \in \mathcal{X}$$

5) There exists a convex objective f_ℓ for which the contribution $\alpha_\ell(\mathbf{x}, W(\omega))$ in the construction of the common descent vector is almost certainly not zero. In addition:

$$\begin{cases} \forall i \in [1,m], f_i(\mathbf{x},W) - f_i(\mathbf{x}^\star,W) \geq 0, \; p.s. \\ \exists \ell \in [1,m], f_\ell(\mathbf{x},W) - f_\ell(\mathbf{x}^\star,W) \geq \dfrac{C}{2}\|\mathbf{x}-\mathbf{x}^\star\|^2, \; p.s. \end{cases}$$

6) The sequence $(\varepsilon_i)_{i\in\mathbb{N}}$ is a σ-sequence.

IMPORTANT.– Assumption (5) implies that the point $\mathbf{x}^{\star}$ almost certainly dominates the point $\mathbf{x}$.

A first theorem proves the convergence in root mean square of the process $(\mathbf{X}_k)_{k\in\mathbb{N}}$ toward a Pareto solution $\mathbf{X}^{\star}$ of the optimization problem.

THEOREM 8.7.– Under the set of assumptions $\mathcal{H}$, the stochastic process constructed by the SMGDA iterations converges in a L^2 norm to a Pareto solution point $\mathbf{X}^{\star}$:

$$\lim_{k\to+\infty} \mathbb{E}\left[\| \mathbf{X}_k - \mathbf{X}_k^{\star} \|^2\right] = 0$$

It is possible to obtain information on the speed of convergence when considering a particular σ-sequence:

THEOREM 8.8.– Let $\mathbf{x}_0$ be a point of $\mathcal{X}$, if we use the σ -sequence defined by:

$$\varepsilon_k = \frac{1}{C/2\cdot k + \dfrac{M^2}{\mathbb{E}[L_0 C/2]}}$$

where:

$$L_0 = \mathbb{E}\left[\| \mathbf{x}_0 - \mathbf{x}_0^{\star} \|^2\right]$$

then:

$$\mathbb{E}\left[\| \mathbf{X}_k - \mathbf{X}_k^{\star} \|^2\right] \leq \frac{1}{\dfrac{C/2}{M^2} k + \dfrac{1}{\mathbb{E}\left[\mathcal{L}_0^{\star}\right]}}$$

As mentioned previously, there is no stopping criterion for the algorithm. There are criteria in the literature that are based on the result of speed of convergence (Dvoretzky 1956; Bouleau *et al.* 1986). However, such criteria presuppose knowledge of all the solutions to the problem: the previous theorem involves the unknown projection $\mathbf{x}_0^{\star}$ of the Pareto front of the initial point $\mathbf{x}_0$. On the other hand, the optimality conditions of the theorem (Mäkelä *et al.* 2014) may make it possible to give a stopping criterion relating to the value of the norm of the descent vector associated with the problem in expectation, in a way equivalent to the criterion proposed for the deterministic algorithm MGDA. As a matter of fact, at a Pareto solution point $\mathbf{x}^\star$ of the optimization problem:

$$\min_{\mathbf{x}\in\mathcal{X}}\left\{\mathbb{E}[f_1(\mathbf{x},W(\omega))],\ldots,\mathbb{E}[f_m(\mathbf{x},W(\omega))]\right\}$$

there exists a convex combination of elements of the Clarke subdifferential expectations that cancel each other out in $\mathbf{x}^\star$:

$$\exists\{\alpha_i\}_{i\in[1,m]},\sum_{i=1}^{m}\alpha_i\mathbf{s}_i(\mathbf{x}^\star)=0$$

where $\mathbf{s}_i\in\mathbb{E}\left[\partial^C f_i(\mathbf{x}^\star,W(\omega))\right]$ and the sum of the α_i positive or null equals to one $\sum_i\alpha_i=1$. The direct extension of the criterion proposed in the MGDA is thus equivalent to calculating the minimal norm element in the expectation of the convex envelope of the union of the Clarke subdifferentials:

$$\zeta=\operatorname{argmin}\left\{\|\mathbf{u}\|,\mathbf{u}\in\mathbb{E}\left[\operatorname{Conv}\left(\bigcup_{i=1}^{m}\partial^C f_i(\mathbf{x}^\star,W(\omega))\right)\right]\right\}$$

The algorithm can then be stopped when the norm of ζ reaches a certain threshold. This stopping criterion is necessarily costly since it requires the calculation of expectations, which is contrary to the philosophy of the SMGDA.

THEOREM 8.9.– The stochastic process described by the sequence of random variables $(\mathbf{X}_k)k\in\mathbb{N}$ almost surely converges to a Pareto solution $\mathbf{x}^\star$:

$$\mathbb{P}\left(\left\{\lim_{k\to+\infty}\mathbf{X}-\mathbf{X}^\star=0\right\}\right)=1$$

8.5. Illustrations

8.5.1. *Performance of the SMGDA algorithm*

In order to illustrate the behavior of SMGDA, we shall solve a representative family of classical multiobjective optimization problems used as reference problems in the optimization community (Huband *et al.* 2006). These problems will be turned into stochastic problems by the arbitrary addition of a number of random parameters in the expression of objectives. The performances of the SMGDA will be compared to those of two algorithms: a genetic algorithm and a direct search algorithm. The NSGA-II developed by Deb *et al.* (2002) is the chosen genetic algorithm. It is a genetic algorithm whose solutions are classified at each generation according to a non-dominant criterion. Therefore, successive fronts of equivalent points following Pareto's definition can be plotted at every generation and used to classify individuals. The second algorithm to which the SMGDA will be compared is a direct search algorithm: Direct MultiSearch (DMS). It appears for the first time in Custódio *et al.* (2011). It is in fact the generalization to the multiobjective framework of directional search methods.

These two algorithms were not originally designed to solve stochastic multiobjective problems. A sample average approximation (SAA) approach is then utilized to construct the estimators of the expectations of the objective functions. The optimization problem will then be rewritten, for these last two algorithms:

$$\min_{\mathbf{x}\in\mathcal{X}}\left\{\frac{1}{N}\sum_{k=1}^{N}f_1(\mathbf{x},w_k^{(1)}),\ldots,\frac{1}{N}\sum_{k=1}^{N}f_m(\mathbf{x},w_k^{(m)})\right\}$$

where $w_k^{(i)}, i=1,\ldots,m$ are random draws independent of the random vector $W(\omega)$. The computation of the expectation estimator must be performed at each iteration of the optimization algorithm, which makes the numerical approach costly.

Moreover, the choice of the number N plays a crucial role in SAA-based approaches: a number that is too low will give a very wide confidence interval and a poor estimator of the objectives, while a number that is too high will make the numerical approach much too expensive.

In order to be able to compare the performance of these three approaches, we shall use two conventional performance indicators: one based on the *purity* of solutions (Custódio *et al.* 2011; Riquelme *et al.* 2015), which compares the proportion of nondominated points that an algorithm is able to identify compared to the solutions found by all the competing algorithms; and the other based on the measurement of the hypervolume (Auger *et al.* 2009), which gives an indication of the dispersion of solutions.

We shall first develop the first problem being addressed in a little more detail. It is a randomized version of the MOP2 test case studied in Huband *et al.* (2006). It involves two non-convex objective functions and 15 optimization variables:

$$\min_{\mathbf{x}\in[-4,4]^{15}} \begin{cases} \mathbb{E}[f_1(\mathbf{x},W)] = \mathbb{E}\left[1-\exp\left(-\sum_{i=1}^{15}(\mathbf{x}_i - \frac{1+W_{1i}}{\sqrt{3}})^2\right)\right] \\ \mathbb{E}[f_2(\mathbf{x},W)] = \mathbb{E}\left[1-\exp\left(-\sum_{i=1}^{15}(\mathbf{x}_i + \frac{1+W_{2i}}{\sqrt{3}})^2\right)\right] \end{cases} \quad [8.11]$$

where $\{W_{1i}\}_{i\in\{1,15\}}$ and $\{W_{2i}\}_{i\in\{1,15\}}$ represent 30 independent random variables of uniform distribution over the interval $[-0,7;0,7]$. The random vector $(W_{1,i};W_{2,i})$ $i\in\{1,15\}$ will be denoted by W.

A budget of 10^5 calls to the objective functions was allocated to the NSGA-II and DMS algorithms but this budget was reduced to 10^4 for the SMGDA, which was initialized from 10^2 points. At the end of the iterations of the algorithms, the Pareto front in the domain of the objectives is estimated based on 10^5 samples of the random variable W. The choice of the σ-sequence is crucial to the quality of the stochastic gradient-based algorithms. Here, we choose the sequence $\varepsilon_k = \dfrac{0.1}{k+1}$.

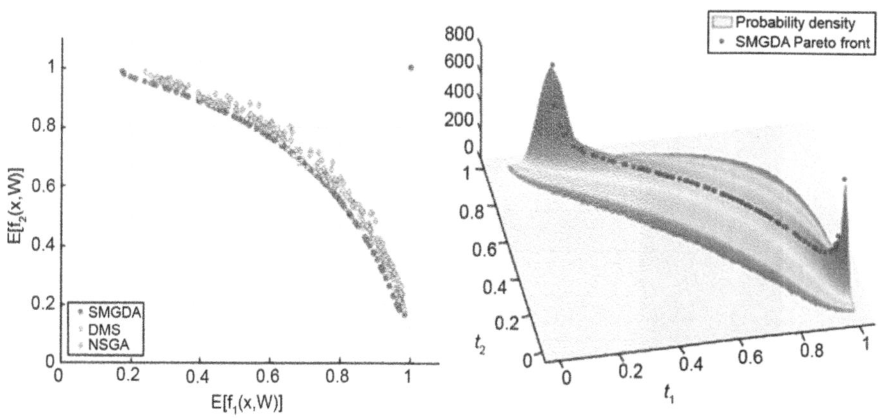

Figure 8.4. *Pareto fronts and dispersion due to uncertainties. For a color version of this figure, see www.iste.co.uk/gogu/uncertainties.zip*

In order to illustrate the impact of hazard on the solution of the optimal problem, we carried out a study of the propagation of uncertainties from 10^5 independent draws of $f_i(\mathbf{x}^*, W)$, for each solution $\mathbf{x}^*$ found on the front, which allowed for building an estimator of the distribution of the Pareto front, plotted in Figure 8.4.

The dispersion of the front is then much greater in the middle of the front than at the extremities, which shows that the effect of randomness is greater for the coupled optimization problem than for each single-objective problem. It can be seen in the figure that the SMGDA gives a more convergent front than the other two approaches; this heuristic finding is confirmed by the performance measures (hypervolume measure shown on the left) (Figure 8.5), a value of 1 designating the best performance.

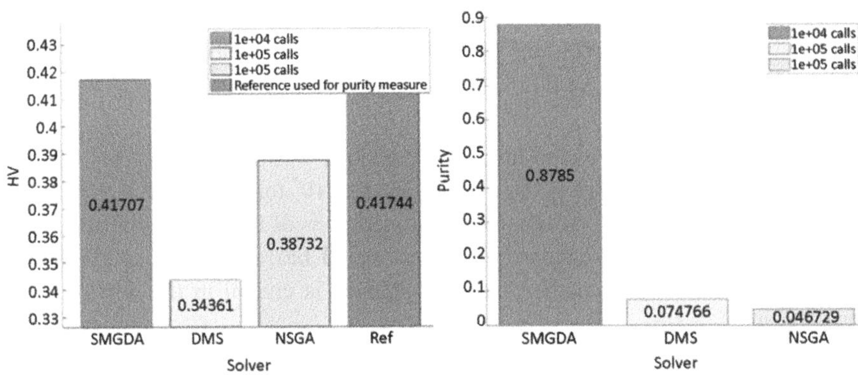

Figure 8.5. *Comparison of performance measures. For a color version of this figure, see www.iste.co.uk/gogu/uncertainties.zip*

Problem	Number of variables		Budget	
	Optimization	Random	SMGDA	NSGA-II, DMS
MOP2	15	30	10^4	10^5
MOP3	2	18	10^3	10^4
MOP6	2	3	10^3	10^4
ZDT{1,2,3}	30	32	5×10^3	5×10^4
JOS1	30	60	5×10^3	5×10^4
JOS2	30	32	5×10^3	5×10^4
SCH1	1	4	10^3	10^4
IM1	2	3	10^3	10^4

Table 8.1. *Test problems (Huband* et al. *2006)*

The performance of the three algorithms was then evaluated with a set of test problems described in Table 8.1. The results obtained for the set of problems are represented in a condensed form by *performance profiles* (Dolan and Moré 2002), which correspond to cumulative distribution functions associated with the performance measures being used. These profiles are represented in Figure 8.6, where a value equal to 1 indicates the best overall performance: for $\tau = 1$, the values plotted on the graph represent the relative number of optimization problems for which an algorithm used is best, for an infinite value of τ corresponding to the relative total number of problems solved by the algorithm. It can be seen that for the test problems under consideration, the SMGDA has better performance than the other two algorithms, and this is true for a budget 10 times less.

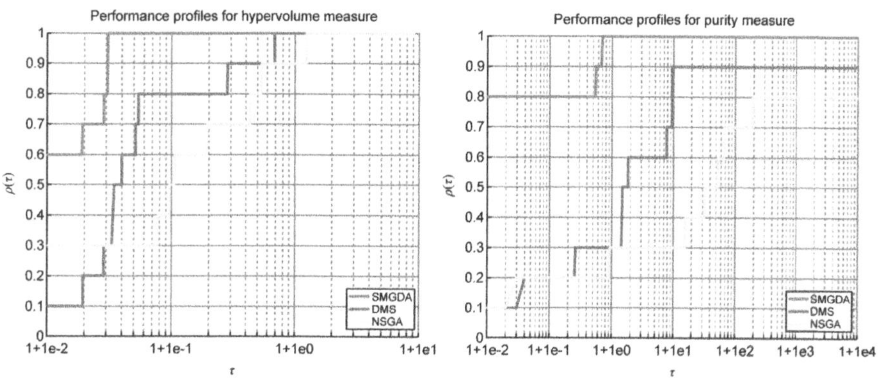

Figure 8.6. *Comparison of performance measures. For a color version of this figure, see www.iste.co.uk/gogu/uncertainties.zip*

8.5.2. *Multiobjective approach to RBDO problems*

In the framework of reliability or risk studies, it is common to consider an optimization problem associated with constraints, expressed in terms of ruin probability. This is, for example, the case in mechanics when the main interest is the reliability of a structure depending on random (uncertain) parameters that ought to be optimized: it would desirable to control the probability that the structure remains reliable during the optimization process. This probability will then become a constraint of the problem. This probability constraint describes the part of the risk that the constructor is willing to accept in the specifications defining the structure.

Consider the following optimization problem:

$$\underset{x \in X^{\mathrm{ad}}}{\operatorname{argmin}} \{\mathbb{E}[f(x,\xi(\omega))]\} \text{ under the constraint } P[g(x,\xi(\omega)) \geq 0] \leq \alpha \qquad [8.12]$$

where f, the cost function, is a locally Lipchitz function defined on a subset X^{ad} of $\mathbb{R}^n$, ξ is a random variable with values in $\mathbb{R}^p$ modeling uncertain parameters, $\mathbb{E}$ designates the mathematical expectation, g represents a limit state associated with the ruin of the system (ruin occurs if this function takes a positive value) and α represents the risk that is accepted (it is generally a very small quantity).

Naturally, other deterministic or probabilistic constraints can be introduced, and similarly, several cost functions to be minimized can be considered. In this chapter, we shall merely address the simple framework presented by equation [8.12].

The intrinsic difficulty of the problem comes from the calculation of the probability of ruin, which is written as:

$$P[g(x,\xi(\omega)] = \int_{g(x,\xi(\omega))\geq 0} P(d\omega) \qquad [8.13]$$

and which requires the analytical definition of the domain of ruin $g(x,\xi(\omega)) \geq 0$. However, this domain is generally unknown because it is implicitly defined. Simplified approaches (FORM/SORM) (Ditlevsen and Madsen 1996; Haldar and Mahadevan 2000) can be used, or this probability can be estimated by a Monte Carlo simulation approach (Royset and Polak 2004). In the latter case, this gives rise to numerical computations that are all the more expensive, as the probability to be estimated is low (which is in principle always the case in terms of reliability).

8.5.3. *Rewriting the probabilistic constraint*

By definition of the mathematical expectation, the ruin probability defined by relation [8.13] can be written as:

$$P[g(x,\xi(\omega)] = \mathbb{E}[\mathbb{I}_{\mathbb{R}^+}(g(x,\xi(\omega))] \qquad [8.14]$$

where $\mathbb{I}_{\mathbb{R}^+}$ designates the indicator function of the set of positive reals. Similarly, the constraint of the optimization problem [8.12] is written by means of an expectation:

$$\mathbb{E}[\mathbb{I}_{\mathbb{R}^+}(g(x,\xi(\omega))-\alpha]\leq 0 \qquad [8.15]$$

the optimization problem is then written as:

$$\arg\min_{x\in X^{\text{ad}}}\{\mathbb{E}[f(x,\xi(\omega))]\};\mathbb{E}[\mathbb{I}_{\mathbb{R}^+}(g(x,\xi(\omega))-\alpha]\leq 0 \qquad [8.16]$$

The solution of such a problem can be addressed using the stochastic Arrow–Hurwicz algorithm, as outlined by Rémi Arnaud in his thesis (Arnaud 2014). This algorithm however is based on the gradients of the objective function and of the constraint. Nonetheless, the indicator function that appears in the constraint is neither differentiable nor locally Lipchitz, and it is necessary to go through constraint regularizing approaches, as proposed by Andrieu *et al.* (2011). The literature in the field of reliability optimization is very rich, for example the following references can be cited: Allen and Maute (2004), Missoum *et al.* (2010), Nikbay *et al.* (2010), Moustapha *et al.* (2016) and Schöbi and Sudret (2017).

Under weak assumptions (Polak *et al.* 2000), it can be shown that the function $x \mapsto P[g(x,\xi(\omega)]$ is differentiable. However, here it is desirable to use a stochastic gradient approach that requires the differential of the function $x \mapsto \mathbb{I}_{\mathbb{R}^+}(g(x,\xi(\omega))$, which is not continuous nor locally Lipchitz. The reliability optimization problem [8.12] is intrinsically linked to the imposed value α of the failure probability. Constructors will surely benefit from keeping it as low as possible. This is the reason why we propose to replace this problem by the following multiobjective optimization problem:

$$\operatorname{argmin}_{x\in X^{\text{ad}}}\{\mathbb{E}[f(x,\xi(\omega))];\mathbb{E}[\mathbb{I}_{\mathbb{R}^+}(g(x,\xi(\omega))]\} \qquad [8.17]$$

It can then immediately be verified that for a given value α, a solution x^* of problem [8.12] belongs, by definition, to the set of optimal Pareto solutions of problem [8.17]. Therefore, by solving problem [8.17], the set of solutions of the reliability optimization problem [8.12] can be directly obtained for all the values of ruin probability $\alpha\in[0,1]$, without having to calculate these probabilities, because the SMGDA algorithm requires knowledge of the gradient or subgradient of the probability of ruin. A notable advantage of the proposed approach is the fact that the optimization process and the process of calculating the probability of ruin are completely decoupled; the latter calculation being done after the algorithm has constructed the Pareto front in the solution space. This approach does not necessitate calculating the probability of failure at each iteration of the algorithm. In particular, it gives access to the set of the best trade-offs for the set of values of the probability of failure $\alpha \in [0,1]$. Any suitable approach can then be employed to construct the Pareto front in the space of objectives, for example subset simulation-based methods coupled with the use of reduced models.

This approach will be compared to a classical approach applied to a very simple example taken from Moustapha *et al.* (2016). The example under consideration is a toy example that would not be properly scaled up with regard to buckling and which, by the choice of probabilistic modeling for uncertain coefficients, cannot represent a real problem: the problem consists of minimizing the area (in order to minimize the mass) $b \times h$ of the section of a rectangular beam under compression subjected to a load F while avoiding its buckling that occurs when the compression load is higher than the critical Euler force. Structural ruin occurs when:

$$F - k\frac{\pi^2 Ebh^3}{12L^2} \geq 0 \qquad [8.18]$$

where E is the Young modulus of the material constituting the beam and L its length, and k is a coefficient modeling different defects (straightness, centering, homogeneity).

Figure 8.7 gives the diagram of the beam that will be studied in this application where b and h are the optimization variables of the problem. The vector containing the two optimization variables is denoted $\mathbf{x} = [b, h] \in [0.15, 0.35]^2$. The cross-section of the beam is expressed as the product of these two variables:

$$M(\mathbf{x}) = b \times h$$

A first deterministic geometric constraint is introduced in this problem so that $b \geq h$. This constraint will be addressed by projection within the context of this application.

Uncertainties are introduced into the model through random variables. The first uncertainty concerns a multiplicative coefficient k of the Euler buckling threshold. The Young modulus E will also be considered as being uncertain, as well as the length of the beam L. The uncertainties follow lognormal laws described in Table 8.2.

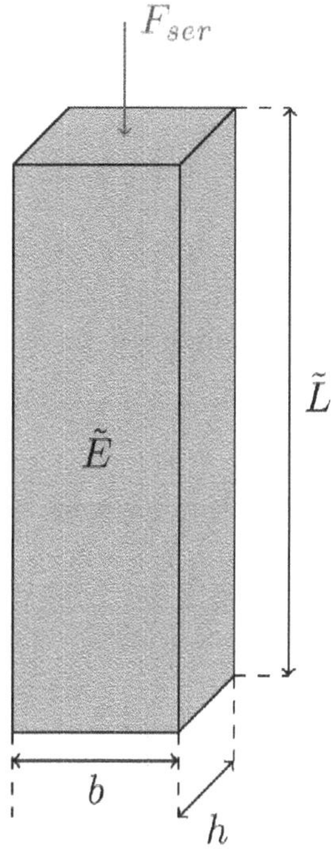

Figure 8.7. *Beam under study under compression*

Parameter	Distribution	Mean: μ	Covariance: δ
k	Lognormal	0.6	10
E (MPA)	Lognormal	10^4	5
L (mm)	Lognormal	3	1

Table 8.2. *Probabilistic uncertainty modeling*

The complete optimization problem can then be written in the form:

$$\begin{cases} \min\limits_{\mathbf{x}\in[0.15,0.35]^2} \quad M(\mathbf{x}, W(\omega)) \\ \text{under the constraints} \begin{cases} b \geq h \\ \mathbb{P}\left(F_{ser} - k(\omega)\dfrac{\pi^2 E(\omega) b h^3}{L(\omega)} \geq 0 \right) \leq \alpha \end{cases} \end{cases} \qquad [8.19]$$

Since uncertainties are modeled by lognormal laws, there is an analytical solution to the optimization problem for a given probability threshold α :

$$b^{\star}(\alpha) = h^{\star}(\alpha) = \left(\frac{12F_{ser}}{\pi^2 \exp(\lambda_{\tilde{k}} + \lambda_{\tilde{E}} - 2\lambda_{\tilde{L}} + \Phi^{-1}(\alpha)\sqrt{\zeta_{\tilde{k}}^2 + \zeta_{\tilde{E}}^2 + \zeta_{\tilde{L}}^2})} \right)^{\frac{1}{4}} \quad [8.20]$$

where $\Phi(\alpha)$ is the cumulative distribution function of the standard Gaussian distribution, $\zeta_j = \sqrt{\ln\left(1+\delta_j^2\right)}$ and $\lambda_j = \ln(\mu_j) - 1/2 \times \zeta^2$.

The optimization problem is then rewritten as a stochastic multiobjective optimization problem:

$$\begin{cases} \min\limits_{x \in [0.15, 0.35]^2} \left\{ M(\mathbf{x}), \mathbb{E}\left[\mathbb{I}_{F_{ser} - \tilde{k}(\omega)\frac{\pi^2 \tilde{E}(\omega) b h^3}{\tilde{L}(\omega)} \geq 0} \right] \right\} \\ \text{under the constraint } b \geq h \end{cases} \quad [8.21]$$

The problem thus formulated corresponds to the formulations adapted to the SMGDA. However, the indicator function does not belong to any class of functions considered here, since it is not locally Lipschitz. In Andrieu *et al.* (2011), the authors propose to approach the indicator function by a convolution product of the indicator with a regular, parameter-dependent function that converges to the Dirac function when the parameter is large. This is the "mollifiers" technique: a regular function $h : \mathbb{R} \rightarrow \mathbb{R}$ is introduced, dependent on an even parameter r, with positive or zero values, of integral equal to 1 with its maximum in 0 and approaching the Dirac function when the parameter r tends to 0.

Consider for example:

$$h(x) = \frac{1}{2}\mathbb{I}_1(x/r) \,; h(x) = (1 - |x/r|)\mathbb{I}_1(x/r) \,; h(x) = \frac{\pi}{4}\cos(\frac{\pi}{2}x/r)\mathbb{I}_1(x/r)$$

The approximation of the indicator function is then constructed on $\mathbb{R}^+$:

$$\mathbb{I}^r_{\mathbb{R}^+}(\mathbf{z}) = \frac{1}{r}\int_{-\infty}^{+\infty} \mathbb{I}_{\mathbb{R}^+}(\mathbf{y}) h(\frac{\mathbf{z}-\mathbf{y}}{r}) dy = \frac{1}{r}\int_0^{+\infty} h(\frac{\mathbf{z}-\mathbf{y}}{r}) dy$$

which can be interpreted as the expectation of the indicator function for a random variable $\mathbf{y}$ of distribution $\frac{1}{r}h(\cdot / r)$ centered around $\mathbf{z}$. This distribution will become increasingly concentrated around $\mathbf{z}$ as r decreases toward 0. Replacing $\mathbf{z}$ with the failure event $g(\mathbf{x}, W(\omega)) \geq 0$ for a fixed draw ω gives:

$$\mathbb{I}^r_{\mathbb{R}^+}(g(\mathbf{x}, W(\omega))) = \frac{1}{r}\int_0^{+\infty} h\left(\frac{g(\mathbf{x}, W(\omega)) - \mathbf{y}}{r}\right) d\mathbf{y}$$

This approximation can then be differentiated with respect to $\mathbf{x}$:

$$\nabla \mathbb{I}^r_{\mathbb{R}^+}(g(\mathbf{x}, W(\omega))) = -\frac{1}{r} h\left(\frac{g(\mathbf{x}, W(\omega))}{r}\right) \nabla g(\mathbf{x}, W(\omega))$$

The resulting final Pareto front is shown in Figure 8.8. A budget of 650 calls to objectives per point was allocated and the final performance of the resulting solutions was evaluated using a SAA method on 10^9 samples, which allowed the estimation of probabilities in the order of 10^{-7}.

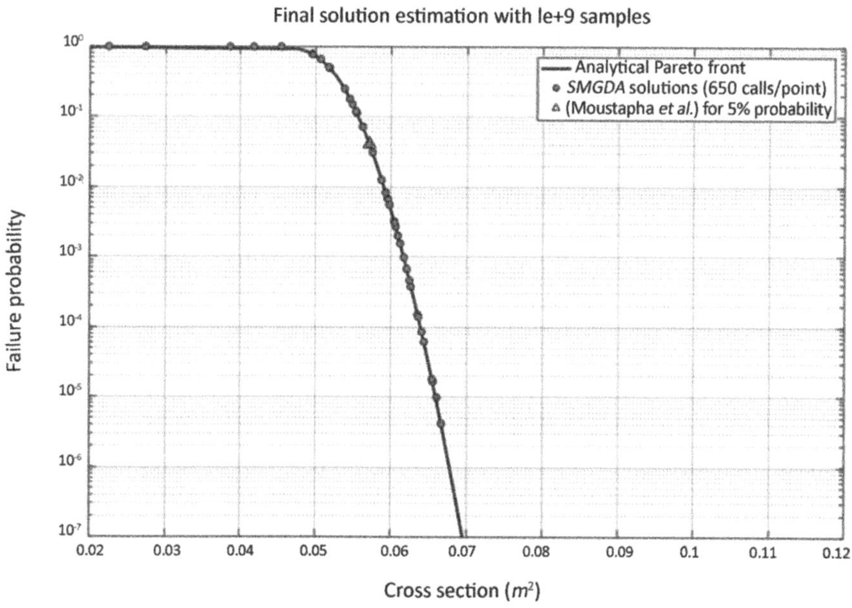

Figure 8.8. *Pareto front in the objective domain for the reliability problem. For a color version of this figure, see www.iste.co.uk/gogu/uncertainties.zip*

The Pareto front obtained by the SMGDA coincides with the analytical Pareto front given by expression [8.20]. The budget allocated to the SMGDA is remarkably low for a stochastically based algorithm.

8.6. References

Allen, M. and Maute, K. (2004). Reliability-based design optimization of aeroelastic structures. *Structural and Multidisciplinary Optimization*, 27, 228–242.

Andrieu, L., Cohen, G., Vàzquez-Abad, F. (2011). Gradient-based simulation optimization under probability constraints. *European Journal of Operational Research*, 212, 345–351.

Arnaud, R. (2014). Optimization of an aeroelastic system with uncertainties. PhD Thesis, CentraleSupélec, Paris.

Auger, A., Bader, J., Brockhoff, D., Zitzler, E. (2009). Theory of hypervolume indicator: Optimal μ-distributions and the choice of the reference point. *Proceedings of the Tenth ACM SIGEVO Workshop Foundations of Genetic Algorithms* 87–102. Association for Computing Machinery, USA.

Bellman, R. and Zadeh, L. (1970). Decision-making in a fuzzy environment. *Management Science*, 17, 141–161.

Bonnans, J.-F., Gilbert, J., Lemarechal, C., Sagastizábal, C. (2006). *Numerical Optimization, Theoretical and Practical Aspects*. Springer-Verlag, Berlin.

Bonnel, H. and Collonge, J. (2014). Stochastic optimization over a Pareto set associated with a stochastic multiobjective optimization problem. *Journal of Optimization Theory and Applications*, 162(2), 405–427.

Bouleau, N., Lamberton, D., Lapeyre, B. (1986). Une méthode numérique adaptée au calcul probabiliste des structures. *Journal de mécanique théorique et appliquée*, 5(5).

Burke, J.V., Lewis, A.S., Overton, M.L. (2002). Approximating subdifferentials by random sampling of gradients. *Mathematics of Operations Research*, 27(3), 567–584.

Clarke, F. (1990). *Optimization and Nonsmooth Analysis*. Society for Industrial and Applied Mathematics, Philadelphia.

Collette, Y. and Siarry, P. (2011). *Optimisation multi-objectif*. Eyrolles, Paris.

Custódio, A.L., Madeaira, F.A., Vaz, A.I.F., Vincente, L.N. (2011). Direct multisearch for multiobjective optimization. *SIAM Journal on Optimization*, 21(3), 1109–1140.

Dantzig, B. (1955). Linear programming under uncertainty. *Management Science*, 1, 197–206.

Deb, K., Pratap, A., Agarwal, S., Meyarivan, T. (2002). A fast and elitist multiobjective genetic algorithm: NSGA-II. *IEEE Transactions on Evolutionary Computation*, 6(2), 182–197.

Désidéri, J.-A. (2009). Multiple-gradient descent algorithm (MGDA). Report RR-6953, INRIA.

Ditlevsen, O. and Madsen, H. (1996). *Structural Reliability Methods*. Wiley-Blackwell, Hoboken.

Dolan, E.D. and Moré, J.J. (2002). Benchmarking optimization software with performance profiles. *Mathematical Programming*, 91(2), 201–213.

Durrett, R. (2010). *Probability: Theory and Examples*. Cambridge University Press, Cambridge.

Dvoretzky, A. (1956). On stochastic approximation. In *Proceedings of the Third Berkeley Symposium on Mathematical Statistics and Probability*, Neyman, J. (ed.). 1, 39–55, University of California Press, Berkeley.

Ehrgott, M. (2006). *Multicriteria Optimization*. Springer, Berlin.

Ermoliev, Y. and Wets, R. (1988). *Numerical Techniques for Stochastic Optimization*. Springer, Berlin.

Eskandari, H. and Geiger, C.D. (2009). Evolutionary multiobjective optimization in noisy problem environments. *Journal of Heuristics*, 15(6), 559–595.

Federer, H. (1996). *Geometric Measure Theory*. Springer, Berlin.

Fliege, J. and Xu, H. (2011). Stochastic multiobjective optimization: Sample average approximations and applications. *Journal of Optimization Theory and Applications*, 151(1), 135–162.

Gabrel, V., Murat, C., Thiele, A. (2014). Recent advances in robust optimization: An overview. *European Journal of Operational Research*, 235, 471–483.

Haldar, A. and Mahadevan, S. (2000). *Probability, Reliability and Statistical Methods in Engineering Design*. John Wiley & Sons, New York.

Huband, S., Hingston, P., Barone, L., While, L. (2006). A review of multiobjective test problems and a scalable problem toolkit. *IEEE Transactions on Evolutionary Computation*, 10, 477–506.

Klamroth, K., Kbis, E., Schbel, A., Tammer, C. (2017). A unified approach to uncertain optimization. *European Journal of Operational Research*, 260, 805–819.

Knowles, J. and Corne, D. (1999). The Pareto evolution strategy: A new baseline algorithm for Pareto multiobjective optimization. In *Proceedings of the 1999 Congress on Evolutionary Computation-CEC99* 98–105, Piscataway, N. (ed.). IEEE Press, Piscataway.

Lelièvre, N., Beaurepaire, P., Mattrand, C., Gayton, N., Otsmane, A. (2016). On the consideration of uncertainty in design: Optimization-reliability-robustness. *Structural and Multidisciplinary Optimization*, 54(6), 1423–1437.

Mäkelä, M.M., Karmitsa, N., Eronen, V.-P. (2010). On generalized pseudo and quasiconvexities for nonsmooth functions. Technical report no. 989, Turku Center for Computer Science, Finland.

Mäkelä, M.M., Eronen, V.-P., Karmitsa, N. (2014). On nonsmooth optimality conditions with generalized convexities. In *Optimization in Science and Engineering*, Rassias, T.M., Floudas, C.A., Butenko, S. (eds). Springer, Berlin.

Miettinen, K. (1999). *Nonlinear Multiobjective Optimization*. Springer, Berlin.

Missoum, S., Dribusch, C., Beran, P. (2010). Reliability-based design optimization of nonlinear aeroelasticity problems. *Journal of Aircraft*, 47(3), 992–998.

Moustapha, M., Sudret, B., Bourinet, J., Guillaume, B. (2016). Quantile-based optimization under uncertainties using adaptive Kriging surrogate models. *Structural and Multidisciplinary Optimization*, 54, 1403–1421.

Nemirovski, A. and Shapiro, A. (2006). Scenarion approximation of chance constraints. In *Probabilistic and Randomized Methods for Design under Uncertainty*, Calafiore, G., Dabbene, F. (eds). Springer, Berlin.

Nikbay, M., Fakkusoglu, N., Kuru, M. (2010). Reliability-based aeroelastic optimization of a composite aircraft wing via fluid-structure interaction of high fidelity solvers. *IOP Conference Series: Materials Science and Engineering*, 10(1).

Polak, E., Wets, R., der Kiureghian, A. (2000). On an approach to optimization problems with a probabilistic cost and/or constraints. In *Nonlinear Optimization and Related Topics: Applied Optimization*, Pillo, G., Giannessi, F. (eds). Springer, Boston.

Riquelme, N., von Lücken, C., Baran, B. (2015). Performance metrics in multi-objective optimization. *2015 Latin American Computing Conference (CLEI)*, Arequipa, 1-11 [Online]. Available at: 10.1109/CLEI.2015.7360024.

Robbins, H. and Monro, S. (1951). A stochastic approximation method. *The Annals of Mathematical Statistics*, 22(3), 400–407.

Roy, R., Hinduja, S., Teti, R. (2008). Recent advances in engineering design optimization: Challenges and future trends. *Manufacturing Technology*, 57, 697–715.

Royset, J. and Polak, E. (2004). Reliability-based optimal design using sample average approximations. *Probabilistic Engineering Mechanics*, 19, 331–343.

Sahinidis, N.V. (2004). Optimization under uncertainty: State-of-the-art and opportunities. *Computers and Chemical Engineering*, 28(6–7), 971–983.

Schöbi, R. and Sudret, B. (2017). Structural reliability analysis for p-boxes using multi-level meta-models. *Probabilistic Engineering Mechanics*, 48, 27–38.

Shapiro, A. (2003). Monte Carlo sampling methods. *Handbooks in Operations Research and Management Science*, 10, 353–425.

Zitzler, E. and Thiele, L. (1999). Multiobjective evolutionary algorithms: A comparative case study and the strength Pareto approach. *IEEE Transactions on Evolutionary Computation*, 3(4), 257–271.

List of Authors

Mathieu BALESDENT
ONERA/DTIS
Paris-Saclay University
Palaiseau
France

Julien BAROTH
Grenoble Alpes University
France

Nathalie BARTOLI
ONERA/DTIS
University of Toulouse
France

Pierre BEAUREPAIRE
SIGMA Clermont
Clermont-Ferrand
France

David BOUHJITI
Grenoble Alpes University
France

Jean-Marc BOURINET
SIGMA Clermont
Clermont-Ferrand
France

Vincent CHABRIDON
EDF R&D
Chatou
France

Sylvain DUBREUIL
ONERA/DTIS
University of Toulouse
France

Frédéric DUFOUR
Grenoble Alpes University
France

Nicolas GAYTON
SIGMA Clermont
Clermont-Ferrand
France

Christian GOGU
Clément Ader Institute
Paul Sabatier University
Toulouse
France

Thierry LEFEBVRE
ONERA/DTIS
University of Toulouse
France

Maurice LEMAIRE
SIGMA Clermont
Clermont-Ferrand
and
Phimeca Engineering
Cournon
France

Cécile MATTRAND
SIGMA Clermont
Clermont-Ferrand
France

Quentin MERCIER
Télécom Paris
Palaiseau
France

Jérôme MORIO
ONERA/DTIS
University of Toulouse
France

Guillaume PERRIN
CEA/DAM/DIF
Arpajon
France

Fabrice POIRION
ONERA
Paris-Saclay University
Palaiseau
France

François WILLOT
Mines ParisTech
France

Index

U, V